Cochlear Mechanics

Hendrikus Duifhuis

Cochlear Mechanics

Introduction to a Time Domain Analysis of the Nonlinear Cochlea

Springer

Dr. Hendrikus Duifhuis
Faculty of Mathematics and Natural Sciences
University of Groningen
Nijenborgh 9,
9749 AG Groningen

BCN-NeuroImaging Center
Antonius Deusinglaan 2,
9713 AW Groningen
H.Duifhuis@rug.nl

ISBN 978-1-4419-6116-7 e-ISBN 978-1-4419-6117-4
DOI 10.1007/978-1-4419-6117-4
Springer New York Dordrecht Heidelberg London

Library of Congress Control Number: 2011944505

Printed on acid-free paper

Springer is part of Springer Science+Business Media (www.springer.com)

Preface

Interest in cochlear mechanics (CM) and particularly in the role of nonlinear cochlear processes expanded significantly during the 1970s. For me it was stimulated particularly through contacts with Jont Allen—a contact that remained important after his visit to Eindhoven—and Egbert de Boer, who was—amongst others—active within the Dutch Auditory Biophysics community.

My move to the University of Groningen in 1980, in combination with international developments, such as the discovery of otoacoustic emissions, led to an immediate increase in interest in CM. We followed up a proposal by Peter Johannesma [International Symposium on Hearing (ISH)—1980] stating that a Van der Pol-oscillator might be a proper model for spontaneous emissions.

At the same time significant theoretical contributions were given by John W. Matthews (1980) and Stephen T. Neely (1981) in their doctoral theses presented at Washington University. They also contributed to the ISH-1980 conference mentioned above. They started to explore nonlinear cochlea models in the frequency and time domain.

As a result, 1980 became a pivotal point in this book!

This book strongly rests on work from the Groningen biophysics department, which was largely performed by graduate students, both at master's (Sietse van Netten, Berk Hess, Johan Kruseman) and PhD levels (Marc van den Raadt, Peter van Hengel) and by postdocs (in particular Peter van Hengel). In addition, national (Max Viergever, Rob Diependaal: Delft University of Technology) and international cooperations have been essential (Bastian Epp: Carl von Ossietzky University of Oldenburg).

Within the University of Groningen we cooperated with mathematicians (Hendrik Hoogstraten, Henk Broer) and with audiologists from the ENT-department at the University Medical Center Groningen (Roel Ritsma, Hero Wit, Pim van Dijk).

Disputes and collaborations with the international community have quite effectively been controlled through the international journals as well as through meetings such as the Mechanics of Hearings conference series (started in 1984).

After formal retirement from the University of Groningen faculty, I was in the position to put our developments together in the underlying book format.[1]

The book is intended for use at the graduate or postgraduate level for students with a background in (bio)physics, (electrical) engineering, applied mathematics, or related specializations and multidisciplinary interest. It is somewhat related to Dallos's *The Auditory Periphery* (1973), but much narrower in scope, and updated with respect to otoacoustic emission data and to nonlinear modeling.

The contents is divided in three parts. Part I contains a historical introduction and deals with developments of linear CMs, up to approximately 1980. Part II presents a selection of experimental nonlinear phenomena, and the time domain study of some global nonlinear models. Part III presents results and open issues. Finally, Part IV contains useful general tools and example results.

The introductory Chap. 1 relies heavily on input from Peter van Hengel. His role in the development of the program and applications is also appreciated.

The final form of this book was significantly improved by the reviewers Bastian Epp, Hero Wit, and in particular Michael Rapson.

Groningen Hendrikus Duifhuis

References

van den Brink G, Bilsen FA (eds, 1980) Psychophysical, physiological, and behavioural studies in hearing (ISH-80), Delft University Press, Delft

Dallos P (1973) The Auditory Periphery. Academic, New York

Matthews JW (1980) Mechanical modeling of nonlinear phenomena observed in the peripheral auditory system. PhD thesis, Washington Univ. Sever Institute of Technology, St. Louis

Neely ST (1981) Fourth-order partition dynamics for a two-dimensional model of the cochlea. PhD thesis, Washington Univ. Sever Institute of Technology, St. Louis

[1] Additional material referred to in the text by unspecified URLs is accessible through the following Springer link: http://www.springerextras.com.

Contents

List of Acronyms and Symbols

ABCD	Residual time technique (Brass, Kemp)
AC	Alternating current
AN	auditory nerve, N VIII (eighth nerve)
ANSI	American National Standards Institute
ASA	Acoustical Society of America
ASR	Automatic Speech Recognition
BCE	Before the Common Era (Before Christ)
BM	Basilar Membrane
BPNL	Bandpass Nonlinear
CB	Critical Band(width)
CDT	Cubic Difference Tone[1]
CEOAE	Click Evoked OtoAcoustic Emission
CP	Cochlear Partition
CT	Combination Tone
CTCF	Constant Tone at CF, level and frequency
DE	Differential Equation
DP	Distortion Product
DPOAE	Distortion Product OtoAcoustic Emission
ECFB	Equivalent LTI Cochlear Filterbank
EM	Electron Microscopy
ENT	Ear- Nose- Throat-
ERB	Equivalent Rectangular Bandwidth
EOAE	Evoked OtoAcoustic Emission
(F)FT	(Fast) Fourier Transform
IHC	Inner Hair Cell
ISH	International Symposium on Hearing
ISI	Inter-Spike Interval
ISO	International Organization for Standardization

[1] A misnomer.

LCR	Inductance, Capacitance and Resistance
LG	Liouville, Green
LTI	Linear Time-Invariant (system)
NL	Nonlinear, Nonlinearity
OAE	OtoAcoustic Emission
OC	Corti's Organ
ODE	Ordinary Differential Equation
OHC	Outer Hair Cell
OW	oval window
PSTH	Post Stimulus Time Histogram
PTPV	Primary-Tone Phase Variation
RK4	Runge-Kutta 4 method
RW	round window
SEM	Scanning Electron Microscopy
SFOAE	Stimulus Frequency Evoked OtoAcoustic Emission
SOAE	Spontaneous OtoAcoustic Emission
SPL	Sound Pressure Level
TEM	Transmisson Electron Microscopy
TeM	Tectorial Membrane
TM	Tympanic Membrane
VDP	van der Pol-oscillator
WKB(J)	Wentzel, Kramers, Brillouin (Jeffreys)

List of Symbols

Symbol	Description	Value	Units	Dimensions
0	As subscript index: initial or average value			
δ	Damping coefficient		–	0
ε	NL-parameter		–	0
η	Dynamic viscosity		Pa s	$L^{-1}MT^{-1}$
γ	Ratio of heat capacities (C_p/C_V)		–	0
ν	Kinematic viscosity		m^2/s	L^2T^{-1}
ρ	Fluid density (water, lymph)	1,000	kg/m^3	ML^{-3}
Φ	Heat source		J	L^2MT^{-2}
a	As subscript: acoustic			
a	Acceleration		m/s^2	ML^{-2}
A	Area		m^2	L^2
d_a	Acoustic damping (resistance)		Pa s/m^3	$ML^{-4}T^{-1}$
e	AC voltage		V	$I^{-1}L^2MT^{-3}$
E	Energy		J	L^2MT^{-2}
F	Force		N	LMT^{-2}
$FT\{s(t)\}$	Fourier transform, operator			$[s(t)].T$
$FT^{-1}\{S(\omega)\}$	inverse Fourier transform, operator			$[S(\omega)].T^{-1}$
g	Gravitational acceleration		m/s^2	ML^{-2}
i	AC current		A	I
I	Sound intensity		W/m^2	MT^{-3}
J	Sound-energy flux		W	L^2MT^{-3}
m	Mass, general physical		kg	M
$m(x)$	Partition mass per area	0.5	kg/m^2	ML^{-2}
m_a	Acoustic mass (inductance)		kg/m^4	ML^{-4}
$p(\mathbf{x}, t)$	Sound pressure at **x** and t		Pa	$L^{-1}MT^{-2}$
$\mathbf{q}$	Heat flux density		Pa	$L^{-1}MT^{-2}$
s	Specific entropy		J/kg.K	$L^2T^{-2}\Theta^{-1}$
S	Entropy		J/K	$L^2MT^{-2}\Theta^{-1}$
s_a	Acoustic stiffness (1/capacitance)		Pa $/m^3$	$ML^{-4}T^{-2}$
$s(x)$	Partition stiffness per area		Pa /m	$ML^{-2}T^{-2}$
t	Time		s	T
T	Temperature		K, (°C)	Θ
$u(t), v(t), w(t)$	x-,y-,z-velocity at t		m/s	LT^{-1}
u	Specific internal energy		J/kg	L^2T^{-2}
U	Internal energy		J	L^2MT^{-2}

List of Symbols (continued)

Symbol	Description	Value	Units	Dimensions
$U(\mathbf{x}, t)$	volume velocity **x**−direction at t		m^3/s	L^3T^{-1}
$U_{st}(t)$	volume velocity stapes at t (positive: inward)		m^3/s	L^3T^{-1}
V, V_0	volume		m^3	L^3
$\mathbf{x}$	3-D x-vector		m	L
x, y, z	length, width, height		m	L
Z, Z_e	electrical impedance V/I		Ω	$L^2MT^{-3}I^{-2}$
Z_m	mechanical impedance F/v		N s/m	MT^{-1}
Z_a	acoustic impedance p/U		Pa s/m^3	$L^{-4}MT^{-1}$
Z_{sa}	specific acoustic impedance p/v		Pa s/m	$L^{-2}MT^{-1}$

Part I
Anatomy and Function of the Linear Cochlea

Part I starts with a short introduction into the historical development of today's views of anatomy and function of the cochlea, and its role as the "front end" in auditory processing, or hearing.

It then presents results of some classical measurements of mechanical parameters, and of dynamical cochlear responses to sound. The relevance of these responses is illustrated with examples from classical auditory psychophysics.

The next step concerns the development of cochlear mechanics analysis and modeling during the 20th century. The first steps appeared around 1925, but the field expanded strongly after 1950. The mathematical—biophysical approach remained mostly linear until the late 1970s, even though several nonlinear auditory phenomena were well known.

The initial study of the linear cochlea is relevant because

- It gives a proper introduction to the mathematical and biophysical concepts that are in use in the field
- Current insight in analysis of complex systems is largely based on tools from linear signal analysis
- And finally, major developments started with the analysis of a linear cochlea.

In the mean time we have learned that linear cochleae only describe cases of severe hearing loss, and that the healthy cochlea is strongly nonlinear. Those properties are treated in Part II.

Chapter 1
Historical Introduction

Abstract After brief examples from Early Greek studies and from the Renaissance of auditory research, developments during the nineteenth and the first half of the twentieth centuries are addressed. This period covers the transition from careful description to formal analysis, and provides the foundation of modern cochlear mechanics, linking medical data to (bio)physical and mathematical concepts.

1.1 The Early Greeks

The history of anatomical research and hypotheses about the functioning of the hearing system can be traced back at least as far as ancient Egypt. The Edwin Smith Surgical Papyrus of 3,000–2,500 BCE includes descriptions of battle injuries to temporal bones, and how they affected hearing. It is likely that the Egyptian knowledge of the ear made its way to Greece as early as 1,000 BCE. The earliest Greek texts on the subject, that we know of, are those of Empedocles, who was born $\approx$500 BCE in Sicily.[1]

Unfortunately, only very few of the original texts of Empedocles have been found, not including those on hearing. We only have the references to his works made by others to try and piece together Empedocles' theory on what was making up the ear. The Doxographi Graeci is a compilation of quotations in ancient texts attributed to various Greek philosophers and writers. It contains these fragments quoting the work of Empedocles:

> Hearing takes place by the impact of wind on the cartilage of the ear, which, [he says], is hung up inside the ear so as to swing and be struck after the manner of a bell.

[1]Empedocles described the world and everything in it in terms of the four elements (he actually never used the term elements, but the word roots).

H. Duifhuis, *Cochlear Mechanics: Introduction to a Time Domain Analysis of the Nonlinear Cochlea*, DOI 10.1007/978-1-4419-6117-4_1,

and

> hearing is the result of noise coming from outside. For when (the air) is set in motion by a sound, there is an echo within; for the hearing is as it were a bell echoing within, and the ear [he calls] an *offshoot of flesh* [2] and the air when it is set in motion strikes on something hard and makes an echo.

We can conclude from this that Empedocles' idea about the process of hearing was that the vibrations in the air somehow caused something inside the head to resonate, and this resonance could be felt. Empedocles was a close friend of the Pythagoreans, and is thought to be heavily influenced by their philosophy.[3]

Aristotle is most likely the first of the Greek scientists to recognize the cochlea as the essential part of the hearing system. He gives the following description:

> Of the ear one part is unnamed, the other part is called the "lobe"; and it is entirely composed of gristle and flesh. The ear is constructed internally like the *trumpet-shell*, and the innermost bone is like the ear itself, and into it at the end the sound makes its way, as into the bottom of a jar. This receptacle does not communicate by any passage with the brain, but does so with the palate, and a vein extends from the brain towards it.

In this text the reference to the trumpet shell appears to indicate the cochlea. The description of the passage to the palate must refer to what we now know as the Eustachian tube which ends in the middle ear. It is conceivable that Aristotle did not differentiate the inner and middle ear, but is describing the two as one.

It is possible that Aristotle may have known what the inside of the mastoid (temporal bone), in which both the middle and the inner ear are located, looked like. The idea that the early Greek philosophers had little anatomical knowledge is incorrect, as is shown by many detailed and highly accurate descriptions of other internal organs in their works. However, the inaccessibility and high vulnerability of the inner ear prevented detailed study with the tools that were available at the time. The cochlea contains some of the most fragile and most sensitive tissues which are encased in the hardest bone found in the human body. It is not surprising that further insight into what made up our hearing organ would have to wait for the development of new technologies. For centuries the best description of the hearing

---

[2]The term *offshoot of flesh*, as translated here, but given by Politzer as *κοχλιοειδήςχόνδρος* has given rise to the idea that Empedocles already recognized the inner structure of the inner ear and the importance of this part to the process of hearing. The term *κοχλιοειδής* refers to *κοχλίας* meaning snail with a spiral shell, or to *κοχλιός* meaning screw.

[3]The Pythagoreans were a "cult" set up by the great mathematician Pythagoras. Unfortunately Pythagoras himself did not believe in writing down any of his teachings, so we only have the texts of some of his followers to get some ideas of the concepts of their philosophy. A central theme was their belief that the world was truly mathematical. So not just was it possible to describe the world using mathematics, but maths was the very fabric of nature. It is interesting to note that an important subject to the Pythagoreans was music, and the harmonicity found in it. It is known that Pythagoras studied sound, especially the relationship between the length of a string and its pitch. Alcmaeon, a member of the Pythagorean colony and physician, is known to have dissected temporal bones and came up with the theory that goats could breathe through their ears. This probably implies he had found what is now known as the Eustachian tube, which connects the nasal cavity to the middle ear.

process was that inside the mastoid was a void filled with air.[4] Movement of air outside—sound waves—would set the air inside this void in motion and that would be sensed somehow.

The knowledge of hearing remained at this level until well into the Renaissance period in Western Europe. Around 1,300 the first challenges to the Aristotlean view of the world had started in the fields of mathematics and physics, and were creating room for different views.

1.2 The Renaissance

In the sixteenth and seventeenth century the intricate accessibility of the inner ear no longer stood in the way of progress. The Italian scientists began to dissect bodies and study them meticulously. Andreas Vessalius (1514–1564) describes the malleus and incus, and from the way he describes these we can infer that both were apparently already known. In 1546, Vessalius's student Philippus Ingrassia found the stapes, the smallest bone in the human body (measuring only $\approx$ mm across). Bartholomeus Eustachius[5] was the scientist who first described the cochlea, although he did not name it. This was done later by Gabriel Falloppio.

In appreciating the contributions made by these early researchers of the ear, we have to take into account, e.g., that the microscope was not invented until the early 1600s and there were no developed methods for fixation and staining of tissues. Considering this it is remarkable what knowledge was collected in those days.[6]

In the seventeenth century the rest of Europe started to get involved in cochlear anatomy, with Thomas Willis (1622–1675) positively identifying the cochlea as the organ where sound was converted into a nerve signal. Antonio Valsalva (1666–1723) discovered that the auditory nerve ended /started somewhere in the basilar membrane, which is one of the structures separating the cochlea into channels over its length. He proposed that the nerve endings were like strings of different lengths

---

[4]Aristotle claimed that the inner air was filled with "purified" air and that hearing problems were due to contamination of this air.

[5]It is interesting to note that Eustachius discovered quite a few details about the cochlea, such as the fact that the cochlea is divided into two channels—as we now know there are actually three, and he was the first to describe the tensor tympani, which is part of the mechanism that protects the cochlea from loud sounds. However, he did not discover the tube connecting the middle ear to the nasal cavity which now bears his name. This connecting tube was already known in Greek times, as we have seen in the texts from Aristotle.

[6]An interesting example is the experiment conducted by the anatomist P.F. Meckel. He set out to demonstrate that Dominicus Cotunnius had been correct in claiming the cochlea was filled with water, thereby implying that Aristotle's assumption (filled with clean air) was false. For his experiment Meckel froze an intact cadaver by leaving it outside on a very cold winter night, and the next morning, when he broke open the ear, found it solidly packed with ice [see, e.g., Beyer (1999), Lustig et al. (2003)].

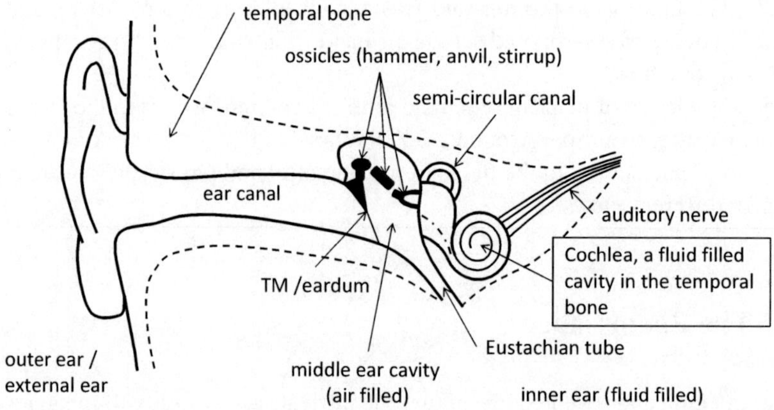

Fig. 1.1 Schematic drawing of the peripheral ear (current view). It is located within the very hard temporal bone. This bone protrudes relatively far into the head. The ear canal and outer ear provide acoustic coupling to the ear drum. The middle ear couples to the cochlea, but it also connects to the environment through Eustachian tube and nasal cavity. This provides the average pressure within the middle ear cavity. The net pressure resulting across the tympanic membrane, the acoustic pressure, is transmitted to the cochlea fluid by the ossicles. The inner ear fluid is practically incompressible; it slides back and force in response to the stimulus, and this requires connections through oval window (stapes footplate) and round window—not indicated. The *dashed–dotted line* indicates the beginning of the cochlear partition. One of the three semicircular canals is indicated schematically as located on the top of the cochlear promontory or inner ear vestibule

stretching across the cochlea (Fig. 1.1). This laid the foundation for the later theories of hearing of Duverney, Willis and later Helmholtz, which will be described in more detail in Chap. 2.

1.2.1 Summarizing the Knowledge at 1800

At the end of the eighteenth century it was known that:

- Perceptible sound is a vibration of air which is channeled to the eardrum through the ear canal.
- The middle ear delivers sound to the cochlea through the chain of middle ear bones: malleus, incus and stapes, the first of which is connected to the ear drum, the last to the cochlea.
- The cochlea is a cavity shaped as a trumpet shell in the temporal bone, the hardest bone in the human body.
- The cochlea is divided over its length into separate channels, called scalae vestibuli and scala tympani; the channels are separated by a structure called BM.
- Within the cochlea, somewhere near the BM, the transmitted vibration is converted into a neural signal which exits the cochlea via the auditory nerve.

INTERMEZZO: This book does not intend to present a scientific study of the history of hearing. Only a few representative cases are presented. Several of excellent treatises on the topic are available, some of which are now accessible in electronic format. The three major categories of study material are (with some subcategories):

1. *books*
   - original studies, from single authors or groups
     Aristotle: De Anima [Eng. transl. by J. A. Smith]
   - historical review books
     Lindsay (1972)[7], Hunt (1978)
   - developments after 1800
     Beyer (1999)
2. *parts or chapters of books*
   Rayleigh (1896)[8], Wever (1949)[9], Lustig et al. (2003)[10], Finger (1994)[11]
3. *(series of-) review articles*
   von Békésy and Rosenblith (1948), Gitter (1990a, 1990b, Hawkins and Schacht: the series of 11 papers "Sketches of Otohistory" published between 2004b and 2006. In 2008 this series was republished as a book, under the same title.

1.3 The Period from 1800 to 1950

During the nineteenth century, scientists generally showed a broad interest, and combinations of activities in arts—in particular music—, physical sciences, mathematics, and medical sciences are frequently encountered.

A real breakthrough in the understanding of sound and hearing was caused by the development of the Fourier-analysis for periodic signals (Fourier 1822). Although presented as a tool for the analysis of temperature patterns, it was quickly picked up by Ohm (1843), who proposed the application to sound as Ohm's law: basically stating that the ear can detect the spectral components of a periodic sound. The idea caught on quickly and convinced most researchers in acoustics and hearing. There

[7]Chaps. 1–18: pre 1800; Chaps. 19–39: 1800–1900.

[8]post 1800.

[9]Chap. 1: pre 1800; Chaps. 2–5: 1800–1950.

[10]Chap. 1: 1800–1900.

[11]Chap. 8, all.

was also opposition: Seebeck (well-known in physics from the Seebeck–effect) proposed that the pitch[12] of a periodic sound depends on the period (Seebeck 1841). This led to a lively discussion in Poggendorfs Annalen.

THE MISSING FUNDAMENTAL Seebeck generated periodic click sounds, for instance with a repetition interval of $T = 5$ ms. The spectrum of such a sound has a 200 Hz fundamental, and many harmonics of a slowly decaying strength (depending on the shape of the click). Seebeck could formally suppress the fundamental, by shifting every other click by ΔT of say 1 ms, so that the successive intervals became $T + \Delta T$ and $T - \Delta T$. He noted that the pitch of the signal did not change strongly (as long as ΔT was small enough), but in a direct comparison with the unaltered series revealed that a weak second harmonic (corresponding to the interval $2T$) became audible. The octave jump to 400 Hz was expected from Ohm's law, but the unaltered $\approx$200 Hz component was not. The experiment, although published, was not taken very seriously and it took a century before Schouten (1938, 1940) carefully repeated the experiments, and confirmed Seebeck's observations. Claims that the fundamental was generated by distortion could later be falsified. It was clearly demonstrated that in case of frequency shifts the perceived pitch did not follow the difference tone frequency, but some average "missing fundamental" frequency (de Boer 1956; Schouten et al. 1962).

Further development of microscopes led to more detailed anatomical results, and the developments in physics, in particular physical acoustics and electricity, led to better insight in the functioning of the auditory system. Major contributions along these lines have been provided by von Békésy. The general increase in scientific interest also led to new approaches in, and a significant expansion of hearing research.

These developments are addressed in more detail in the following sections.

1.3.1 Development of Tools

Besides the mechanical tools that were used for development of surgical tools, as tools for the controlled generation of acoustic stimuli, and high precision measuring tools, also the optical techniques made considerable progress. Acoustics and optics were understood in more detail, and that led to better tools. It is also important

[12]The auditory percept "pitch" denotes that subjective attribute of a sound that can be used to order sounds on a scale from low to high. For tones with only one spectral component, i.e., for simple tones, pitch depends primarily on the frequency of the component (ANSI S1 2005).

to mention the development of the theory and experimental method of electricity explicitly. This enabled the development of electrophysiology.

1.3.1.1 Light Microscopy

The light microscope had become available in the seventeenth century. Originally it was a smart combination of two lenses—later more—, which led to a marked improvement of the optical resolution. Further development of physical optics led to reduction of optical aberrations, and additional improvements have been achieved up to today, with advancement in lens production as well as in control of light sources.

1.3.1.2 Electron Microscopy

Electron microscopy became available in the 1930s. Both the transmission type (TEM) and scanning type (SEM) electron microscopes were developed in Germany during that period by Max Knoll. He moved to the Electrical Engineering Department at the Princeton University in 1948, and returned to Munich (Technische Hochschule) in 1956, from which position he retired in 1966.

The first prototype of a TEM was built by Max Knoll and Ernst Ruska in 1931, and a resolution greater than achievable with light microscopy was obtained in 1933. The first commercial TEM became available in 1939. During the course of this development the same group also introduced the first SEM prototype. In 1935 they obtained an image of silicon steel showing electron channeling contrast. The SEM was further developed by several others, and the first appearance on the market came a number of decades later in 1965.

The techniques found a quick application in auditory research, and the improved resolution helped experimenters to develop new insights in anatomy and function of the auditory system.

In 1986 Ruska was awarded with the Nobel Prize in Physics—with Gerd Binnig and Heinrich Rohrer (for the scanning tunneling microscope)—for the development of the electron microscope and of electron microscopy. He died in 1988.

1.3.2 Developments in Anatomy

As the Italians were the first to start anatomical research, it is only fair that one of the greatest discoveries in cochlear anatomy should be made by one of their countrymen. The marchese Alfonso Giacomo Gaspare Corti was the first to describe the structure on the basilar membrane between scala media and scala tympani. This structure holds the sensory cells which convert the mechanical energy into electrical energy and make the initial neural encoding of sound. This complex of cells now carries his name: the organ of Corti (abbreviated: OC).

It was common for inquiring scientists to travel around between the European universities or courts and to absorb and communicate knowledge. (Of course, budget problems and scientific competition were also known.) Corti did his research in Würzburg (Germany), he learned his craft in Vienna (Austria), picked up additional skills in London and Paris, and was shown how to make his preparations in Utrecht University (Netherlands). Finally, the instrument he used—the microscope—was first developed in England and produced in Germany.

In 1851, just after publication of his famous paper "Recherches sur l'organe de l'ouïe des mammifères," describing his studies on over 200 cochleae (from cattle, swine, sheep, cats, dogs, rabbits, moles, mice and man), Corti received word of his father's death. This forced him to return to Italy to divide his father's legacy and, in spite of many promises to do so, he never returned to the study of the ear, leaving it to Otto Friedrich Deiters, Arthur Böttcher, Friedrich Matthias Claudius, Viktor Hensen, and Magnus Gustav Retzius to identify and lend their names to the various cell types in the organ of Corti [see, e.g., Wever (1949), Hawkins (2004a), for more details on the life and works of Corti].

These investigators not only described the various cell types, they also performed countless measurements quantifying the dimensions of various (sub)structures. It was Hensen who wrote that the increase in BM width from the base to the apex would probably be the cause of the frequency separation of tones in the cochlea, following the initial idea of strings resonating posed by Valsalva. He also found that the auditory nerve runs up to the hair cells, identifying them as the most likely candidates for the conversion of the mechanical energy (movement of the cochlear fluid) into electrical energy (the nerve signal). New techniques such as methylene blue and silver staining exposed the innervations of the hair cells in greater detail. Held found that one nerve fiber connects to multiple hair cells, and that one hair cell is innervated by many nerve cells. Kolmer was the first to suggest (in 1926) that the cochlea was not only innervated by afferent nerve fibers, leading the information of sound up to the brain, but also by efferent fibers leading some sort of information from higher brain regions back into the cochlea. This was confirmed by Rasmussen who described the pathway of the olivocochlear bundle.

In 1831 Ernst Reissner found the membrane—now carrying his name—separating the upper channel into two channels (*scala vestibuli* and *scala media*). Guilbert Breschet (1784–1845) showed that scala media is filled with a different fluid than the other two scalae: endolymph as opposed to perilymph, the difference being the ionic content. He also found the helicotrema connecting scala vestibuli and *scala tympani* at the far end—the *apex*—of the cochlea.

1.3.3 The Emergence of Auditory Neurophysiology

At the turn of the century the development of electric and electronic equipment began to meet the requirements necessary for the measurement of electric responses of the nerve.

The first published paper on auditory nerve responses to sound (Buytendijk 1911) in guinea pig and rabbit may have been missed originally by most colleagues. But the publications by Forbes et al. (1927) in the American J. of Physiology, and a little later by Wever and Bray (1930b) e.g., in Science and the PNAS, about two decades later evoked a lively discussion. For example, Hallpike and Rawdon-Smith (1934)'s paper started as follows:

> In 1930, as is now well known, Wever and Bray [1930a,b] first described certain electrical phenomena occurring in the trunk of the acoustic nerve during the reception of auditory stimuli. This effect, which has since come to be known as the Wever and Bray phenomenon, consists essentially of potentials alternating at a frequency accurately corresponding to that of the stimulating sound. Thus, with suitable amplification, speech is clearly reproduced and the voice of the speaker recognized. After excluding certain artifacts, Wever and Bray satisfied themselves that the effect was a vital one, in that it was dependent upon the functional integrity of the cochlea; and further, concluded that it was attributable to true action potentials produced in the fibers of the auditory nerve.

The first electrodes were not suited for the recording of responses from single nerve fibers, they picked up the combined electrical response from the nerve, with a tip distance of a few mm. Whereas Buytendijk used a sensitive string-galvanometer (sensitive down to $\approx 10\,\text{nA}$) and a photo camera, Wever and Bray had better amplifiers and could use oscilloscopes and therefore had disposition of more straightforward means of measuring currents and potentials.[13] Initially there was doubt that the measured potentials and currents represented nerve responses. Derbyshire and Davis, e.g., first hypothesized that the responses showed recordings which originated directly from the sensory hair cells and disputed the nerve source interpretation (Derbyshire and Davis 1935).

Measurement techniques progressed with improved electronics, and the beginning of the rise of computing power. In the 1940s Galambos and Davis (1943, 1944) published results from single auditory nerve fibers of the cat, using micropipette electrodes. They reported response characteristics such as response areas and inhibitory areas that set a standard for the decades to come. They also report the adaptation of spike rate for a continuous stimulus, reaching a stable response in "a few tenths of a second." The summary of their 1943 paper ends with:

> At constant frequency an increase in sound intensity causes an increase in rate of discharge by the single fiber (Fig. 8). Most fibers reach a maximum of 450 discharges per second after an intensity increase of about 30 db.
>
> The frequency band capable of exciting a given fiber increases markedly as the intensity level is raised (Fig. 5). At levels about 100 db above threshold tones as far away as 3 octaves below and $\frac{1}{2}$ octave above may be adequate.
>
> The auditory-nerve fiber discharges in synchronism with a definite part of the stimulating sound-wave cycle (Fig. 12).

[13] At this time the occurrence of aural combination tones (CTs) was also well known, but the origin was not so clear (Wever and Bray 1938; Wever et al. 1940a). These studies (including Wever et al. 1940b) also used whole nerve electrical responses. The conclusion was that overtones are generated "at a point beyond the stapes" and that CT generation may have a middle ear component, but that the middle ear is not the sole seat of CT distortion, as proposed by Helmholtz.

> The results are held to support a place theory of hearing according to which pitch is a function of where, and loudness a function of how much of, the BM is disturbed.

conclusions which appear to be quite up-to-date.[14]

Several of the studies mentioned above discuss the relevance of the findings for the theory of hearing. In particular, the observation that temporal structures are retained in neural responses up to a few kHz, has been proposed to be supportive of the place theory for pitch perception. On the other hand, neural synchronization also led to the Volley theory, with the interpretation that neural time coding would allow central pitch recognition without a precise frequency map.

1.3.4 Studies of Auditory Function

Over the centuries, the link of anatomical data to auditory perception has led several times to the hypothesis of a cochlear frequency-to-place map, although an alternative hypothesis, viz. that sensory hair cells basically receive the complete sound signal and the analysis occurs in the brain, persisted for a rather long time (cf. Wever 1949). During the transition from the nineteenth to the twentieth century the development on the European mainland was probably dominated by Hermann Von Helmholtz. He was a German ENT specialist and a physicist, and author of the influential book: *Die Lehre von den Tonempfindungen*, first published in 1863, which was translated to: *On the Sensation of Tones* by A.J. Ellis. Helmholtz supported the frequency-to-place map, and related the mechanism of the map to the systematical change of the width of the basilar membrane.

Two years earlier he had described the function of the middle ear as a transformer between the air outside the head and the fluid-filled cochlea. The difference in size of the ear drum and the stapes footplate, the curvature of the ear drum and the lever action of the ossicles, all combined to transform the sound waves in air (small pressure variations, large volume displacements) into sound waves in fluid (large pressure variations, small volume displacements).

Helmholtz based his description of the function of the cochlea largely on the observations of Hensen, who had shown that the width of the basilar membrane varied along the length of the cochlea. Combining this result with Ohm's description of sound as a combination of tones of different frequencies, led Helmholtz to the view that the cochlea performed a sort of Fourier analysis of the incoming sound waves. The physical process by which this Fourier analysis operated was the resonance of BM fibers (across the membrane width) with different resonance frequencies (varying with the lengths of the fibers). A sound wave would set all the fibers of the BM in motion, but the ones which had resonance frequencies closest to the frequencies present in the sound signal would respond maximally. This would create a pattern over the length of the cochlea describing the frequency content of the signal: the frequency map, or place coding.

---

[14]The dB had not yet made it to the standard abbreviation.

In short, the theory of Helmholtz states that the BM fibers can be regarded as strings, where the tension and the mass of the string cause it to resonate at a certain frequency (see Sect. 8.5.1). Since the frequency of resonance of a string is related to the string's length, Helmholtz concluded that the variation of the BM width—actually the variation in the length of the transverse BM fibers—would lead to a systematic variation in the resonance frequency of the fibers. A quantitative explanation and prediction of these properties is necessary to evaluate the applicability to human (and in general: mammalian) hearing. What are the string properties and how much will it vibrate for frequencies close to and further away from the resonance frequency?

We start with simpler case: a mass attached to a spring. The string can be "translated" to that problem. Assume a structure in which a certain mass m (kg) is connected to a spring with a certain stiffness s (kg/s^2). The equation of motion for such a system driven by a sinusoidal force $F(t) = F_0 \cos(\omega t)$ is:

$$m \frac{\partial^2 x}{\partial t^2} + d \frac{\partial x}{\partial t} + sx = F_0 \cos(\omega t) \tag{1.1}$$

where d (kg/m) is a damping term, damping the oscillations, and $x(t)$ the excitation or response to the driving force $F(t)$. It has a driving amplitude F_0 and (angular) frequency ω respectively. The solution to this equation is a motion of the form:

$$x(t) = \frac{F_0}{|Z_m|\omega} \sin(\omega t - \phi) \tag{1.2}$$

where

$$Z_m = d + \mathrm{i}\left[\omega m - \frac{s}{\omega}\right] \tag{1.3}$$

and

$$\phi = \arctan\left[\frac{1}{d}\left(\omega m - \frac{s}{\omega}\right)\right] \tag{1.4}$$

The function Z_m is called the mechanical impedance of the resonator (see Sect. 8.3). The behavior of the resonator can be derived by considering this function for different values of ω.

It is clear that Z_m is a complex function and that its size $|Z_m|$ is minimal when the imaginary part $\omega m - s/\omega$ vanishes. This happens for the (angular) frequency $\omega_0 = \sqrt{s/m}$. This frequency is called the resonance frequency of the resonator. At this frequency, the impedance Z_m equals the damping d, and the imaginary part is 0. For a lossless system ($d = 0$), the amplitude of the motion $x(t)$ would be infinity at resonance. For frequencies other than ω_0, the amplitude behaves as $1/[1 - (\omega/\omega_0)^2]$.

For $d > 0$ the amplitude of $x(t)$ depends on the frequency ω, and shows a maximum near the resonance frequency. When d is small, this maximum occurs close to ω_0. For increasing d the maximum moves toward smaller values of ω. For $\omega = 0$ the impedance Z_m is undefined, but the product $|Z_m|\,\omega$, which occurs in the denominator of (1.2) remains defined and the response amplitude approaches F_0/s. In order to determine whether a value of d is small it needs to be compared to some reference. It seems natural to compare the value of d to those of m and s and the combination $\sqrt{ms}$ has the correct dimension kg/s. The *damping factor* $\delta = d/\sqrt{ms}$ is used to indicate whether d is relatively small or not, and therefore describes the amplitude curve. A related, and widely used, way to describe the relative size of the damping is by giving the so-called quality factor of the resonator:

$$Q_{3\mathrm{dB}} = \frac{\omega_0}{\Delta\omega} = \frac{\sqrt{ms}}{d} = \frac{1}{\delta} \tag{1.5}$$

In this formula, $\Delta\omega$ is the width of the amplitude curve. This width is measured where the amplitude of the motion is $1/\sqrt{2}$ times the peak value of the amplitude curve. This definition is most commonly used for the quality factor Q. Because a factor $1/\sqrt{2}$ corresponds fairly closely to 3 dB, this value is also referred to as $Q_{3\mathrm{dB}}$. In some cases, the width is measured at $1/\sqrt{10}$ times the peak amplitude. In this case, the resulting quality factor is denoted by $Q_{10\mathrm{dB}}$.

The frequency-to-place map is often characterized as a *Place Theory*. The specific interpretation by Helmholtz is one of the subset of *Resonance Theories*, which contained *tuned elements*, *membrane resonance*, or *tube resonance*. There was also a number of *Nonresonance or Wave Theories* in the Place Theory category. The opposing category—at the time—was the group of *Frequency Theories* either with a nonanalytic (telephone) or analytic further specification. Or schematically:

Frequency-to-Place Map Theories around 1950

- Resonance Theories
  - tuned elements
  - membrane resonance
  - tube resonance
- Frequency Theories
  - analytic
  - nonanalytic (e.g., telephone theory)

An excellent overview of these theories has been presented by Wever (1949).

The alternatives to the Place Theory generally assumed traveling or standing waves on the BM or even on Reissner's membrane. In the standing wave theory, the number of loops and/or nodes that would form on the BM is related to the frequency of the signal and the total pattern of vibration of the organ of Corti would be interpreted at some higher level in the brain. The Fourier analysis would thus not take place in the cochlea but at some higher level. The hair cells would only serve to encode the temporal signal (time coding).

Experimental support for the frequency mapping essential to the Place Theory came from the experiments of Georg von Békésy, in the late 1920s. After receiving his doctorate in physics von Békésy went to work for the Hungarian Telephone Company and the University of Budapest. His first study concerned the effect of nonlinear properties of iron in telephone receivers on the quality of reception. From there, he turned to the study of the mechanics of the inner ear and developed ingenious methods to observe the motion of the real basilar membrane, as well as a sophisticated mechanical model of the cochlea which he used both for demonstrations and for study.

Commenting on his first view of the organ of Corti, von Békésy wrote:

> I found the inner ear so beautiful under a stereoscopic microscope that I decided I would just stay with that problem. It was the beauty and the pleasure of beauty that made me stick to the ear.

His work on checking the telephone line quality for the Hungarian Telephone company led him to look for the weakest part—in terms of reception quality—in the system. After concluding it was the receiver, the next question was how much it needed to be improved. After all, improving it beyond the perceptual limits of human hearing was futile. This led von Békésy to study the properties of the ear drum and the middle ear and finally the inner ear.

Von Békésy decided to start with the study of a scale model of the human inner ear, which he had to construct first. Unlike earlier attempts, von Békésy was very careful to properly scale the model, which means that he performed the necessary scaling of all parameters, such as the density of the fluid. In his models he found a behavior that he could not adjust to either of the two theories (von Helmholtz's bank of filters, and the standing wave model), but could not find any mistake in his model. Thereupon, he decided to attempt something which had never been done before: actually measure the motion of the BM in response to sound in an (almost) intact cochlea. In his 1942 paper,[15] he describes carefully shaving off the material of the temporal bone of cadavers, and visualizing the basilar membrane with a microscope. By using very high sound pressure levels (over 120 dB SPL) he was able to elicit a motion of the membrane large enough to be measurable. The motion patterns he described on the one hand supported the place theory of Helmholtz, while simultaneously contradicting the resonator filterbank model.

[15] Von Békésy's papers—English translations—are available in *Experiments in Hearing*, published in 1960.

Von Békésy observed and described a pattern of wave motion that started off with a very small amplitude at the stapes. While the amplitude increased as the wave moved along the length of the basilar membrane, its traveling speed decreased, until the wave almost came to a halt, showed a maximum amplitude and then quickly decreased in size and vanished before reaching the helicotrema.[16] The location where the wave reached its maximum clearly depended on the stimulus frequency. High frequencies resulted in maxima near the stapes, low frequencies had their maxima nearer the helicotrema, with a monotonic mapping of frequency to place over the cochlear length. This mapping seemed to follow a roughly logarithmic trend, in accordance with the logarithmic frequency scale, or octave scale known from music. Such a mapping was in full support of the place coding of frequency following from Helmholtz's resonator filterbank theory. However, the wave nature of the motion pattern was in clear contradiction with this. The resonators in Helmholtz's filterbank react to the sound pressure stimulus independently, and a single oscillator can at most be 180° out of phase with the stimulus.[17] The wave behavior described by von Békésy implies that moving away from the stapes the phase of the BM motion shows an increasing phase lag relative to the stimulus. This phase lag accumulated to at least 1,080° as three full cycles of the wave could be seen on the BM.

The term "traveling wave" was proposed by von Békésy himself to describe he observations, and it has been a cause of controversy ever since. Right at the publication of the English version of the manuscript objections were made against the use of the term, as it suggested that a wave was traveling along the BM. For instance, the experiments of Held and Kleinknecht had shown that lesions in the basilar membrane only destroyed hearing at the frequencies corresponding to the lesions, implying that the BM was in motion beyond the lesion. This would be impossible if the wave traveled along the membrane (Held and Kleinknecht 1927). Instead, it was argued, the motion of the basilar membrane actually was more like a surface wave, reflecting motion of the fluid in the cochlear channels. Von Békésy argued that he did not specify in which medium the traveling was done, and that he used the term to describe the appearance of the motion pattern. So the name persisted, as did the controversy. In 1954 Wever, Lawrence and von Békésy jointly agreed that the patterns of motion described by von Békésy

> can be referred to as ... a traveling wave, provided that ... nothing is implied about the underlying causes (i.e. how any given segment of the BM gets the energy that makes it vibrate)

Presently Wever and Lawrence's note seems to be forgotten.

---

[16] Actually, von Békésy could not visualize the entire cochlear length and he filled in the patterns near the stapes and near the helicotrema based on the motion he observed in his models.

[17] Over the frequency range from $\omega = 0$ to $\omega \rightarrow \infty$ the phase in the response of s single oscillator drops by 180° (or π). The phase range depends on the precise input and output properties. In the given example, at $\omega = 0$ the phase starts with a 90° lead, reaching 0 at $\omega = \omega_0$ and a 90° lag if $\omega \rightarrow \infty$.

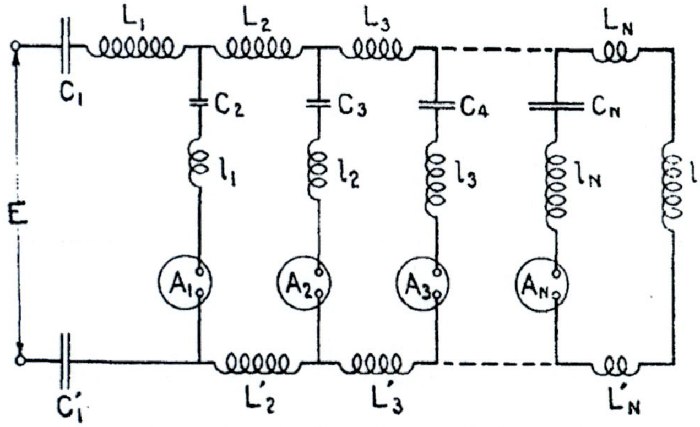

FIG. 7b. Electrical analogue of the cochlea.

Fig. 1.2 Wegel and Lane's *Fig. 7b* (1924), the first transmission line model of the cochlea. C_1 and C_1' represent elasticity at OW and RW; L_i and L_i' represent fluid mass coupling in scala vestibuli and tympani; l_i and C_i mass and elasticity parameters of the BM and coupled structures, and the ammeters A_i represent the connection to nerve terminals. Reprinted figure with permission from RL Wegel and CE Lane, Physical Review 23, p.266, 1924. Copyright 1924 by the American Physical Society

In 1962 von Békésy was awarded the Nobel Prize in physiology and medicine for his work.

Another important development that had considerable impact on the study of hearing—as well as on the study of other sensory modalities—was the quantification of experimental psychology through the introduction of psychophysical methods. The definition of subjective quantities along with the development of subjective scales opened the possibility to quantify subjective measures as pitch and loudness and relate those to the physical parameters frequency and amplitude or sound pressure level (e.g. Stevens and Davis 1936; Stevens et al. 1937; Stevens 1951). The methods developed by this group, at the time centered at Harvard, provided a solid basis for reproducible measurements of absolute and differential thresholds.

During the same period (the 1920s), the interest in acoustics and hearing increased in industry (telephone companies), as well as in the academic institutions: academic hospitals, schools of science, and schools of engineering. It also led to the establishment of new professional societies, where people from these different backgrounds met. The Acoustical Society of America (ASA) was founded in 1929, and the first volume of its journal appeared in the same year.

One of the key contributors to hearing was Harvey Fletcher (1884–1981). After an academic education in physics, and a brief postdoctoral period in his home state Utah, he took a position at the Western Electric Company in New York. From this laboratory Wegel and Lane (1924) had just presented their tone-on-tone masking results,[18] and "its probable relation to the dynamics of the inner ear." Figure 7b from their paper (reprinted in Fig. 1.2) can be considered *the* precursor of the later

[18]These tone-on-tone masking data are discussed in Chap. 4, Sect. 4.2.1, Fig. 4.3.

transmission line models. In that environment, Fletcher developed a now famous audiometer and was also involved in the development of the early electronic hearing aids. In the mid-1920s, when the Bell Telephone Laboratories where founded, he was appointed director of auditory research (1928), and by 1935 he was promoted to director of the Physical Research. His contributions to speech and hearing, many of which came though the ASA, have been summarized and reviewed by Jont B. Allen (1996).

Laboratories of the larger telephone—and electronics companies in Europe, such as Philips and Siemens—joined the active interest in hearing research, as reflected in contributions by Schouten (Philips) in his studies on the "Residue" (e.g., 1938, 1940), and by Trendelenburg (Siemens) in his studies on sound perception, sound analysis, and sound transmission (1935).

References

Allen JB (1996) Harvey Fletcher's role in the creation of communication acoustics. J Acoust Soc Am 99:1825–1839

ANSI S1 (2005) Acoustical Terminology. American National Standards Institute, Inc., including 2005 updates

Aristotle (∼350 BCE) Sense Perception (DA II.5), in: Greek: $\Pi\varepsilon\rho\iota\ \Psi\upsilon\chi\tilde{\eta}\zeta$, or Latin: De Anima. English transl. by J. A. Smith, publ. Oxford, 1928. Ebook: e.g., MIT the Internet Classics Archive

von Békésy G (1960) Experiments in Hearing. McGraw-Hill, New York

von Békésy G, Rosenblith WA (1948) The early history of hearing—Observations and theories. J Acoust Soc Am 20:727–748

Beyer RT (1999) Sounds of our times: two hundred years of acoustics. Springer-Verlag, New York

de Boer E (1956) On the 'residue' in hearing. PhD thesis, Univ. of Amsterdam, Netherlands

Buytendijk FJJ (1911) On the negative variation of the nervus acousticus caused by a sound. Proceedings KNAW 13(II):649–652

Derbyshire AJ, Davis H (1935) The action potential of the auditory nerve. Unknown Journal 113:476–504

Finger S (1994) Origins of Neuroscience. Oxford University Press, Oxford, ch.8 The Ear and Theories of Hearing. 9 Audition and the Central Nervous System

Forbes A, Miller RH, O'Connor J (1927) Electric responces to acoustic stimuli in the decerebrate animal. Amer J Physiol 80:363–380

Fourier JBJ (1822) Théorie Analytique de la Chaleur. Firmin Didot, Paris

Galambos R, Davis H (1943) The response of single auditory-nerve fibers to acoustic stimulation. J Neurophysiol 6:39–57

Galambos R, Davis H (1944) Inhibition of activity in single auditory nerve fibers by acoustic stimulation. J Neurophysiol 7:287–303

Gitter AH (1990a) Eine kurze Geschichte der Hörforschung - Teil 1: Antike. Laryngorhinootologie 69:442–445

Gitter AH (1990b) Eine kurze Geschichte der Hörforschung - Teil 2: Renaissance. Laryngorhinootologie 69:495–500

Hallpike CS, Rawdon-Smith AF (1934) The "Wever and Bray Phenomenon." A study of the electrical response in the cochlea with especial reference to its origin. J Physiol 81:395–408

Hawkins JE (2004b) Sketches of otohistory. Part 1: Otiprehisotry: How It All Began. Audiol Neurootol 9:66–71

Hawkins JE (2004a) Sketches of otohistory. Part 3: Alfonso Corti. Audiol Neurootol 9:259–264
Hawkins JE, Schacht J (2008) Sketches of Otohistory. Karger, Basel
Held H, Kleinknecht F (1927) Die lokale Entspannung der Basilarmembran und ihre Hörücken. Pflüg Arch ges Physiol 216:1–31
von Helmholtz HLF (1863) Die Lehre von den Tonempfindungen, 1st edn. Vieweg und Sohn, Braunschweig, english edition: On the Sensations of Tone, transl. by A.J. Ellis (1885) of 4th German edition (1877), publ. by Dover in 1954
Hunt FV (1978) Origins in Acoustics. Yale University, 2nd edition: 1992 by the Acoustical Society of America
Lindsay RB (ed) (1972) Acoustics: Historical and Philosophical Development. Benchmark Papers on Acoustics, Dowden, Hutchinson and Ross, Stroudsburg, PA
Lustig LR, Niparko JK, Minor LB, Zee DS (2003) Clinical neurotology: diagnosing and managing disorders of hearing, balance and the facial nerve. Martin Dunitz Ltd, London, (collection of review papers)
Ohm GS (1843) Ueber die Definition des Tones, nebst daran geknüpfter Theorie der Sirene und ähnlicher tonbildender Vorrichtungen. Poggendorff's Annal Phys Chem 59:513–565
Politzer A (1907) Geschichte der Ohrenheilkunde. I. Band. F. Enke, Stuttgart
Politzer A (1913) Geschichte der Ohrenheilkunde. II. Band. F. Enke, Stuttgart
Rayleigh JWS (1896) Facts and Theories of Audition, vol II, 2nd edn, MACMILLAN, London, chap XXIII. [Dover edition: 1945]
Schacht J, Hawkins JE (2006) Sketches of otohistory. Part 11: Ototoxicity: drug-induces hearing loss. Audiol Neurotol 11:1–6
Schouten JF (1938) The perception of subjective tones. Proc Koninklijke Nederlandse Akademie van Wetenschappen 41:1086–1093
Schouten JF (1940) The residue, a new concept in subjective sound analysis. Proc Koninklijke Nederlandse Akademie van Wetenschappen 43:356–365
Schouten JF, Ritsma RJ, Cardozo BL (1962) Pitch of the residue. J Acoust Soc Am 34:1418–1424
Seebeck A (1841) Beobachtungen über einige Bedingungen der Entstehung von Tönen. Ann Phys Chem 53:417–436
Stevens SS, Volkmann J, Newman ER (1937) A scale for the measurement of the psychological magnitude pitch. J Acoust Soc Am pp 185–190
Stevens SS (ed) (1951) Handbook of Experimental Psychology. Wiley, New York
Stevens SS, Davis H (1936) Psychophysiological acoustics: Pitch and loudness. J Acoust Soc Am 8:1–13
Trendelenburg F (1935) Klänge und Geräusche. Springer, Berlin
Wegel RL, Lane CE (1924) The auditory masking of one pure tone by another and its probable relation to the dynamics of the inner ear. Phys Rev 23:266–285
Wever EG (1949) Theory of Hearing. Wiley, [Dover edition: 1970, isbn: 486623998]
Wever EG, Bray CW (1930a) Action currents in the auditory nerve in response to acoustical stimulation. P N A S 16:344–350
Wever EG, Bray CW (1930b) Auditory nerve impulses. Science 71(1834):215, [Feb. 21]
Wever EG, Bray CW (1930c) Present possibilities for auditory theory. Psych Rev 37:365–380
Wever EG, Bray CW (1938) Distortion in the ear as shown by the electrical responses in the cochlea. J Acoust Soc Am 9:227–233
Wever EG, Bray CW, Lawrence M (1940a) The origin of combination tones. J Exp Psychol 27(3):217–226
Wever EG, Bray CW, Lawrence M (1940b) A quantitative study of combination tones. J Exp Psychol 27(5):469–496
Wever EG, Lawrence M, von Békésy G (1954) A note on recent developments in auditory theory. P N A S 40:508–512

Chapter 2
Developments from 1950 to 1980

Abstract This chapter presents some details of technical and experimental progress that was made during the period 1950–1980

2.1 Developments of Techniques and Tools for Vibration Measurements

2.1.1 Electronics

The advancement of electronic equipment in the 1950–1980 period has been overwhelming. Initially the improvement focused primarily on analogue equipment. Signal generators as well as meters and analyzers were developed to more demanding specifications, aiming at better controllable and reproducible experiments. For instance, a high quality oscillator which tended to be somewhat sensitive to temperature variation—and therefore providing a slightly shifting frequency during warm-up—produced very little distortion or background noise. In contrast, the first digital signal generator produced extremely stable frequencies, but it took decades of additional developments to achieve the same precision in amplitude and signal purity as had been available in the analogue equipment.

Analogue set-ups could already be automated using electronic control techniques and analogue computing equipment. However, analogue control techniques began to lose impact in the late 1960s and 1970s, when digital control started to take over.

High quality equipment became available in laboratories for biophysical, electrophysiological, and psychophysical experiments, often with specialized fine-tuning for specific tasks. The benefit of these developments was not limited to vibration measurement. It was essential for all branches of hearing research. This also applies to laboratory computers, which appeared around 1970, and rapidly became indispensable, slightly after the step from slide rules to calculators.

H. Duifhuis, *Cochlear Mechanics: Introduction to a Time Domain Analysis of the Nonlinear Cochlea*, DOI 10.1007/978-1-4419-6117-4_2,

Miniaturization of electronic equipment started gradually with the invention and application of transistors. These replaced vacuum tubes almost completely over a short period, and further miniaturization of chips—including electrode arrays—has not yet stopped.

2.1.2 The Mössbauer Technique

Innovative as the experiments of von Békésy were, the excitation patterns he had obtained were far too wide to be related to the frequency tuning observed in psychophysics. In order to make the motion of the BM visible, von Békésy had to apply a driving force to the stapes which would correspond to sound levels of up to 150 dB SPL, well above the pain threshold at approximately 120 dB. It was therefore conceivable that his measurements destroyed part of the mechanical processes which he attempted to measure. Moreover, the measurements were performed on cadaver cochleae. If the mechanics of the intact cochlea would depend on, e.g., blood supply this also would have been affected in his measurements. Extrapolating his excitation profiles to lower levels and ultimately to the threshold of hearing convinced von Békésy and others that something was missing, because the results indicated membrane motions less than 1 pm at the threshold of hearing (at $\approx$20 μPa).

For the explanation of the problem of frequency selectivity, von Békésy started investigating the neural mechanisms of lateral inhibition, known from neurons in the visual system. A different solution, which was proposed by Gold in 1948, might solve both the problem of the frequency selectivity and the higher motion amplitudes needed at the hearing threshold. With a background in electrical acoustics, Gold proposed an active amplification mechanism which would counteract the damping at low levels. He even suggested that such an amplification mechanism might suffer from slight mistuning which would lead to a feedback loop producing sound spontaneously [see, however, footnote 10 in Sect. 3.4 and comments in Sect. 7.3].

In 1967 another new technology was applied to hearing research by Johnstone and Boyle: the Mössbauer technique. With this technique, data could be obtained at significantly lower sound levels, viz. within the normal range of hearing–although, at its upper end.

In 1957 Mössbauer discovered resonant and recoil-free emission and absorption of gamma rays in solids. This causes very narrow line widths in the generated gamma spectrum and allows the measurement of a Doppler effect when the source and detector move relative to one another. A small source emitting narrow band gamma radiation is placed on the cochlear partition and a detector is tuned to the radiated energy when the source is at rest. With this configuration, the rate of detected photons becomes a function of the velocity of the source relative to the detector. Unfortunately, this function shows strong compressive nonlinear behavior. This limits undistorted measurements of velocities to a limited dynamic range.

Applying the technique to the cochlea is not straightforward. The gamma ray sources have to be prepared and carefully placed in an intact, living cochlea, and the measurement system must be accurately calibrated. But it allows much smaller motions to be measured than can be achieved by light microscopy. Rhode perfected the technique and in 1971 he was able to show measurements from cochleae of 20 squirrel monkeys at stimulus levels from 70 dB to 100 dB SPL. These measurements were the first to show that the cochlea clearly behaved nonlinearly at these, still relatively high, levels.

From 1971 on, several researchers started measuring the vibrations of the cochlear partition using the Mössbauer technique, showing the motion of the cochlear partition to be a nonlinear function of stimulus level. At first, attempts to reproduce the measured nonlinearity, even by Rhode himself, failed to reproduce the data. In hindsight, this was probably due to the vulnerability of the nonlinearity, which, as we know now, rapidly decreases after death of the animal or any damage to the structure of the cochlea.

2.1.3 Optical Techniques: Application of Lasers

The possibility to use (phase-) coherent light sources allowed a big step in the improvement of optical resolution, both for still and moving pictures. The spatial resolution was no longer limited to the order of magnitude of the wavelength of the source, but became dependent on the accuracy with which the phase of the light wave could be controlled and measured.

The principle of interference methods was recognized at the end of the 1950s, and patented by Minsky in 1961, but application of laser interferometry and confocal laser scanning techniques in biophysical experiments had to wait until the 1990s.

In laser interferometry, a laser bundle is split into two coherent beams. One beam—the reference beam—is sent to a detector as directly as possible. The other is focused on a moving reflector, the target, and the reflected light is sent to the detector. By studying the pattern created by the interference of the reference beam and the reflected beam, the velocity of the target can be deduced. Applications to cochlear and lateral line hair cell research were developed adjacently by Khanna (1986); Khanna et al. (1986) and van Netten (1988).

2.1.4 Pressure Measurement

Two lines of development of pressure measurements are important for cochlear mechanics (CMs). The first involves the outer and middle ear, the second the cochlea.

Middle ear pressure measurements are measurements of acoustic pressure (in air) within the middle ear cavity. The employed microphones require proper coupling to the middle ear structure. Basically, they are specializations of normal audio-equipment (e.g., Puria et al. 1997). The most accurate microphones tend to be of the condensor type. They have an impressive dynamical ranges, and a rather wide spectral sensitivity. Both the low-level boundary and the frequency characteristic improve with increasing microphone size. However, over the period 1950–1980 the common diameter of high-quality microphones reduced from 1 to $\frac{1}{4}''$ without significant reduction of quality.

The intracochlear pressure is a basic experimental variable in cochlear mechanics, but experimental accessibility is demanding. Results of reliable pressure measurements within the (cat) cochlear were first presented by Nedzelnitsky (1974, 1980). The measurements were difficult. Within the scala vestibuli (SV), data could be obtained over a 40–105 dB SPL range (measured at the tympanic membrane); for the scala tympani (ST), this range reduced to 75–105 dB SPL. The experimental difficulty is probably one of the reasons that the method did not become very popular, although it reappears from time to time, with slightly improved techniques. Dancer and Franke presented guinea pig results in 1980. Like Nedzelnitsky, they used a fluid filled thin probe connected to a sensitive sensor, in this case a piezoresistive transducer. The outer tip diameter was 0.25–0.35 mm.

Pressure measurements are important components of reliable power flow measurements, which require the independent measurement of, e.g., local pressure and local volume velocity. The majority of techniques in current use focuses successfully on velocity measurements, and associated pressure data are usually lacking. They might be estimated on the basis of a local impedance, but that assumes a sufficiently valid linear analysis and associated data. The independent measurement is more reliable. However, it also runs into the problem of measurements at the edge of what is fundamentally possible, viz. the fundamental uncertainty principle. In the present case the product $\Delta p\,\Delta U$ is bounded (p = local pressure, U = local volume velocity). But there is also uncertainty along the time dimension: energy and power are not instantaneous quantities, but they imply averaging across some time window. The points are addressed in more detail in Sect. 5.1.3.1.

2.1.5 Development of Computing Power

Around 1950 the slide rule and computation tables were still very much in use for the analysis of numerical problems. These provided the only computation option for cases where the analytic approach or approximation was not feasible. Computer centers started to develop, both in industrial and academic environments. Initially, however, these were primarily tied to (applied) mathematics departments. It took more than a decade before lab computers became available and another before they became powerful enough to gradually become indispensable. Among the first minicomputers used for data processing are the TX-0 and the LINC at MIT (e.g., Kiang et al. (1965), Sachs and Kiang (1968), and Goblick and Pfeiffer (1969)), predecessors of the Digital Equipment Corporation (DEC) PDP-line. The PDP-1 was produced in 1963, and in 1965 the PDP-8 systems became one of the first successful minicomputers. Around 1970 many PDP-4 and PDP-8 systems were used in many auditory research laboratories. There was some competition from systems from Honeywell (DDP), and Data General (Nova, Eclipse). Personally I have been interested in this development using minicomputers since the early 1970s for simple modeling (DDP, DEC, and Data General).

To conclude the period covered in this chapter: around 1980, the use of minicomputers in auditory research labs had become common. They were employed for experiment control, data processing, and simple modeling. It took another 15 years before these were replaced by still rather expensive workstations, and then by the gradually more powerful, and economical, modern laptops and workstations.

At the same time, another line of computing power virtually disappeared. Based on the properties of op-amps (operational amplifiers), economical analog computers were developed and applied in signal analysis and control theory. The advantage of these setups is that they operate using continuous time (no time sampling). However, the amplitude range has natural upper and lower bounds, determined by power supply and internal noise, respectively. Second order differential equations, as used throughout this book, can be represented by a three op-amp circuit. (The interested student should be able to find the relevant literature in electrical engineering textbooks as referenced in the 2005 review paper by Lundberg).

2.2 Anatomical and Physiological Progress

2.2.1 Cochlea: Application of Electron Microscopy

In 1955 the TEM (see Sect. 1.3.1.2) was first introduced into the field of otology (the study of the anatomy and physiology of the ear) by Engström. It meant a marked improvement of the amount of detail that could be observed. Studies by, e.g., Spoendlin and Flock now showed the structure of the hair bundles on top of the hair cells and showed these to contain actin, leading to their official name of stereovilli (from the Greek *stereos* = stiff and *villi* as the official term for actin-filaments attached to a cell). Kimura showed in 1965 that the longest outer hair cell stereovilli are firmly connected to the tectorial membrane by showing the imprints left on the underside of the TM. The TEM technique operates on thin slices of tissue, and the high optical sensitivity of this microscope implies that it is very sensitive to the slice preparation techniques. These have been improved and (semi-) automated over time.

In 1969 the scanning electron microscope was used by Lim for a detailed study of the structure of the organ of Corti (e.g., Lim and Melnick 1971). The greater depth of view of the SEM pictures allowed a good representation of its complicated 3-dimensional structure. Although SEM works on surfaces instead of on slices, the surfaces needed are too small to allow access in intact preparations, but structure within a preparation can be conserved much better than for TEM. On the other hand, in order to receive significant reflection from the surface, this has to reflect electron beams, which requires proper coating of the surface. In other words, the natural surface is potentially distorted both by properties of the coating, and by the electron beam.

Nevertheless, both TEM and SEM have provided marvelous images of the static structure.

2.2.2 Outer Hair Cells

One of the benefits of the introduction of EM into the lab was that they provided convincing evidence that both inner and outer hair cells are directionally sensitive (Flock et al. 1962; Engström et al. 1962). They confirmed that the structure of the hair bundle, and its orientation are relevant, and Engström et al. reported that mammalian cochlear hair cells no longer have a kinocillium but only a remaining basal body.

Flock et al. confirmed that the hair bundle structure and the location of the basal body of the kinocillium indicate the directional sensitivity. This was found for guinea pig OHCs, and for vestibular hair cells. Up to 1956 microphonic potential data from lateral line had been interpreted as indicators of bi-directional sensitivity of cupular hair cells (e.g., Kuiper 1956), but now that hypothesis was replaced by the notion that two sub-populations with opposite orientation can account for the microphonic data—as long as there is a nonlinear stimulus–response relation, because otherwise the microphonic response would disappear. These ideas were confirmed and accepted almost immediately.

Up to 1970 the difference in function between IHCs and OHCs remained a matter op speculation. The morphological differences had been established convincingly, and also the results of the first extensive studies of the afferent (and also the efferent) innervation had been presented (Spoendlin 1970). At the time there seemed to be no relevant role for the OHCs, in particular because of the distribution of afferent nerve fibers: Spoendlin estimated that at least 90% of the approximately 50.000 fibers innervate IHC and that 5–10% might innervate OHCs. At first, this was a matter of serious dispute (e.g., Eldredge 1967), but later studies confirmed the basic distinction and disproved the objections.

During the 1970s and 1980s, it became clear that there was a correlation between HC-damage caused by noise exposure and auditory nerve response. More in particular: the amount of damage depended on exposure level and duration, and damage started at the OHCs, beginning at the outer row, and finally reaches IHCs. As long as IHCs are (almost) intact, the auditory nerve fibers do give a response, but the tip of the tuning curve deteriorates with increasing damage level. IHC damage, finally, leads to deterioration of the cell and of the nerve fibers. Active nerve fiber labeling techniques have been used to confirm the nerve fiber—IHC connections (e.g., Liberman 1984; Liberman and Dodds 1984a,b; Liberman and Kiang 1984).

This conclusion was reached at the time that the source(s) of cochlear nonlinearity and a cochlear amplifier were still unknown. Now the pieces of the puzzle fell into place, and the role of intact OHCs appeared to become clear, or at least to possibly fill the gap. Detailed study of the OHC cell body indicates that its cytoskeleton is involved in electromotile responses of the cell (e.g., Santos-Sacchi 1992; Kalinec et al. 1992; Frank et al. 1999). Although this tells us in what direction progress can be made, it has not yet provided answers to all open questions.

2.2.3 Auditory Nerve and Beyond

Progress in auditory neurophysiology over the 1950–1980 period was also impressive. The basic techniques had just started (see Chap. 1), but biophysical understanding of the basics of the neural action potentials is attributed to Hodgkin et al. (1952); Hodgkin and Huxley (1952b,a,c) and did not develop before the early 1950s.

The primary auditory nerve fibers can be considered the output channels of the cochlea, and in this view the nerve response is the relevant output response of the cochlear system. However, the nerve response is not a simple analogue response, but a spike (=neural action potential) train, and the relation between stimulus parameters and response characteristics is not trivial. Although the response remains a deterministic, causal process, it becomes complicated enough to be characterized more efficiently in stochastic terms. Details of the transform of the continuous stimulus to discrete neural response can be avoided by using threshold detection techniques and constant response techniques. With the threshold technique, which requires the detection of a signal response in comparison to the background noise (spontaneous activity) it was possible to measure tuning curves–band-filter transfer functions—of single auditory nerve fibers (e.g., Kiang et al. 1965). It was immediately clear that the results were much more in accordance with psychophysical tuning than with the BM tuning observed by von Békésy. The same group also measured a large set of click responses, using positive and negative clicks, termed condensations click with inward stapes motion and rarefaction for the opposite. The Post Stimulus Time Histograms (PSTH) in response to clicks show a response profile that is equivalent to a half-wave rectified impulse response; click potential reversal evokes the other half-wave rectified part.

Response strength increases with stimulus strength, but saturates at some spike rate. The response area, used as a term for the CP range innervated by responding nerve fibers, also increases with level. In other words, as the level increases, the number of responding nerve fibers increases. This might seem a solution to the limitation caused by the saturation of the responses in single fibers, but, unfortunately, this only works for narrow band signals. The saturation properties are not analyzed in detail; they are addressed briefly in Sect. 7.1.1.

At the same time, other groups were studying time domain properties of the neural responses: in particular, the group of Rose and Hind focused on questions of waveform representation in the precise timing of the spikes. They found—for sufficiently low frequencies—a good synchronization to periodic stimulus properties. For a single tone, the response follows the half-wave rectified sine wave, both in synchronized period histograms (PSTH) and in interspike interval histograms (ISI) (e.g., Rose et al. 1967).

But besides the interest in these linear properties, their was also increasing interest in aspects that appeared to be in conflict with a linear interpretation. Without aiming to be complete, we refer the broad interests in:

- Studies of 2-tone suppression, apparently a nonlinear 2-tone interaction rather than an adapting feedback sharpening process (e.g., Nomoto et al. 1964; Sachs and Kiang 1968; Hind et al. 1970; Arthur et al. 1971; Abbas and Sachs 1976);
- CT phenomena, where auditory nerve data gave the first strong indication that the origin of the phenomenon had mechanical origin (i.e., prior to the nerve) in the cochlea (e.g., Goldstein and Kiang 1968; Dallos 1969; Smoorenburg et al. 1976);
- Changes in shape of tuning curves and excitation patterns with level (e.g., Goblick and Pfeiffer 1969; Anderson et al. 1971; Kiang and Moxon 1974).

More limitations of the linear cochlea are listed in the next chapter in Sect. 3.8. They are discussed in Chap. 4.

2.3 Auditory Perception

Auditory perception involves the entire auditory system. Here we focus on some specific topics, which traditionally have been assumed to depend largely on primary auditory system properties.

Firstly, this concerns studies of temporal and spectral sensitivity and selectivity, such as the relation between critical bands and effective bandwidths of tuning curves, between auditory masking patterns and cochlear excitation patterns, and the time–frequency uncertainty relation $\Delta f \Delta t \geq 1$ in relation to sharpness of tuning ($Q_{3\mathrm{dB}}$) and time resolution.

Although the tools and theories largely originate from linear systems analysis, it had been clear from the beginning of masking studies that the data showed unmistakable nonlinear effects. Wegel and Lane's classical study clearly showed that the growth of masking with masker level is nonlinear, especially if probe frequency and masker frequency are different (Wegel and Lane 1924). Moreover, there was a difference in the masking behavior above and below the masker frequency. In short, the masking pattern broadens significantly with increasing masker level, in particular on the hf-side of the masker.

The shape of the masking pattern of a fixed, narrow band masker has the characteristics of a band-filter. Its bandwidth became subject to several discussions. Several definitions are used, and sometimes not clearly separated: the Q_{3dB} and Q_{10dB} are defined as the ratio of the center frequency and the bandwidth determined at, respectively, 3 dB and 10 dB below the peak; the *ERB*—the equivalent rectangular bandwidth—is the width of the rectangular filter covering the same spectral area $\int H(f)\,\mathrm{d}f$, using the same peak level.

The critical band, CB, was introduced as by Fletcher (1940) a subjective effective bandwidth. It was introduced as the maximum bandwidth of a narrow band of noise over which all noise power integrates when masking a tone. Later, it played an important role in loudness computations (Zwicker). Some controversy remained between European (+Canada) and US laboratories about the width of the CB, where a difference by a factor of about 2.5 remains (Swets et al. 1962). The greater CB made it to a standard for loudness computation (ISO R 532, 1975 [to be revised status], in the US recently replaced by ANSI S3.4-2005).

Another line with a historical basis was picked up again in the 1960s, and concerns the study of aural CTs (e.g., Plomp 1965; Goldstein 1967). Since the description of the effect by the Italian violinist Tartini (1692–1770) in 1754—although von Helmholtz (1863, Chap. VII) mentions that the earlier inventor (1745) was the German organist Sorge—the phenomenon of the perception of combination tones had been known in the musical environment. Both Sorge and Tartini reported CTs at frequencies below the primaries or differential tones, Helmholtz also claims to have perceived summational tones.

Meanwhile the perception of the lf intermodulation frequencies (i.e., $f_{CT} = m\,f_1 \pm n\,f_2$, where f_1 and f_2 are the primary components, $f_{CT} \ll f_1 \ll f_2$) has been well documented and quantified. A difference has sometimes been reported and claimed between even-order and odd order ($m + n$ = even or odd) CTs. The difference tone $f_2 - f_1$ is of the even-order type, the most common $2f_1 - f_2$ is odd. Unfortunately, the latter is also often called a cubic difference tone, because a cubic nonlinearity can generate this type of distortion component. Hence the suggestion that there is a cubic nonlinearity in the system. However, there is ample evidence that the nonlinearity cannot be cubic; in fact it appears to be much closer to a cubic root, or a power of 1/3 rather than 3 (e.g., Smoorenburg 1972). The perception of hf intermodulation products or overtones has not been confirmed. Aural CTs are generated only if the primary components are within a certain proximity. The effect is optimum at a primary frequency ratio of about 6/5.

A different line of research addressed psychophysical 2-tone suppression, which appears to be related to neurophysiological 2-tone suppression. This phenomenon was studied extensively during the 1970s. One of the breakthroughs was the introduction of the "pulsation threshold" technique by Houtgast (1972). The basic result, similar in psychoacoustics and in neurophysiology, is that apparently a strong enough tone can not only mask a tone, or generate a CT, but can also suppress the response to an adjacent (in frequency) simultaneously presented tone.

References

Abbas PJ, Sachs MB (1976) Two-tone suppression in audtory-nerve fibers: extension of a stimulus response relationship. J Acoust Soc Am 59:112–122

Anderson DJ, Rose JE, Hind JE, Brugge JF (1971) Temporal position of discharges in single auditory nerve fibers within the cycle of a sine-wave stimulus: frequency and intensity effects. J Acoust Soc Am 49:1131–1139

Arthur RM, Pfeiffer RR, Suga N (1971) Properties of 'two-tone inhibition' in primary auditory neurones. J Physiol 212:593–609

Dallos PJ (1969) Combination tone $2f_l - f_h$ in microphonic potentials. J Acoust Soc Am 46:1437–1444

Dancer A, Franke R (1980) Intracochlear pressure measurements in guinea pig. Hear Res 2:191–205

Eldredge DH (1967) Review of: The Organization of the Cochlear Receptor, by h spoendlin. J Acoust Soc Am 41:1386–1388, bookreview

Engström H, Ades HW, Jr JEH (1962) Structure and function of sensory hairs of the inner ear. J Acoust Soc Am 34:1356–1363

Fletcher H (1940) Auditory patterns. Rev Mod Phys 22:47–65

Flock Å, Kimura R, Lundquist PG, Wersäll J (1962) Morphological basis of directional sensitivity of the outer hair cells in the organ of Corti. J Acoust Soc Am 34:1351–1355

Frank G, Hemmert W, Gummer AW (1999) Limiting dynamics of high-frequency electromechanical transduction of outer hair cells. P N A S 96:4420–4425

Goblick TJ, Pfeiffer RR (1969) Time-domain measurements of cochlear nonlinearities using combination click stimuli. J Acoust Soc Am 46:924–938

Goldstein JL (1967) Auditory nonlinearity. J Acoust Soc Am 41:676–699

Goldstein JL, Kiang NYS (1968) Neural correlates of the aural combination tone 2f1-f2. Proc IEEE 56:981–992

von Helmholtz HLF (1863) Die Lehre von den Tonempfindungen, 1st edn. Vieweg und Sohn, Braunschweig, english edition: On the Sensations of Tone, transl. by A.J. Ellis (1885) of 4th German edition (1877), publ. by Dover in 1954.

Hind JE, Rose JE, Brugge JF, Anderson DJ (1970) Two-tone masking effects in squirrel monkey nerve fibers. In: Plomp R, Smoorenburg GF (eds) Frequency Analysis and Periodicity Detection in Hearing, Sijthoff, Leiden

Hodgkin AL, Huxley AF (1952a) The components of membrane conductance in the gaint axon of *loligo*. J Physiol 116:473–496

Hodgkin AL, Huxley AF (1952b) Currents carried by sodium and potassium ions through the membrane of the gaint axon of *loligo*. J Physiol 116:449–472

Hodgkin AL, Huxley AF (1952c) The dual effect of membrane potential on sodium conductance in the gaint axon of *loligo*. J Physiol 116:497–506

Hodgkin AL, Huxley AF, Katz B (1952) Measurement of the current-voltage relatons in the membrane of the gaint axon of *loligo*. J Physiol 116:424–448

Houtgast T (1972) Psychophysical evidence for lateral inhibition in hearing. J Acoust Soc Am 51(1885–1894)

Kalinec F, Holley MC, Iwasa KH, Lim DJ, Kachar B (1992) A membrane-based force generation mechnism in auditory sensory cells. P N A S 89:8671–8675

Khanna SM (1986) Homodyne interferometer for basilar membrane measurements. Hear Res 23:9–26
Khanna SM, Johnson GW, Jacobs J (1986) Homodyne interferometer for basilar membrane measurements. II. Hardware and techniques. Hear Res 23:27–36
Kiang NYS, Moxon EC (1974) Tails of tuning-curves in auditory nerve fibers. J Acoust Soc Am 55
Kiang NYS, Watanabe T, Thomas EC, Clark LF (1965) Discharge patterns of single fibers in the cat's auditory nerve. The M.I.T. Press, Cambridge, Mass.
Kuiper JW (1956) The microphonic effect of the lateral line organ. PhD thesis, University of Groningen, The Netherlands
Liberman MC (1984) Single-neuron labeling and chronic cochlear pathology. I. Threshold shift and characteristic-frequency shift. Hear Res 16:33–41
Liberman MC, Dodds LW (1984a) Single-neuron labeling and chronic cochlear pathology. II. Stereocilia damage and alterations of spontaneous discharge rates. Hear Res 16:43–53
Liberman MC, Dodds LW (1984b) Single-neuron labeling and chronic cochlear pathology. III. Stereocilia damage and alterations of threshold tuning curves. Hear Res 16:55–74
Liberman MC, Kiang NYS (1984) Single-neuron labeling and chronic cochlear pathology. IV. Stereocilia damage and alterations in rate- and phase-level functions. Hear Res 16:75–90
Lim DJ, Melnick W (1971) Acoustic damage of the cochlea. Arch Otolaryng 94:294–305
Lundberg KH (2005) The history of analog computing. IEEE Contr Syst Mag 25(3):22–28
Minsky M (1961) Microscopy apparatus. Patent 3,013,467
Nedzelnitsky V (1974) Measurement of sound pressure in the cochleae of anesthetized cats. In: Zwicker E, Terhardt E (eds) Facts and Models in Hearing, Springer, Berlin, pp 45–53
Nedzelnitsky V (1980) Sound pressures in the basal turn of the cat cochlea. J Acoust Soc Am 68:1676–1689
van Netten SM (1988) Laser interferometer microscope for the measurement of nanometer vibrational displacements of a light-scattering microscopic object. J Acoust Soc Am 83:1667–1674
Nomoto M, Suga N, Katsuki Y (1964) Discharge patterns and inhibition of primary auditory nerve fibers in the monkey. J Neurophysiol 27:768–787
Plomp R (1965) Detectability threshold for combination tones. J Acoust Soc Am 37:1373–1378
Puria S, Peake WT, Rosowski JJ (1997) Sound pressure measurements in the vestibule of human-cadaver ears. J Acoust Soc Am 101:2754–2770
Rose JE, Brugge JF, Anderson DJ, Hind JE (1967) Phase-locked response to low-frequency tones in single auditory nerve fibers of the squirrel monkey. J Neurophysiol 30:769–793
Sachs MB, Kiang NYS (1968) Two-tone inhibition in auditory-nerve fibers. J Acoust Soc Am 43:1120–1128
Santos-Sacchi J (1992) On the frequency limit and phase of outer hair cell motility: effects of the membrane filter. J Neurosci 12:1906–1916
Smoorenburg GF (1972) Audibility region of combination tones. J Acoust Soc Am 52:603–614
Smoorenburg GF, Gibson MM, Kitzes LM, Rose JE, Hind JE (1976) Correlates of combination tones observed in the response of neurons in the anteroventral cochlear nucleus of the cat. J Acoust Soc Am 59:945–962
Spoendlin H (1970) Structural basis of peripheral frequency analysis. In: Plomp R, Smoorenburg GF (eds) Frequency Analysis and Periodicity Detection in Hearing, AW Sijthoff, Leiden, pp 4–36
Swets JA, Green DM, Tanner WP (1962) On the width of the critical band. J Acoust Soc Am 34:108–113
Wegel RL, Lane CE (1924) The auditory masking of one pure tone by another and its probable relation to the dynamics of the inner ear. Phys Rev 23:266–285

Chapter 3
Emerging Cochlear Mechanics

Abstract Cochlear mechanics is a field that relies strongly on fluid mechanics, linear and nonlinear signal processing, and additional mathematical tools. This is applied to a biological structure. A selection of useful—and possibly superfluous—prerequisites is presented in Appendix 8. Throughout, terminology and unit definitions follow the ANSI (2005) standard acoustical terminology.

3.1 1-D Cochlea, Long Wave vs. Short Wave

Although the notion of CMs was by no means new, the impact of a number of papers presented at the Speech Communication Conference at M.I.T. (May 31–June 3, 1950), plus the paper by Peterson and Bogert that appeared earlier that year, laid the foundation for modern CMs. The conference brought the latest experimental results together, from psychophysics to neurophysiology, and from signal analysis to speech communication proper. The Proceedings appeared in the November issue (# 6) of the Journal of the Acoustical Society of that year. Of special interest for CMs are the papers by Petersen and Bogert's in issue # 3, and by Ranke and Zwislocki, both in # 6. All three were well aware of von Békésy's results, and aimed to optimally incorporate these results in their models.

Von Békésy, Ranke and Zwislocki all had been in the field for quite some time. They started their research on the European mainland around 1930. Many of their initial publications, which contain important details, appeared in German in European journals. The 1950 papers by Ranke and Zwislocki are reviews of their previous work, and remain primarily descriptive, whereas the more recent study by Petersen and Bogert goes into detail. Although Ranke and Zwislocki describe a very similar objective, their basic assumptions start to deviate almost from the beginning, and they dispute each other's hypotheses rather unmistakably.

We start with the points on which there was good agreement. All three papers start with a 3-D cochlear structure. This structure consists of fluid-filled ducts within the temporal bones. The auditory nerve is the output channel of the cochlea. It

H. Duifhuis, *Cochlear Mechanics: Introduction to a Time Domain Analysis of the Nonlinear Cochlea*, DOI 10.1007/978-1-4419-6117-4_3,

carries a neural code of the cochlear response to the brain. Fluid moves in 3-D ducts, but its input stems from a 1-D stapes motion (at least almost 1-D at levels below $\approx 80\,\text{dB}$). There is also general agreement about boundary conditions at stapes, round window, and helicotrema, and about the irrelevance (negligibleness) of the curvature of the cochlea for the mechanical behavior.

Now we arrive at the intellectual bifurcation: Zwislocki assumes, and defends, that the resulting acoustic wave in the cochlear fluid is a plain wave, moving in the longitudinal direction, because its wavelength is always large compared to the cochlear cross-sectional sizes (height and width, or radius of the ducts). If this is true, then it would fulfill a sufficient condition for the plain wave hypothesis. Peterson and Bogert follow the same practical line and also select the 1-D approximation for a plain wave within the ducts, but they leave more room for modification, in particular at and beyond the point of resonance, where the wave speed decreases noticeably, which implies a similar decrease of the wavelength ($f = c/\lambda$).

Ranke, on the other hand, considered the whole duct as a shallow water case where the water is covered by the cochlear partition that has very different parameters than the fluid. The fluid itself being practically incompressible, moves in the direction of the partition, and that gives an effect similar to shallow water waves. The classical analysis of surface waves in both deep shallow waters, however, had been carried out on the more common water–air interface (Lamb 1895 Chap. 19, and Rayleigh 1896 Sect. 353) and the effect of the fluid motion in one duct on the motion in another duct, through the cochlear partition interface, is quite different from the effects of the water–air surface tension. Hence, the direct comparison might have been misleading, but the consideration that the fluid motion near the point of maximum response at the partition has slowed down significantly, and, thus, its wavelength must have decreased accordingly, remains highly relevant: it invalidates the common long-wave assumption! This issue has been brought up again by Siebert (1974) and by Lighthill (1981), and several investigators thereafter.

A general consequence of the above consideration is that a 1-D approximation of both long wave and short wave analysis should be considered a "contradiction in terminis." For the long wave, approximation problems arise near resonance where the resulting cochlear partition response requires a transpartition behavior, and for the short wave, which would allow this transpartition effect, the problem arises at the basal end, where the shallow water approach is too extreme.

3.2 Mechanical Properties of the Cochlear Structure and the System Equations

Cochlear mechanics starts with a specification of the relevant fluid mechanics parameters and the mechanical structure parameters of the cochlea. This regards the cochlear fluids, the cochlear shape, and properties of the moving structures.

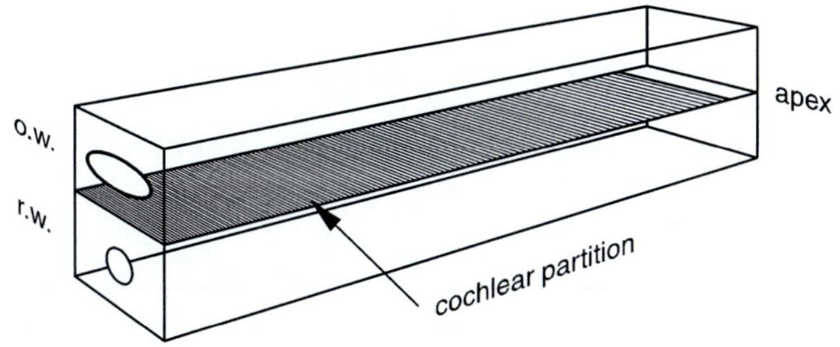

Fig. 3.1 Box-model of the 3-D cochlea (partly transparent). The scala media is embedded within the cochlear partition. The width of the basilar membrane is not indicated. Abbreviations o.w. and r.w. represent the *oval window* and *round window*. The box is oriented in a Cartesian coordinate system: length $\Rightarrow x$, width $\Rightarrow y$, and height $\Rightarrow z$. The stiffness s of the partition is supposed to depend on longitudinal position x: $s = s(x)$

In practice, several simplifying assumptions are made in order to make the system mathematically tractable. A major simplification that has been used in all first studies [and remains to be used frequently] is the *linear* approximation of the system. It comes with the very useful consequence that, insofar as this approximation is correct, all tools from linear system analysis can be applied without any loss of information. Therefore, the cochlear response can be analyzed using the stapedial motion as the input parameter. The outer and middle ear interfaces can be treated as linear black boxes which contain the correct filter–transformer information. It also implies that it is not essential which of the stapes variables: deflection, volume deflection, velocity, or volume velocity is employed, because all of these are simply related. The stapedial volume velocity U_{st}, which is taken positive when the velocity is inward (i.e., in the positive x-direction), will be employed most frequently.

The acoustic parameters of the cochlear fluid are such that for almost the entire frequency range the wavelength is large compared to the cochlea dimensions (see Sect. 3.1, and Chap. 8). Therefore, details of the structure of the cochlear ducts may be ignored. Most theoretical studies have investigated an unrolled model cochlea, either as a box (as in Fig. 3.1) or as some sort of tube (e.g., Peterson and Bogert 1950; de Boer 1980). Peterson and Bogert (1950) and Fletcher (1951) did analyze the fluid compressibility and viscosity (Fletcher) in some detail, but later studies tend to start with the approximating assumption of incompressibility of an inviscid fluid. Obviously this implies that the sound velocity in that medium is practically infinite.

As long as we realize that for the upper range of auditory frequencies only (above ~8 kHz) the above approximation may need significant corrections, we can conclude that the approximation appears to be valid over an important range of human hearing. A straightforward criterion for the overestimation of the sound speed concerns the actual travel times of cochlear waves. The actual speed in the cochlear fluid is ≈1,500 m/s (similar to the speed in water), which implies a wavelength of about 190 mm, or about five times the length of the human cochlea. Phase differences computed with infinite speed are considered acceptable at this point for the frequency range mentioned above.

As indicated in Sect. 8.1, the mass conservation equation now reduces to (8.4), or, when expressed in the velocity potential $\phi(\mathbf{x})$, to the Laplace equation:

$$\nabla \cdot \mathbf{v}(\mathbf{x}) = \nabla \cdot \nabla \phi(\mathbf{x}) = \Delta\, \phi(\mathbf{x}) = 0. \qquad (3.1)$$

The cochlear walls are also taken to be solid (practically incompressible). Consequently, the cochlear fluid velocity into the walls is zero. The cochlear partition, however, has mechanical properties which enable it to move, but its volume is also supposed to be constant: its structure is also assumed to be incompressible. The apical connection between scale vestibuli and scale tympani also allows fluid flow, but its dimensions imply some mechanical resistance. Acoustic effects of endolymphatic and perilymphatic connections with the vestibular system and beyond (endolymphatic sac) are assumed to be negligible.

Under these assumptions a volume influx at the stapes has to be equal to an efflux at the RW, and this flux must be crossing the cochlear partition and/or the helicotrema. These 3-D considerations simplify the transition to the 1-D approach.

An input stimulus delivers a sound-energy flux input J. The size of J is determined by the product of stapes-velocity u_{st} and stapedial driving force $p_{\mathrm{st}} \times A_{\mathrm{st}}$, or of volume velocity U_{st} and the stapedial pressure p_{st}. The quotient of p_{st} and u_{st} is the cochlear input impedance, or more specifically: the specific acoustic input impedance. The quotient of p_{st} and U_{st} defines the acoustic input impedance (see Sect. 8.3). If the volume velocity is not associated with fluid compression, it must imply fluid movement. In other words, the input impedance is determined by the summed movement of the cochlear fluid. This consists of inertia effects (fluid motion) in combination with the parallel effects of all responding cochlear partition elements and the helicotrema. The oval and RW connections are assumed to be almost lossless. This means that they are assumed to be elastic, with negligible mass and damping.[1]

Under the specifications defined above it is useful to split the cochlear response to the input stimulus into two separate components: a fast component which is due to the (in)compressibility of the fluid, and a slow component which involves interaction between fluid and cochlear partition. We first consider the fluid pressures associated with these components. If the stapedial input pressure is $p_{\mathrm{st}}(t)$ and the volume velocity $U_{\mathrm{st}}(t)$, then the input sound-energy flux is

$$J_{\mathrm{st}} = \langle p_{\mathrm{st}}(t) \cdot U_{\mathrm{st}}(t) \rangle \qquad (3.2)$$

where the brackets indicate time averaging over at least one cycle of the input signal. The dot product represents the effect of the phase angle between p and U, and U

[1] The OW mass- and damping parameters can be incorporated in the stapes footplate properties. For both OW and RW, the primary elastic properties regard attachment at the window edges. Usually, they are included in a middle ear model (Sect. 5.3.2.2). In simpler models, the RW-stiffness is neglected altogether (Sect. 5.3.2.1).

is supposed to be perpendicular to the footplate—OW surface A_{st}, otherwise the spatial input angle has to be accounted for as well.

If a stimulus is simply sinusoidal, and represented by the real part of the complex exponential

$$p_{\text{st}}(t) = \hat{p}_{\text{st}}\,\Re\{e^{-i\omega t}\} \qquad \text{and} \qquad U_{\text{st}}(t) = \hat{U}_{\text{st}}\,\Re\{e^{-i(\omega t+\phi)}\}$$

then the average dot product is $\frac{1}{2}\,\hat{p}_{\text{st}}\,\hat{U}_{\text{st}}\cos\phi$, the absolute value of which maximizes for $\phi = 0 \pm n\,\pi$ and minimizes for $\phi = \pi/2 \pm n\,\pi$.

Let the pressure in the total scala vestibuli be denoted as $p_v(x,t)$ and the pressure in the scala tympani as $p_t(x,t)$. Then the fast pressure wave is defined as the average, and the slow pressure wave as the difference:[2]

$$p_f(t) = \frac{p_v(x,t) + p_t(x,t)}{2} \tag{3.3a}$$

$$p_s(x,t) = \frac{p_v(x,t) - p_t(x,t)}{2}. \tag{3.3b}$$

At $x = 0$, the force balance at the oval window implies $p_{\text{st}}(t) = p_f(t) + p_s(0,t) = p_v(0,t)$, and if no acoustic energy leaves at the RW, then $0 = p_f(t) - p_s(0,t) = p_t(0,t)$. The last condition (zero pressure across round window) leads to

$$p_s(0,t) = p_f(t) = \frac{1}{2}p_{\text{st}}(t). \tag{3.4}$$

If we combine this result with that of the incompressibility, which leads to $U_{\text{st}}(t) = U_v(0,t) = -U_t(0,t)$, we observe that the sound-energy influx following the slow wave divides equally over the 2 scalae:

$$\begin{aligned} J_{\text{st}} &= \langle p_{\text{st}}(t) \cdot U_{\text{st}}(t)\rangle = \langle p_v(0,t) \cdot U_v(0,t)\rangle + \langle p_t(0,t) \cdot U_t(0,t)\rangle \\ &= \big\langle \{p_f(t) + p_s(0,t)\} \cdot U_{\text{st}}(t)\big\rangle + \big\langle \{p_f(t) - p_s(0,t)\} \cdot -U_{\text{st}}(t)\big\rangle \\ &= 2\,\langle p_s(0,t) \cdot U_{\text{st}}(t)\rangle\,. \end{aligned} \tag{3.5}$$

The fast wave, however, produces a sound-energy flux in the $+x$-direction in the scala vestibuli, simultaneously with an equal flux in the $-x$-direction in the scala tympani. The net sound-energy flux distributed over the 2 scalae by the fast wave is therefore equal to zero (except for transients), and the energy that is dissipated within the cochlea is exclusively described by the slow wave (see, e.g., Peterson and Bogert 1950; Lighthill 1981; Duifhuis 1988).

[2]Since the fast wave travels at a very high speed in the x-direction, we can drop—in first approximation—the x-dependence in $p_f(t)$.

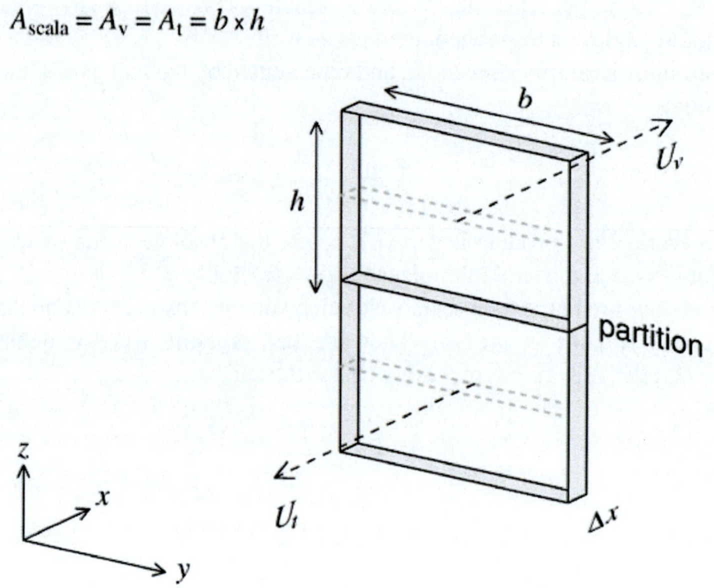

Fig. 3.2 Cross-section through box model at x. The scala area $A = b \times h$ for both scalae. Volume velocity in the scala vestibuli is $U_v(x,t)$ at point x in the x-direction. It equals the average particle velocity ($u_v(x,t)$, in the same direction) times the cross sectional area A

The effect on a partition element is analyzed further, starting from connecting fluid slices in the scalae and taking Δx for the slice thickness (Fig. 3.2). The input volume flux in the scala vestibuli is $U_v(x,t)$ at x, and the output flux at $x + \Delta x$ is $U_v(x + \Delta x, t)$. The difference must be compensated by flow through the partition section between x and $x + \Delta x$, and/or an associated pressure change within the slice if the fluid is compressible. In first approximation, the compressibility is neglected and this pressure change is ignored. The transpartition volume velocity is represented by the average point velocity $w(x,t)$, which is taken positive in the direction from scala tympani to scala vestibuli. This gives:

$$w(x,t)\,b\,\Delta x = U_v(x + \Delta x, t) - U_v(x,t), \tag{3.6}$$

or, if $\Delta x \to 0$

$$w(x,t)\,b = \frac{\partial U_v(x,t)}{\partial x} = A_v\,\frac{\partial u_v(x,t)}{\partial x}. \tag{3.7}$$

For the scala vestibuli, the simplified form of Euler's equation (8.5) for the 1-D model is:

$$A_v\,\frac{\partial p_v(x,t)}{\partial x} = -\rho\,\frac{\partial U_v(x,t)}{\partial t} \quad \text{or} \quad \frac{\partial p_v(x,t)}{\partial x} = -\rho\,\frac{\partial u_v(x,t)}{\partial t}. \tag{3.8}$$

Next the mechanical properties of the cochlear partition are introduced. It should be obvious that an implicit simplification has been introduced by describing the scala media structure, including BM and organ of Corti (OC) with tectorial membrane (TeM), as the cochlear partition (CP). The structure has elastic properties (proposed by von Helmholtz and demonstrated by von Békésy), also some mass, and a sensory system also must pick up signal power, which implies that there also must be a damping term. This can be combined in local second order mechanical structure properties, where local should be taken as a macroscopic specification. Micromechanical properties at single cell level—or smaller—are neglected insofar as they do not arise in the macromechanics. Note that around 1950 CMs started basically with the same parameters, but with a slightly different argumentation.

The force balance at the partition gives

$$p_v(x,t) - p_t(x,t) = 2\,p_v(x,t) = -w(x,t)\,Z_{\mathrm{CP}}(x) \tag{3.9}$$

where $Z_{\mathrm{CP}}(x)$ represents the local specific acoustic impedance of the partition (Pa.s/m). This quantity is equivalent to the mechanical impedance (force / velocity) per unit area (see Sect. 8.3). Usually properties of a partition slice Δx are considered to act as if coupled in series, which yields

$$w(x,t)\,Z_{\mathrm{CP}}(x) = m(x)\,\frac{\partial w(x,t)}{\partial t} + d(x)\,w(x,t) + s(x)\int w(x,t)\,\mathrm{d}t \tag{3.10}$$

and where mass, damping and stiffness ($m(x)$, $d(x)$, and $s(x)$) are taken per unit area.

Additional common simplifying approximations are

1. The partition mass per unit area is constant over x,
2. The cross-sectional areas of scala vestibuli and scala tympani are equal,
3. The cross-sectional areas change very slowly over the cochlea length (this was used above in (3.7) when applying $\frac{\partial U_v(x,t)}{\partial x} = A_v\,\frac{\partial u_v(x,t)}{\partial x}$, and the same for the time derivative in (3.8)), and
4. Boundary layers due to fluid viscosity are neglected.

These equations can be combined to give local equations in either fluid pressure p, fluid velocity u or volume velocity U, or partition velocity (or deflection, or acceleration). We start with the pressure equation. Substituting (3.9) in (3.7) allows elimination of the partition velocity $w(x,t)$. Next we take the time derivative on both sides, to obtain:

$$-\frac{2b}{Z_{\mathrm{CP}}(x)}\frac{\partial p_v(x,t)}{\partial t} = A_v\frac{\partial^2 u_v(x,t)}{\partial x\,\partial t}. \tag{3.11}$$

Then the u-term is eliminated using the spatial derivative of (3.8), and $A_v = bh$ is applied, reducing the scala size parameters to a single h. This yields for $p_v(x,t)$:

$$\frac{\partial^2 p_v(x,t)}{\partial x^2} - \frac{2\rho}{h\,Z_{\mathrm{CP}}(x)}\frac{\partial p_v(x,t)}{\partial t} = 0 \tag{3.12}$$

or, taking Fourier transforms[3] and thereby replacing $\frac{\partial}{\partial t}$ by $\mathrm{i}\omega$:

$$\frac{\partial^2 p_v(x,\omega)}{\partial x^2} - \frac{2\,\mathrm{i}\omega\rho}{h\,Z_{\mathrm{CP}}(x,\omega)} p_v(x,\omega) = 0. \tag{3.13}$$

Equation (3.9) gives a simple linear relation between p and w, because at a fixed point x at the partition the impedance Z_{CP} is assumed to be depending on ω only. Since linear factors can be dropped after substitution, the same equation is obtained for w:

$$\frac{\partial^2 w(x,\omega)}{\partial x^2} - \frac{2\,\mathrm{i}\omega\rho}{h\,Z_{\mathrm{CP}}(x,\omega)} w(x,\omega) = 0, \tag{3.14}$$

and, similarly, also for u, as long as exchange of order of spatial and temporal differentiation is allowed, a condition that has been assumed tacitly so far. Details and proof are left to the reader.

Equations (3.13) and (3.14) would be rather simple if—and only if—the parameters would be constant, i.e., independent of x. Dependency on ω would still provide straightforward solution to monochromatic stimuli, or pure tones. Note that a constant partition impedance Z_{CP} would make (3.12) a diffusion equation. However, the structure is more complicated, because relevant parameters do depend on x. The equations do represent the behavior of the response at different positions along the partition, and the complete system also needs the boundary conditions at $x = 0$, the base, and at $x = x_{\mathrm{apex}}$.[4] The condition at the base was discussed above (3.2–3.5), the condition at the apex need more discussion, and will be continued in Sect. 3.5, and again in more detail in Sect. 5.3.3. At this point, we remark that the frequently used condition that the apex is a shortcut is questionable. We will follow the alternative assumption, that was proposed by Dallos (1970, 1973), and later by Lynch et al. (1982) and Puria and Allen (1991). It involves an estimate of the fluid mass and resistance (due to viscosity), which would provide a better termination than the shortcut.

Although the CM papers from the early 1950s gave an excellent start of this field, they were hampered by the fact that the data on which they based the estimations

[3]The Fourier transform of the pressure $p_v(x,t)$ is written as $p_v(x,\omega)$. The rather common capital notation for the transforms is not used because the capital symbols are reserved for other variables. Although some duplications will be unavoidable, we try to minimize the occurrence, and use P for power and U for volume velocity.

[4]In human the apex occurs at $x \approx 35$ mm.

of the parameter values were premature. Results were obtained at cadaver samples and the stimulus levels that had to be used to provide detectable responses were significantly above the hearing range. Therefore, although the data presented by von Békésy showed the proper global response behavior, the quantitative results were not directly applicable to the normal human cochlea, a point that took a few decades to be unraveled convincingly. Most of the early studies aimed at a low frequency selectivity, which led to the growth of several hypothetical sharpening mechanisms, and although von Békésy's original "lateral inhibition" hypothesis had largely disappeared during the 1970s, several alternatives were still in serious scientific competition around 1980.

As long as the local partition impedance is modeled as a simple second order mechanical system, with mass, damping, and stiffness, it seems clear that one of the objectives: understanding the cochlear frequency map, can be met readily. Local tuning depends on local resonance, or primarily on the interaction of local mass and stiffness. The BM stiffness was measured directly, and estimates of the mass could be made and checked to fit the expected range from about 20 Hz to 20 kHz. However, the resonance frequency depends on the ratio of stiffness and mass,[5] not on the individual values. That left sufficient room for uncertainty for quite some time. Precise information about the phase of the response, or of the impulse response, could have settled this point, but hardly did. The other parameter, damping, was even more difficult to estimate.

Here we have reached a common problem in the development of theories: should a theory be optimizing output results by "curve fitting" through parameter optimization, or should the theory describe the underlying biophysical processes optimally and relying completely on parameter data, measured independently as accurately as possible? If a system is very complex, i.e., if there are many parameters, then it can take a long time to obtain the required estimates; if the number of free (unknown) parameters is too large, then mathematical optimizing of estimations quickly becomes curve fitting, because identical results can be obtained with different parameter sets. Cochlear mechanics still has not reached the stage where the properties of the elements of the structure have been determined completely. Our approach to the analysis is to use available data as accurately as possible, and reduce the number of remaining free parameters to a minimum number of representative global parameters.

---

[5]The angular resonance frequency of a mechanical mass–spring system is $\omega_r = \sqrt{s/m}$, with mass is m and the stiffness s.

It is commonly assumed that the CP-mass per unit area is constant, based on the observation that the BM thickness hardly changes over x. However, the CP-structure size does change, and the effective mass that includes (part of) the relevant boundary layer is also changing from base to apex. Therefore, *constant mass per area* should be regarded as a first approximation.

This approach was also taken in the early 1950s. In their original cochlear mechanics study, Peterson and Bogert (1950) set up the basic equations without the incompressibility assumption. The authors also discussed effects of viscosity, and noted that the dynamic) viscosity measured by von Békésy ($\eta = 0.002\,\text{Pa s}$) was approximately 3 times the value of water at body temperature ($\eta = 0.0007\,\text{Pa s}$). Viscosity effects had been studied by Zwislocki's (e.g., 1948, 1950, 1953) and it was also discussed extensively by Fletcher in 1951. A detailed study of compressibility and viscosity was given by Dallos (1973) in his textbook. We present the effects of fluid compressibility and viscosity quantitatively, following his analysis.

Dallos' general equation for the *FT* of the pressure (his (4.82)) can be written in the form

$$\frac{1}{p_v(x,\omega)}\frac{\partial^2 p_v(x,\omega)}{\partial x^2} = \frac{2\,\mathrm{i}\omega\rho}{h\,Z_{\mathrm{CP}}(x,\omega)}\left[1+\frac{\mathrm{i}\omega\,h\,Z_{\mathrm{CP}}(x,\omega)}{2\,\rho\,c^2}\right]\left[1+\frac{R(x,\omega)}{\mathrm{i}\omega\rho}\right]$$
$$= \frac{2\,\mathrm{i}\omega\rho}{h\,Z_{\mathrm{CP}}(x,\omega)}\;[1+\text{compression correction}]\;[1+\text{viscosity correction}] \tag{3.15}$$

where we use our parameter definitions. The first bracket term on the right represents the effect of compressibility, and is completely equivalent to the term obtained by Peterson and Bogert.[6] The second one describes viscous effects. It contains a new parameter $R(x,\omega)$ which is defined as:

$$R(x,\omega) = \sqrt{4\pi\mu\omega/A_v} \tag{3.16}$$

containing the dynamic viscosity μ. For a quantitative estimate, we employ both the value for water (at 40°C) and the value for perilymph reported by von Békésy. Note that the two correction terms depend on both frequency and place (ω and x). The compressibility term is even more complex because it contains the partition impedance completely, which means that the frequency effect depends strongly on place, due to the x-dependence of partition damping and stiffness. Estimates of the corrections of the two terms are presented in Table 3.1 for the frequencies $f = 100\,\text{Hz}$ and $f = 1\,\text{kHz}$. The compressibility terms are computed at $x = 5\,\text{mm}$, 17.5 mm, and 30 mm.

> The verification of the values presented in Table 3.1 is left to the readers as a valuable training exercise. Dallos (1973) presents very useful explanations and discussion points, as well as links to earlier studies.

[6] Peterson and Bogert (1950) use a differently defined Z, viz. $Z(x,\omega) = \mathrm{i}\omega\,Z_{\mathrm{CP}}(x,\omega)$ (their (14)), and their "bracket term" γ^2 (their (16–18)) looks different because they extracted ω^2/c^2 instead of $2\,\mathrm{i}\omega\rho/h\,Z_{\mathrm{CP}}(x,\omega)$. They did not find an analytical solution for this equation, but presented numerical results.

Table 3.1 Correction terms for the 1-D linear cochlea model

Frequency	Compressibility factor at $x =$ (mm)			Viscosity factor	
↓	5.0	17.5	30.0	Water	Perilymph
	$\times 10^{-6}$	$\times 10^{-6}$	$\times 10^{-6}$		
100 Hz	$0.28 - 465.7i$	$0.15 - 13i$	$0.075 - 0.13i$	$-0.05i$	$-0.09i$
1 kHz	$0.28 - 461.3i$	$0.15 - 8.6i$	$0.075 + 4.2i$	$-0.016i$	$-0.03i$

The table gives the computed fractions for compressibility and for viscosity at 100 Hz and at 1 kHz; the compressibility terms are given at $x =$ 5 mm, 17.5 mm, and 30 mm from the base. The cochlear partition parameters are: $m = 0.5\,\text{kg/m}^2$, stiffness follows the corresponding Greenwood-map, and the damping is chosen to provide a quality factor that is proportional to the square root of the frequency. These values are discussed in Sects. 5.3.5.1 and 8.5.1

Checking the entries of the table we note that the correction for compressibility appears to be less than 0.05% at the basal part of the cochlea and even less apically. In general the imaginary part of the term dominates the real part, but the net effect on the real 1-vector remains small. The viscosity factor is stronger, especially for low frequencies but the imaginary terms of 5% for water or 9% for perilymph according to von Békésy are still significantly below 1, and since the effects are purely imaginary, they will have relatively small phase effects (approximately 0.05 or 0.09 rad, respectively), and even smaller effects on the amplitudes of the responses. These effects are considered sufficiently small compared to general consequences of the 1-D approximation to justify the common neglect.

The box-model simplification that omits the tapered shape of the cochlea also received attention in the first generation CM models. The anatomical structure clearly shows a general decrease of the cross-sectional areas A_v and A_t from base to apex. Again several of the earlier studies addressed the point in one form or other (e.g., Dallos 1973). The anatomical data are usually described with 1 parameter (e.g., (4.92) in Dallos' book proposes $A_v + A_t = 5.10^{-6}\,e^{-50x}$ with x in m and S in m^2). For the 1-D linear model the impact of this parameter remains also small. Note, however, that several bat families have very typical specializations of their cochlear duct cross section profiles, in which also the shapes of A_v and A_t can differ substantially.

The formulation given in (3.15) for the spectral profile of the pressure ($p_v(x, \omega)$) can readily be adapted to discard either compressibility or viscosity, simply by dropping the relevant bracket term. Dropping both terms brings us back at the solution that was obtained earlier: in that case the result is equal to (3.13), which is equivalent to (14) in Zwislocki (1953), and also to (6.j) in de Boer (1980). The solution to this equation is treated next.

3.3 Solutions for the 1-D Linear Cochlea

Note: Equations in this subsection are solutions to the frequency domain representation of the cochlea equations. Time domain versions are obtained by inverse *FT*. For a simple tone stimulus this can be obtained by taking the real part of $e^{i\omega t}$, and adjusting amplitude and phase.

The main differential equation (3.13) that describes the CMs variables $p(x, \omega)$, $u(x, \omega)$, and $w(x, \omega)$ can be rewritten as

$$\frac{\partial^2 \psi(x,\omega)}{\partial x^2} = \frac{2\,i\omega\rho}{h\,Z_{\mathrm{CP}}(x,\omega)}\,\psi(x,\omega)$$

$$= -k^2(x,\omega)\,\psi(x,\omega), \tag{3.17}$$

where ψ represents any of the variables p, u, or w. As indicated in the previous section, the difficulty of the equation is that the term $k(x, \omega)$ is not a constant, but does depend on x and ω; it is a complex variable, and real and imaginary parts play different roles in the wave propagation. Let those parts be defined as $k(x, \omega) = k_{\mathrm{r}}(x, \omega) + i\,k_{\mathrm{i}}(x, \omega)$ where the indices "r" and "i" denote the real and imaginary components.

For constant k, the general solution of (3.17)

$$\psi(x,\omega) = C_1\,e^{i\,k\,x} + C_2\,e^{-i\,k\,x} \tag{3.18}$$

$$\psi(x,\omega) = C_1\,e^{+i\,k_{\mathrm{r}}\,x}\,e^{-k_{\mathrm{i}}\,x} + C_2\,e^{-i\,k_{\mathrm{r}}\,x}\,e^{+k_{\mathrm{i}}\,x} \tag{3.19}$$

where two standard cases can be recognized. If the real part of k dominates ($k_{\mathrm{r}} \gg k_{\mathrm{i}}$) then (3.17) is a simple wave equation, and k_{r} represents the wavenumber (see Sect. 8.2). If, on the other hand, the imaginary part dominates, and $k^2 = -k_{\mathrm{i}}^2$, then (3.17) becomes a standard Helmholtz equation with the general solution:

$$\psi(x,\omega) = C_{11}\,e^{k_{\mathrm{i}}\,x} + C_{12}\,e^{-k_{\mathrm{i}}\,x} \tag{3.20}$$

where the additional first indices of C are meant to stress that the integration constants in (3.20) differ from those used in (3.18) and (3.19).
In general, the integration constants C_{in} have to be adjusted to match the boundary conditions at base and apex.

The difficulty of assessing the behavior of k originates from the quotient $\omega\,/\,Z_{\mathrm{CP}}(x,\omega)$, in which the partition impedance reflects the x-dependencies [$Z_{\mathrm{CP}}(x,\omega)$ was introduced in (3.9)]. It is useful to analyze three conditions for the cochlear responses as special cases in more detail:

1. The high-frequency asymptote, effective where $\omega \gg \omega_r(x)$, or apically from x_r, and where $Z_{\mathrm{CP}}(x,\omega) \Rightarrow i\omega m$;
2. The region around resonance, at $x_r = x_r(\omega)$, where $Z_{\mathrm{CP}}(x,\omega) \Rightarrow d(x)$;

3. The low-frequency asymptote, effective where $\omega \ll \omega_r(x)$ or basalward from x_r, and $Z_{\mathrm{CP}}(x, \omega) \Rightarrow s(x)/\mathrm{i}\omega$.

Case 1, the high-frequency asymptote, implies that $k^2 = -2\rho/(hm)$. Consequently, $k = \mathrm{i}\, k_\mathrm{i}$, independent of x, and $k_\mathrm{r} = 0$. This leads to the solution presented above (3.20). Both terms represent standing waves, the first with the amplitude factor C_{11} growing with x, and the second term with the factor C_{12} has an exponentially decreasing amplitude. The value of k_i is $2.10^3\,\mathrm{m}^{-1}$, which amounts to a change by a factor of about 1,000 over a 7 mm x-shift. Thus, the increasing term grows very fast and would tend to explode. Therefore, the decreasing second term is assumed to dominate the response; only for very low frequencies the increasing first term cannot be ruled out completely.

Case 2, near resonance, where stiffness and mass terms are in balance. In that case the damping component dominates the impedance, and the solution for k^2 will be purely imaginary: $k^2 = 2\mathrm{i}\omega\rho/\{h\, d(x, \omega)\}$. The roots of k^2 are complex conjugates for which $k_\mathrm{r} \equiv k_\mathrm{i}$, and either $k = k_\mathrm{r} + \mathrm{i}k_\mathrm{i}$, or $k = -k_\mathrm{r} - \mathrm{i}k_\mathrm{i}$. Note that the last solution represents a decreasing amplitude with x, in combination with propagation in apical direction. In order to evaluate the value ($|k|$) in more detail we have to explore the damping term, which has not yet been specified. In many CM studies, authors have selected a damping profile that produces tuning with a constant quality factor $Q = f_\mathrm{r}/\Delta f$. However, experimental data on auditory frequency selectivity, from classical psychophysics, neurophysiology, and ultimately also from CMs data, show that the selectivity at the high-frequency end is sharper than at the low-frequency end. Further quantitative discussion is postponed to the end of this section, after discussion of the full response. If the damping is low enough, the cochlear partition response will show a maximum near the point of resonance; for increasing damping the maximum will begin to shift basalward. In case of a very high damping $Q \ll 1$, the maximum disappears altogether.

Case 3, the low-frequency asymptotic behavior, characterizes the response if the stimulus frequency is below the local resonance frequency. This occurs basal-wards of the point of resonance. The cochlear partition impedance is now dominated by the stiffness term, which means that k^2 becomes a positive constant. This leads to two real roots $\pm k_\mathrm{r}$, and it follows that now (3.17) is a lossless wave equation, with the solution:

$$\psi(x, \omega) = C_{31}\, \mathrm{e}^{\mathrm{i}\, k_\mathrm{r}\, x} + C_{32}\, \mathrm{e}^{-\mathrm{i}\, k_\mathrm{r}\, x} \tag{3.21}$$

showing two waves traveling in opposite directions. Continuing the reasoning above, and realizing that the solutions for k are supposed to change gradually from one to the other, it follows that the second term, the apically traveling wave will be the prominent solution. Boundary conditions may involve a second term. In that case, please note that the sum of two waves traveling in opposite directions is equal to the sum of a standing wave and a traveling wave. The latter reflects the excess of the strongest of the initial components. Note also that only the traveling wave transfers power along the CP. In a smooth linear cochlea, i.e., without any reflections of the traveling wave, and without internal sources, the only source is the external source driving (at) the stapes. This is consistent with the above conclusion that only the wave traveling toward the apex (with amplitude C_{32}) is relevant. The traveling wave is not simple because k_r is not a constant, but it depends on x and ω as follows: $k_\mathrm{r}(x,\omega) = \omega\sqrt{2\rho/\{hs(x)\}}$. The stiffness $s(x)$ decreases with increasing x, and consequently k increases, and the wavelength $\lambda = 2\pi/k$ decreases. Hence, for a constant stimulus frequency, the phase velocity (ω/k) also decreases.

The x-dependence of the stiffness follows the Greenwood-map properties, based on its estimate of the local resonance frequencies and the constant partition mass (per unit area). Greenwood (1961, 1990) proposed the equation

$$\omega_r(x) = 2\pi\, f_r(x) = 2\pi\, 165.4\,\left(10^{2.1-60\,x} - k_G\right) \tag{3.22}$$

where x, in m, is the distance to the base.[7] The value of the parameter k_G is assumed to be between 0.8 and 1. We will use $k_G = 0.85$. This sets the most apical resonance frequency to 24.81 Hz, which might be somewhat too low: the fact that the lower frequency limit of hearing is at about 20 Hz does not necessarily require a cochlear filter tuned to that frequency; it only implies that the most apical filters must respond to that frequency.

The constant k_G is significant at the apical end of the cochlea only. At the basal end, the formula is approximated closely by the exponential form

$$\omega_r(x) = 2\pi\, f_r(x) = \omega_0\, 10^{-60\,x} = \omega_0\, \mathrm{e}^{-138.2\,x} \tag{3.23}$$

where $\omega_0 = 130832$ Hz. For the stiffness we obtain $s(x) = m\,\omega^2(x)$ or $s(x) = s_0\, 10^{-120\,x} = s_0\, \mathrm{e}^{-276.4\,x}$. Hence, the basal part of Z_CP tends to $s_0\,\mathrm{e}^{-276.4\,x}/(\mathrm{i}\omega)$. Rewriting the traveling wave equation for this condition one obtains:

$$\frac{\partial^2 \psi(x,\omega)}{\partial x^2} = -k^2(x)\,\psi(x,\omega) = -k_0^2\,\mathrm{e}^{276.4\,x}\,\psi(x,\omega) \tag{3.24}$$

which is a differential equation of the Bessel family. The results are discussed in detail in Dallos (1973) and de Boer (1980), and summarized in Sect. 3.3.1. They provide the analytical, close approximations, to the solutions sufficiently basalward from the resonance point x_r. In this basal part of the cochlea, the waves propagate practically lossless as plane waves in apical directions. As discussed above, the input power is split into equal energy fluxes in the two scalae.[8] The effect of damping is to draw power toward the partition, where it is dissipated. The physics of this effect can be analyzed realistically in 2-D or 3-D, and is discussed in more details in Sect. 3.6.

---

[7] Greenwood's original formula uses the distance from the apex, which is replaced here by $(0.035 - x)$. The partition length of 0.035 m generates the $10^{2.1}$ term.

[8] Energy flux is a less common concept in acoustics. It is discussed in more detail in Sect. 3.6.1.

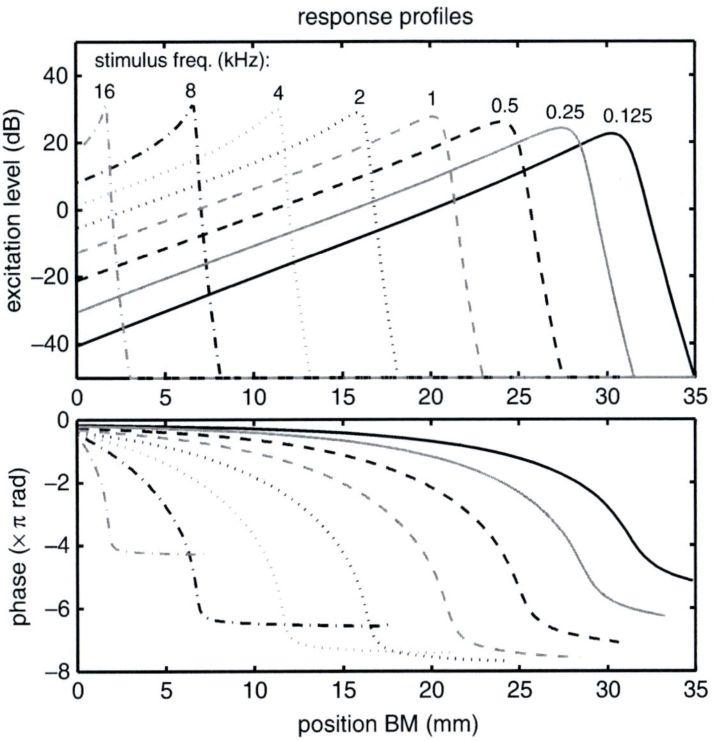

Fig. 3.3 Response profiles for single tones (0.125–16 kHz) with level response in the *top panel* (partition velocity in dB re stapes velocity) and related phases in the *lower panel*, presented from base (0 mm) to apex (35 mm). The frequency map follows Greenwood's formula (e.g., 1990). Phase information is omitted where the response amplitudes vanishes. Note that we do not use a constant Q_{3dB} for the partition m-d-s systems, as has been done in many studies, but use a more realistic changing selectivity, by taking Q_{3dB} proportional to $\sqrt{f}$. Note also that the responses differ markedly from the response of a second order system, although we have to use some caution: for a direct comparison of the result with a filter characteristic, it is necessary to obtain the local transfer characteristics at CP points. The result will look similar (e.g., de Boer 1980, Figs. 6.7 and 6.8). In particular, the apical slopes beyond the peaks are steeper. Also the total phase roll off would be only π for a second order system, here it is obviously significantly more

Other properties of the solution that are tied to the changing partition properties are the values of the stiffness and damping variables. Due to these changes, the traveling wave velocity decreases as it approaches the point of resonance. Initially (i.e., at the base) the slow down of propagation is much stronger than the increase of dissipation in the partition damping. This means that the vibration amplitude increases. Near the point of resonance dissipation exceeds propagation, and the response decreases rather sharply. The maximum response is obtained before (basalward of) the point of resonance. The precise location depends on the damping. Examples for evoked response profiles for sinusoidal stimuli are shown in Fig. 3.3.

3.3.1 Application of the WKBJ Method

The difficulty of the Fourier transformed wave equation (3.17) is that the term $k^2(x)$ is not a constant. However, as long as $k^2(x)$ varies slowly enough, an analytical approximation is available, which was first developed in theoretical physics. The method is named after its developers: Wentzel, Kramers, Brillouin, and Jeffreys. The reference "J" for Jeffreys is sometimes missing. Therefore, the interested reader should also check information on the WKB method. Note that mathematicians denote the same method as the LG method, after Liouville and Green, nineteenth-century.

Application to CMs was first proposed by Zweig et al. (1976) and followed up by de Boer (1980). We follow de Boer's approach.[9] Starting from (3.17) and dropping ω from the notation:

$$\frac{\partial^2\psi(x)}{\partial x^2} + k^2(x)\,\psi(x) = 0$$

we start from the asymptotic basal (Case 3) situation, where k^2 is real and positive, and varies relatively slowly over x. Assuming a nonzero solution

$$\psi(x) = \psi_0 \exp\left(\int_0^x \phi(\chi)\,\mathrm{d}\chi\right) \tag{3.25}$$

one finds

$$\frac{\partial\psi(x)}{\partial x} = \psi_0 \exp\left(\int_0^x \phi(\chi)\,\mathrm{d}\chi\right)\phi(x) = \psi(x)\,\phi(x)$$

$$\frac{\partial^2\psi(x)}{\partial x^2} = \psi(x)\,\phi^2(x) + \psi(x)\,\frac{\partial\phi(x)}{\partial x} \qquad \text{or,}$$

$$\psi(x)\,\phi^2(x) + \psi(x)\,\frac{\partial\phi(x)}{\partial x} + k^2(x)\,\psi(x) = 0 \qquad \text{and,}$$

$$\phi^2(x) + \frac{\partial\phi(x)}{\partial x} + k^2(x) = 0. \tag{3.26}$$

This solution is developed further by iteration. The first approximation assumes that the variation in $\phi(x)$ is small enough, or

$$\phi_1^2(x) = -k^2(x) \qquad \text{or} \qquad \phi_1(x) = \pm\,\mathrm{i}k(x). \tag{3.27}$$

---

[9] Note that De Boer's function $g(x)$ is equal to our function $k(x,\omega)$.

Choose the—sign solution, for the solution in the apical direction, and put this back in (3.26) in the form:

$$\phi(x) = \left[-k^2(x) - \frac{\partial\phi(x)}{\partial x}\right]^{1/2} = -\mathrm{i}\left[k^2(x) + \frac{\partial\phi(x)}{\partial x}\right]^{1/2} \quad \text{to obtain,}$$

$$\phi_2(x) = -\mathrm{i}\left[k^2(x) + \frac{\partial\phi_1(x)}{\partial x}\right]^{1/2} = -\mathrm{i}\left[k^2(x) + \frac{\partial \mathrm{i}k(x)}{\partial x}\right]^{1/2} \tag{3.28}$$

$$\approx -\mathrm{i}k(x) - \frac{1}{2k(x)}\frac{\partial k(x)}{\partial x} \tag{3.29}$$

where the square root is approximated by the first two terms of its series expansion, assuming that $|\partial k(x)/\partial x| \ll |k^2(x)|$. Substituting ϕ_2 in the assumed solution gives:

$$\int_0^x \phi(\chi)\,\mathrm{d}\chi = \int_0^x \left\{-\mathrm{i}k(\chi) - \frac{1}{2k(\chi)}\frac{\partial k(\chi)}{\partial x}\right\}\mathrm{d}\chi \tag{3.30}$$

$$= -\mathrm{i}\int_0^x k(\chi)\mathrm{d}\chi - \frac{1}{2}\left[\log\left(k(x)\right) - \log\left(k(0)\right)\right] \quad \text{or,}$$

$$\psi(x) = \psi_0\left[\frac{k(x)}{k(0)}\right]^{-1/2} \exp\left(-\mathrm{i}\int_0^x k(\chi)\mathrm{d}\chi\right). \tag{3.31}$$

The solution meets the accuracy condition, $|\partial\phi(x)/\partial x| \ll |\phi^2(x)|$ up to and slightly over the point of maximum response. Beyond that point, the condition "Case 1" applies. This implies that (a) there is no energy flux toward the apex and (b) the remaining standing wave decays quickly. An impression of this behavior can be obtained from Fig. 3.7, which shows a flow diagram for the upper duct (scala vestibuli), but the figure shows the result for a 2-D cochlea model. The energy flux pattern shows that the 1-D approximation of a plane wave is appropriate up to a few millimeters before the point of resonance x_r for the given stimulus.

3.4 Filterbank and Other Signal Analysis Issues

Since the original solutions to the cochlear wave equation could only be evaluated analytically for asymptotic cases, the practical responses have mostly been evaluated numerically. Also, the Fourier transformation has generally been used (time $\Rightarrow$ frequency) to remove the time derivative. These methods are suited to completely describe the characteristic of a linear time invariant system (LTI), as long as one realizes that the frequency domain variables are complex, or that both amplitude and

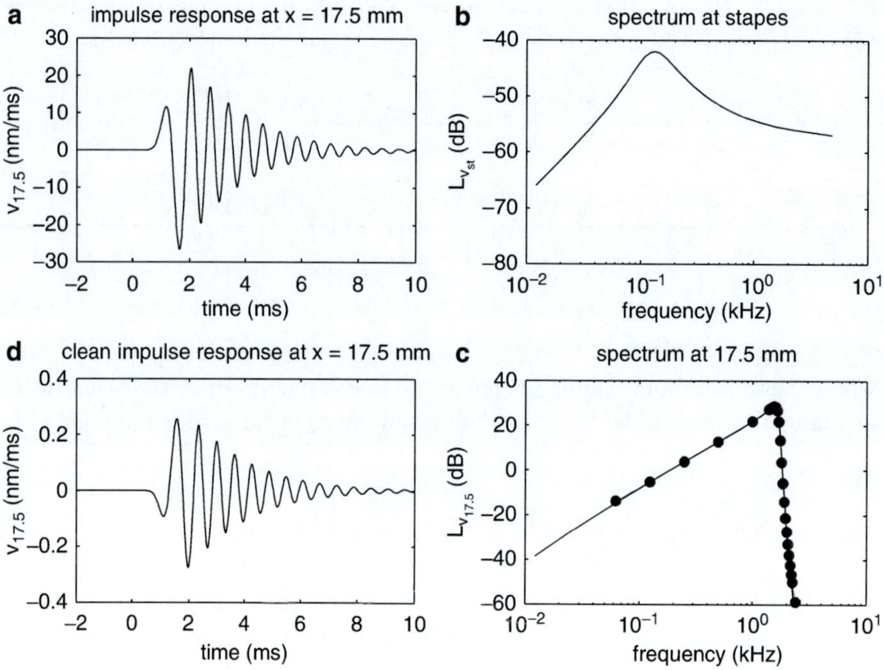

Fig. 3.4 Panel A shows a click response at $x = 17.5$ mm. A brief (50 μs) click was presented at the entrance of the ear canal at 80 dB SPL (peak level). Hence the response (*solid line*) includes the transmission through the middle ear. Panel B shows the click response spectrum at the stapes, and panel C show the cochlear transfer to point x, obtained by separating middle ear $H_{\mathrm{ME}}(\omega)$ and cochlea $H(x, \omega)$. The *dots* give single points obtained from an alternative narrow-band (pure tone) analysis. Finally, panel D gives the *FT* of the cochlear transfer function as the clean cochlear impulse response

phase data are required for complete responses. For a linear cochlea, the response behavior along the partition can be represented equally well by response profiles in the frequency domain (transfer functions $H(x, \omega)$, as in the time domain by the relevant impulse responses $h(x, t)$.

In Fig. 3.4, we show the transfer function and the impulse response for the point halfway the cochlea, at $x = 17.5$ mm from base and apex. The local resonance frequency is 1.7 kHz. Panel A shows the response to 50 μs click presented at the ear canal; B gives the response amplitude spectrum at the stapes; C shows the transfer function $|H(\omega)|$ at this point. The drawn line represents the impulse response data, and the dots are results of single tone tests. D is the inverse *FT* of C, and gives the click response of the cochlea from stapes to $x = 17.5$ mm. The minor deviations at the edges have two different sources: at the lf-end the impulse response is not very accurate because of limitation of the time window length (100 ms), and at the hf end, we run into the limited dynamic range of the *FT* (computed with an *FFT*).

Note that the zero crossings of the click response are not equidistant: they appear to converge toward $1/f_{\mathrm{R}}$, but they start with greater values. The first cycle, measured from the first +to- zero crossing, is about 0.848 ms, the subsequent five values are 0.676 ms, 0.641 ms, 0.624 ms, 0.614 ms, and 0.605 ms, which are still greater than $1/f_R = 0.588$ ms. Apparently, this phenomenon arises as the consequence of the complexity of the linear system, and in itself is not an indicator of nonlinear behavior. It was mentioned in the Fig. 3.3 caption that the phase roll off of the cochlear response is relevant. For a simple linear system, the phase lag is $\pi/2$ per order of the system. So in terms of systems analysis, it is clear that the transfer function from stapes to a point on the CP represents a higher order system than a single mass–spring resonator with damping would generate. This is due to the fact that many of such CP elements are present, and that they are coupled. In the case of the cochlea, the coupling of two adjacent oscillators depends on structural mechanical properties within the partition, and on constraints of the driving properties of the fluid environment. The 1-D transmission line models of the cochlea, basically following Zwislocki and Peterson & Bogert, focus on the fluid coupling. The transmission line interpretation of the coupling is presented in Fig. 3.6. A force (or pressure) stimulus at the stapes "finds" an approximately resistive input impedance, which means that the velocity response—and not the displacement—follows the force immediately. The fluid mass movement is constrained by its inertia. Despite the incompressibility, which gives $c \Rightarrow \infty$, separate points along the partition are limited in mechanically responding to the applied force, because the conveying fluid can not move freely. The fluid mass conservation, in combination with its inertia, provides complex constraints to the driven partition elements. They are no longer representable by second order mechanical systems. Therefore, Helmholtz' theory of the frequency selectivity of the cochlea misses an important point, and does not cover the data (see von Helmholtz 1863, Chap.VI and App. XI, and Fig. 3.5 this Section). His interpretation that the radial fiber structure within the BM dominates the longitudinal coupling remains. It stimulated development of the theory of hearing considerably—but simultaneously suppressed conflicting ideas for several decades.[10]

A general choice that has been made in the electrical analogues for the acoustic network is that we have chosen to map pressure and volume velocity on voltage (potential difference) and current. The product of p and U has the dimension of power, just like the electrical[11] $e \times i$ and the mechanical $F \times v$ products. The selection of p and U implies that the network elements have the dimension of acoustic impedance (Pa.s/m^3). More specifically, the dimension of acoustic mass is $\mathrm{M\,L^{-4}}$.

---

[10] It is noted in passing that the theoretical arguments proposed by Gold (1948) suffer from exactly the same limitation: he also tries to model separate points of the partition as independent second order systems. Moreover, he tries to match a selectivity strength characterized by a $Q_{3\mathrm{dB}}$ between 60 and 250 for the frequency range from 1 kHz to 10 kHz, a value which is nowadays considered unrealistic. Lastly, his constraints are based upon limitations of independently oscillating strings in water, but it is very doubtful that they can move independently. (The limitations predicted by him appear to match the values used in Fig. 3.3.) Although the BM properties play a significant role, the total partition stiffness contains additional contributing elements.

[11] Here e is the AC-voltage and i the AC-current.

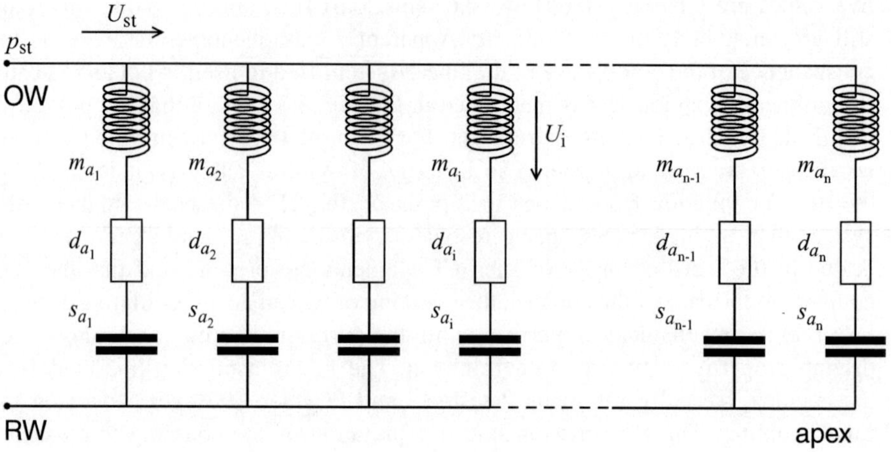

Fig. 3.5 Network representation of a 1-D cochlea model as a filterbank, as proposed by Von Helmholtz. Each section represents a filter element from the bank as an LRC line, with acoustic partition mass given by the parallel inductance $L(x) \Rightarrow m_{a_i}$, where the resistance represents the acoustic damping term $R(x) \Rightarrow d_{a_i}$, and the capacitance models the acoustic stiffness of the section $1/C(x) \Rightarrow s_{a_i}$. All elements are directly linked to the stapes pressure, and the stapes volume velocity is distributed over the branches as determined by the line impedances. U_i represents transpartition volume velocity at branch i. The property characterizing the network as filterbank is provided by the parallel wiring of all LRC-elements

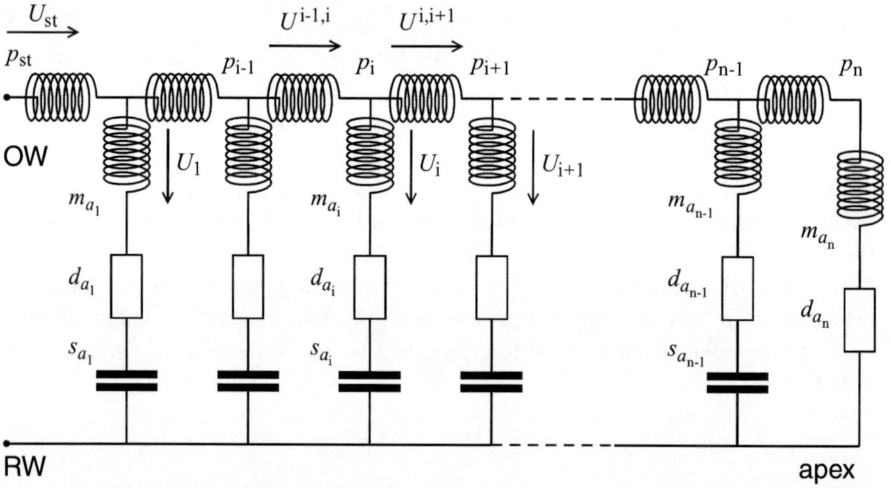

Fig. 3.6 Network representation of a 1-D cochlea model as a transmission line, following Zwislocki, and Peterson and Bogert. Each section from the transmission line represents an acoustic fluid mass, represented by the series inductance $L(\Delta x) \Rightarrow m_a^{i-1,i}$, and a partition section given as an LRC line, with acoustic partition mass given by the parallel inductance $L(x) \Rightarrow m_{a_i}$, where the acoustic resistance represents the damping term $R(x) \Rightarrow d_{a_i}$, and the capacitance models the acoustic stiffness of the section $1/C(x) \Rightarrow s_{a_i}$. Local pressure is then equivalent to voltage and volume velocity to current. Volume velocities $U^{i,j}$ apply to the duct, U_i represents trans partition volume velocity. Indicated pressures p_i represent trans partition pressure at i. The property characterizing the network as transmission line is provided by fluid mass coupling between adjacent LRC-elements

Summarizing this section, and thereby making use of some important properties of linear signal analysis:

- Yes, a linear cochlea is fully equivalent to a filterbank where the individual elements reflect all properties of cochlea sections *plus* the effects of the cochlear fluid coupling. This filterbank will be termed the equivalent LTI cochlea filterbank, or shortly: ECFB.
- No, a cochlea cannot be modeled properly with any bank of independent second order filters (or even fourth order).
- A gamma-tone filterbank (not discussed in detail [12]) approximates an ECFB.

3.5 The Impedance Concept

In Sect. 3.2 we used the mechanical impedance Z_m, and the specific acoustic impedance Z_{sa}. In Sect. 8.3 these terms are specified in detail, together with the acoustic impedance Z_a. These concepts, which are part of the standard acoustics domain, are comparable to the impedance concept defined in electricity: $Z_e = e/i$, where it has been defined as quotient of electric potential (voltage) and current.

The use of the impedance concept relies strongly on the validity of the use of (complex) spectral analysis. This means that time effects have to be analyzed using the inverse *FT*.

The analogy between the acoustic impedances and electric impedance has been used to apply electric network analysis results to CMs, both in theoretical analysis and in hardware modeling. Hardware models imply discretization of the problem, and a leading question then becomes: *What is the minimum number of discrete sections* n *that is required for a realistic description of the mechanical response of the cochlea to sound?*

A related question is: would there be an upper boundary to this number? This last question is simpler: the anatomical structure itself is discretized by the cellular constituting parts. Hair cell, supporting cells, and pillar cells, e.g., have a longitudinal size of about 10 μm. This is in reasonable accordance with a maximum of about 3,200 elements. In practical implementations, smaller numbers are easier to build in hardware, and to analyze in software (although this problem hardly exists with today's computing power, it was definitely a point before 1980, or even up to 2,000). The numbers reported in the studies up to 1980 are: Bogert (1951) used $n = 175$ in a hardware model, and the same was used in software by Hall (1974). Hubbard and Geisler (1972) based their model on the same parameters, but their version was limited to $n = 31$. Numbers of around $n = 100$ have also been used by Flanagan (1962).

---

[12] The gammatone filters were introduced by Aertsen and Johannesma (1980) and promoted as elements of a practical auditory filterbank by Patterson et al. (e.g. 1991).

For cochlea models that use global parameters, the combination of a number of coupled cells is useful. This suggest n-values between 200 and 800. Hence, Bogert's choice remains noteworthy.

The value of n defines the CP step size, Δx, which value is necessary to define impedance values of sections of the structure.

3.6 Extension to 2-D and 3-D

Although even the 1-D cochlea models usually evolve from a 3-D starting point, not many 3-D and only a few 2-D models have been analyzed during the 1950–1980 period. The major reason is that the analytical solution of the 3-D fluid mechanics was practically impossible; even more important, the 3-D dynamics of the fine structure of the OC was impossible because of lack of accurate data. Although computer power was increasing, it took additional time to achieve a reasonable 3-D computation of the fluid mechanics.

The short-wave—long-wave dispute originated from different opinions about how to reduce the 3-D structure to a 1-D model description. Although details of his model may be questionable, it seems to be justifiable to state that the first 2-D analysis had been proposed by Ranke (1950) (e.g., Siebert 1974). After a numerical analysis of an improved 2-D model by Lesser and Berkley (1972) several other studies followed during the 1970s (e.g., Lien 1973; Viergever 1977; Allen 1977; Allen and Sohndi 1979; Steele and Taber 1979a; de Boer 1979) and 1980s (e.g. Holmes and Cole 1984; Chadwick 1985; Diependaal and Viergever 1989).

The 2-D approach already gives a better physical description of the flow moving to the CP than can be obtained form the 1-D model. For the incompressible fluid approximation, mass conservation in the $x - z$-domain is expressed by Laplace's equation (3.1), or

$$\frac{\partial^2 \phi(x,z)}{\partial x^2} + \frac{\partial^2 \phi(x,z)}{\partial z^2} = 0. \tag{3.32}$$

where ϕ, introduced in Sect. 3.2 represents the commonly used potential functions, which relate the fluid particle velocity components (u in the x-direction, and w in the z-direction) as:

$$u(x) = \frac{\partial \phi(x,z)}{\partial x} \quad \text{and} \quad w(z) = \frac{\partial \phi(x,z)}{\partial z}.x \tag{3.33}$$

Now we have additional boundary condition for the z-dimension, viz., no fluid flow through the (hard) walls. This means that $w(z)$ is 0 at the walls. (The similar condition at the apex does apply to the 1-D case already.) Equation (3.32) also applies directly to the pressure (consider the proof as an exercise):

$$\frac{\partial^2 p(x,z)}{\partial x^2} + \frac{\partial^2 p(x,z)}{\partial z^2} = 0. \tag{3.34}$$

Next, combine (3.34) with the CP wave equation (3.13) and use the function $k^2(x,\omega)$ defined in (3.17), to obtain:

$$\frac{\partial^2 p(x,z)}{\partial x^2} = -k^2\, p(x,z) \qquad \text{and,} \tag{3.35a}$$

$$\frac{\partial^2 p(x,z)}{\partial z^2} = k^2\, p(x,z). \tag{3.35b}$$

In general, k is complex, or $k(x,\omega) = k_{\mathrm{r}}(x,\omega) + \mathrm{i}k_{\mathrm{i}}(x,\omega)$, as stated before.

The considerations given in Sect. 3.3 with regard to the different cases of k are still applicable, but there is an additional boundary condition for the z-direction. A real positive k^2 is equivalent to real wave numbers in the x-direction, which implies imaginary wave numbers, or attenuation or amplification[13] in the z-direction and vice versa.

The solution for the pressure wave in the z-direction at $x = x_1$ is straightforward:

$$p(x_1,z) = C_{x_1,1}\mathrm{e}^{k(x_1)z} + C_{x_1,2}\mathrm{e}^{-k(x_1)z} \tag{3.36}$$

The boundary condition at $z = h$, at the wall, must imply that at that point the pressure gradient in the direction of the wall vanishes: $\partial p(x,z \to h)/\partial z = 0$, and at $z = 0$ the pressure at the CP is part of the trans-partition pressure. There is only 1 solution that meets the condition of the vanishing derivative. This is the solution

$$p(x_1,z) = C_{x_1,3}\cosh\{k(x_1)\,(h-z)\} \tag{3.37}$$

in which case the derivative, which contains $\sinh\{k(x_1)\,(h-z)\} = 0$ if its argument approaches 0, or if $z \to h$, and $C_{x_1,3}$ together with $\mathrm{e}^{(\pm kh)}$ represent $C_{x_1,1}$ and $C_{x_1,2}$. The remaining factor matches the condition at $z = 0$, or

$$p(x_1,0) = C_{x_1,3}\cosh\{k(x_1)h\} \qquad \text{and} \qquad C_{x_1,3} = \frac{p(x_1,0)}{\cosh\{k(x_1)h\}}. \tag{3.38}$$

Given this profile we can also compute the average pressure over the height at position x_1:

$$\begin{aligned} p(x_1,\bar{z}) &= < p(x_1,z) >_{\mathrm{av}(z)} = \frac{1}{h}\int_0^h p(x_1,z)\,\mathrm{d}z \\ &= \frac{p(x_1,0)}{\cosh\{k(x_1)h\}}\frac{1}{h}\int_0^h \cosh\{k(x_1)h\}\,\mathrm{d}z \\ &= p(x_1,0)\frac{\tanh\{k(x_1)h\}}{\{k(x_1)h\}} = \frac{p(x_1,0)}{\alpha(x)}. \end{aligned} \tag{3.39}$$

[13]The possibility of expanding solutions was discussed in Sect. 3.3, and can be neglected most of the time.

In a study that aimed to address nonlinear properties, Duifhuis (1988) introduced the parameter $\alpha(x)$ to insert a 2-D compensation term in the cochlear wave equation. While still neglecting significant rotation effects it is possible to introduce a correction factor in Euler's equation (3.8) where the volume acceleration in the x-direction is related to the pressure gradient in that direction. If there is not a pure plane wave, i.e., if the pressure varies over z, then a first approximation of a better solution is to use the average pressure value for the horizontal fluid, whereas for the trans partition equation the pressure at $z = 0$ is relevant. For the plane wave these are equal, in the 2-D approximation they differ by the factor $\alpha(x)$ that was derived above. Taking this into account in the wave equation just requires the modification of k^2 term to

$$k^2(x) = -\frac{2\mathrm{i}\,\omega\,\rho\,\alpha(x)}{h\,Z_{\mathrm{CP}}(x)}. \tag{3.40}$$

In the long-wave approximation the product $k(x)h$ will be much less than 1, and $\alpha(x) \approx 1$. In the short-wave case, when $k(x)h \gg 1$ the tanh-function will approach 1, so that $\alpha(x) \approx k(x)h$. The combination of (3.40) and (3.39) leads to the following equation for $k(x)$:

$$k(x)h\ \tanh\{k(x)h\} = -\frac{2\mathrm{i}\,\omega\,\rho\,h}{Z_{\mathrm{CP}}(x)}. \tag{3.41}$$

This equation is equivalent to the eikonal equation in optics. It does not have an analytic solution, but current tools in mathematical packages provide means for efficiently finding numerical answers.[14] It is used in the next section to compute the energy flux through the cochlea.

The general solution for the pressure wave traveling in the positive x-direction is:

$$p_v(x,z) = p_0\,F(x)\,\cosh\{k(x)\,(h-z)\}\,\exp\left(-\mathrm{i}\int_0^x k(\chi)\,\mathrm{d}\chi\right). \tag{3.42}$$

Note that z occurs only in the cosh term. The term $F(x)$ is relatively slowly varying scaling factor. Finally, for a purely traveling wave, where $k = k_{\mathrm{r}}$ represents the wave number in the x-direction, the argument of the exponent reflects the phase delay. The x-factor can be expressed in several equivalent forms, none of which becomes very simple, e.g.,

$$F(x) = \frac{k(0)h}{\tanh(k(0)h)}\left[\frac{k(0)h/\cosh^2(k(0)h) + \tanh(k(0)h)}{k(x)h/\cosh^2(k(x)h) + \tanh(k(x)h)}\right]^{1/2}\frac{1}{\cosh(k(x)h)}. \tag{3.43}$$

This solution is the WKB solution for the 2-D cochlea presented by Steele and Taber (1979a) and by Viergever (1980).

---

[14] A simple Matlab© script, named "eikonal.m" is available on the URL.

In the 1-D long wave approximation $k(x) \ll 1$ for all x, and in that case the z-factor approaches 1, and $F(x)$ reduces to $\sqrt{k(0)/k(x)}$, which gives for the 1-D long wave solution (3.31):

$$p_v(x) = p_0 \left[\frac{k(0)}{k(x)}\right]^{1/2} \exp\left(-\mathrm{i}\int_0^x k(\chi)\,\mathrm{d}\chi\right). \tag{3.44}$$

as was first shown by Zweig et al. (1976).

The 2-D approach is sometimes compared to the behavior of surface waves in shallow water. The similarity is a relevant profile is apparent at a "surface," and that the scala wall has some similarity to the shallow bottom. Fundamental differences are that the CP is not simply comparable to a water–air interface, and neither is the CP impedance comparable to the water–air surface tension. A basic difference is that the CP properties depend on place. Also, for CMs we usually neglect the effect of gravity, which is crucial in surface waves. What remains is similarity in the fluid motion. That is why a general application of the Navier–Stokes formulation remains powerful, and the simplifications should be used with care.

One of the first 3-D studies, by Steele and Taber (1979b), presents a 3-D analysis of the WKB approximation, but compares their results primarily to results of hardware fluid mechanical models. Two conclusions of their parameter analysis are that (1) the tapering of the scalae has little effect, and that (2) damping has also little effect. Moreover, they can conclude from a comparison of 1-D, 2-D and 3-D results that

> the popular 1-D "long wavelength" mathematical model is virtually of no value for quantitative investigation of cochlear function.

For the same parameters, the 3-D model gives a sharper peak. This is consistent with the notion of the energy flux in the cochlea: for a narrow band signal, this flow focuses at (or before) the point of resonance, and when it focuses more sharply in the 3-D case—longitudinal and radial sharpening—it will generate a sharper peak.

3.6.1 Energy Flux and Dissipation

Before going into more detail about the flow of acoustic power or energy through the cochlear fluid, we will take a closer look at these concepts. We start from the well defined concepts of power P, expressed in W, the energy E, expressed in Ws, J or Nm, and sound intensity I, expressed in W/m^2. This is a good starting point because power and energy are supposed to be common knowledge, and acoustic intensity is presented in most acoustic textbooks. Much less common are the acoustic definitions of *sound energy* and *sound energy density*, and of *sound-energy flux*, *sound-energy flux density* and *sound-power density*. Proper definitions are given in the acoustic standard ANSI S1.1-1994 (including 2005 erratum).

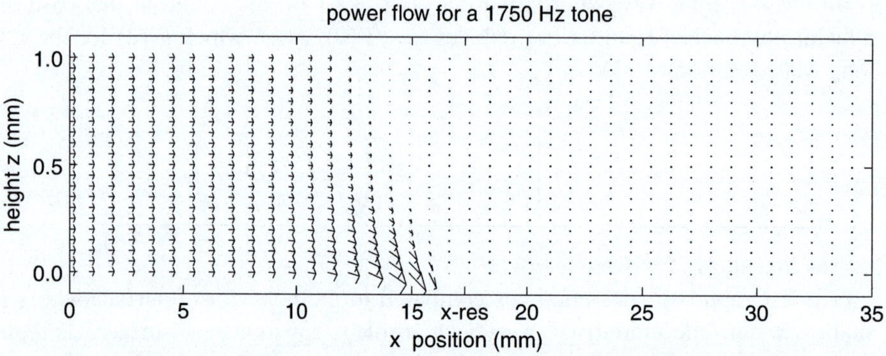

Fig. 3.7 Energy flux profile in the upper scala computed for a 1750 Hz tone. The flow vectors are scaled linearly, and matched to the horizontal and vertical length scales. The height scale is expanded. Near the base ($x = 0$) the profile is that of a plane wave. At $x = 10$ mm, it becomes apparent that the flow near the CP begins to increase, and such at the cost of the flow at greater heights. The associated change in direction becomes clear at 12 mm, and now the flow into the CP (at $z = 0$) shows increasing dissipation. All power has vanished before it could reach $x - res$, the 1750 Hz resonance point, and no transport occurs beyond that point

The important characteristics are:

sound energy:	amount of energy exclusively attributable to sound (J)
sound energy density:	sound energy per volume unit (J/m^3)
sound-energy flux:	flow of sound energy per unit of time through a specified area (W); symbol, J
sound-energy flux density:	= sound intensity I (W/m^2)
sound-power density:	= sound intensity I (W/m^2)

Even though a full 3-D analysis will give a more precise result, the somewhat simpler 2-D analyses already provides a reasonable insight in the energy flux trough the cochlea. The result of the analysis is, as mentioned in Sect. 3.2, that the energy flux starts as a laminar scalar fluid flow, carried by a plane wave in apical direction. As the point of resonance approaches, part of the flow tends to be directed toward the CP, where it is dissipated. Figure 3.7 shows the energy flux profile, as computed using (3.42) and (3.43), computed for a sinusoidal stimulus of 1750 Hz at the arbitrary stapes pressure of 1 Pa. The pressure value is arbitrary because the system is linear, and the response profile is scaled linearly to provide a reasonable flow profile.

The energy flux in, e.g., the upper duct is $\mathbf{J} = \mathbf{J}(x, w)$ with the components:

$$\mathbf{J}(x) = < p(x,z) \cdot u(x,z) >$$
$$\mathbf{J}(w) = < p(x,z) \cdot w(x,z) > \tag{3.45}$$

where the $\cdot$ implies a dot product. The particle velocity at (x, z) has the components u in the x-direction and w in the z-direction. The particle velocity $u_v(x, t)$ was given in (3.8). Dropping the index for scala vestibuli, and taking the *FT*, we are left with

$$u(x) = -\frac{1}{\mathrm{i}\omega\rho}\frac{\partial p(x, z)}{\partial x}, \qquad \text{and} \tag{3.46}$$

$$w(z) = -\frac{1}{\mathrm{i}\omega\rho}\frac{\partial p(x, z)}{\partial z}. \tag{3.47}$$

We obtain the derivatives from (3.42), the WKB approximation for the pressure. The z-derivative is simple because z occurs only in the cosh function; x, however, occurs in the factor, and in k, which appears in two places.

With some effort, it should be possible to verify that

$$\frac{\partial p(x, z)}{\partial x} = p(x, z)\left[\frac{F'(x)}{Fx} - \mathrm{i}\,k(x) + (h - z)\frac{\partial k(x)}{\partial x}\ \tanh\{k(x)(h - z)\}\right] \tag{3.48}$$

$$\frac{\partial p(x, z)}{\partial z} = -p(x, z)\,k(x)\ \tanh\{k(x)(h - z)\} \tag{3.49}$$

which implies that the averaged pressure–velocity dot products (3.45) reduces to the form $< p \cdot p\,c >$, with the result $\frac{1}{2}|p^2|\,\Re\{c\}$, or:

$$\mathbf{J}(x) = \frac{|p^2(x, z)|}{2\omega\,\rho}\,\Re\left[\mathrm{i}\left(\frac{F'(x)}{F(x)} - \mathrm{i}\,k(x) + (h - z)\frac{\partial k(x)}{\partial x}\ \tanh\{k(x)(h - z)\}\right)\right]$$

$$\mathbf{J}(w) = \frac{|p^2(x, z)|}{2\omega\,\rho}\,\Re\left[-\mathrm{i}k(x)\ \tanh\{k(x)(h - z)\}\right] \tag{3.50}$$

These results are used to generate Fig. 3.7.

3.7 Micromechanics

The CP contains the sensory cells (IHCs and OHCs) that transmit information to the auditory neural network. One basic question regarding the microstructure concerns the quantitative relation between information transmission and energy dissipation. How much of the acoustic power that enters the cochlea is transmitted (converted) to the brain or dissipated (and lost) in the transmission process, and how much is lost in viscous and other (fluid-) mechanical losses.

Obviously, the microstructure of the cochlear partition must contain the information that is necessary to answer the above questions. Over the last decades some progress has been made into this direction, but almost all of the many studies in this area remain in the initial phase of probing interesting ideas.

The basic reason for this is that the structure is too complex. Selecting a few parameters from the complete set tends to remain arbitrary. On the other hand, the full set is too large. In that case, one is easily fooled by a simplistic model with too many free parameters that cannot be based on proper independent information.

Over the last decades a significant amount of information has become available, in particular, focused on hair cell properties. More detailed structural information has been presented about the organ of Corti structure. Tests have been done concerning the mechanical role of the tectorial membrane, or just a partial hair cell fine-structure. Partial, because so far, all studies that consider the cross-sectional OC and scalae seem to limit analysis of cell structure to connections within that plane, disregarding, e.g., the spiral structure in the longitudinal direction (Corti's tunnel).

In other words, to make progress, work needs to be done in order to be able to estimate a sufficiently accurate structure based on verified data.[15] The situation is very similar to the one described by Holmes and Cole (1984), from the introduction of which I take two quotes:

> The problem is made more difficult by the current state of the experimental evidence that is available.

and

> It is with this rather unsettling state of affairs that we consider the three-dimensional hydroelastic model of the cochlea that is outlined below.

Almost all modeling and analysis that has been done along this line used linear analysis tools that are not very helpful for the further analysis of the fundamental nonlinear properties.

3.8 Problems with the Linear Cochlea

Some nonlinear hearing phenomena had been know for centuries, but mostly they tended to be seen as by-products of auditory processing. However, claims emerged that they probably were more than just that. Moreover, additional nonlinear phenomena were observed, all pointing to a cochlear source. In this section, we list the most relevant phenomena that arose before 1980. They will be discussed in more detail in Chap. 4

- *Dynamic Range*
  The dynamic range of an intact human ear is between 100 dB and 120 dB. Sensitivity along the scale is relative ($\Delta I/I$ is approximately constant). At the output, the auditory nerve fibers, the dynamic range is probably between 30 dB and 50 dB. The range is compressed by a factor between 2 and 4.

---

[15] During the 2008 International Workshop on Mechanics of Hearing about half of the contributions related to microstructure or substructure properties.

- *CTs*
  Aural CTs, known to be audible for centuries, have now been proven to originate within the intact cochlea. They clearly are (by)products of a nonlinear process.
- *Emissions*
  Toward the end of the 1970s, the first data on evoked (Kemp 1978) and later on spontaneous auditory emissions were presented (Kemp 1979). In particular, the presence of spontaneous emissions suggests a cochlear source with nonlinear properties.
- *Linearization through OHC loss*
  A final important finding is that cochlear hearing loss reduces or even eliminates otoacoustic emissions and CTs, and also linearizes the cochlea, thereby limiting the input dynamic range to the output dynamic range. The hearing loss is accompanied by the threshold increase.

3.9 Where to Go from Here?

All observed nonlinear phenomena, some quite old and some very new, indicate at this point that a linear model of the cochlea will not be a reasonable model for the normal mammalian cochlea. Some of the nonlinear effects, like CT generation, and otoacoustic emission persist to very low levels, even below threshold. Hence, modeling the nonlinearities as minor distortion effects of a basically linear system is questionable. Apparently, we are unable to define below what level the system becomes linear. It is not obvious if there is only one major source of nonlinear behavior, or that more independent sources exist.

These considerations led us to a different approach to CMs modeling, viz. to a time domain analysis of a nonlinear model. This approach comes with the disadvantage that the tools for nonlinear system analysis are extremely limited, and that the more common knowledge of linear analysis sometimes can put us on the wrong track.

Part II will address these consequences. Selected implications are worked out in Part III, together with some remaining open issues.

References

Aertsen AMHJ, Johannesma PIM (1980) Spectro-temporal receptive fields of auditory neurons in the grassfrog. Biol Cybernetics 38:223–234, URL http://dx.doi.org/10.1007/BF00337015, 10.1007/BF00337015

Allen JB (1977) Two-dimensional cochlear fluid model: New results. J Acoust Soc Am 61:110–119

Allen JB, Sohndi MM (1979) Cochlear macromechanics: Time domain solutions. J Acoust Soc Am 66:123–132

ANSI S (2005) Acoustical Terminology. American National Standards Institute, Inc., including 2005 updates de Boer E (1979) Short-wave world revisited: resonance in a two-dimensional cochlear model. Hear Res 1:253–281

de Boer E (1980) Auditory physics. Physical principles in hearing theory. I. Physics Reports 62:87–174

Bogert BP (1951) Determination of the effects of dissipation in the cochlear partition by means of a network representing the basilar membrane. J Acoust Soc Am 23:151–154

Chadwick RS (1985) Three dimensional effects on low frequency cochlear mechnics. Mech Res Commun 12:181–186

Cooper NP, Kemp DT (eds) (2008) The Biophysics of Hearing, Mechanics of Hearing, World Scientific, Singgapore

Dallos P (1970) Low-level auditory characteristics: Species dependence. J Acoust Soc Am 48:489–499

Dallos P (1973) The Auditory Periphery. Academic, New York

Diependaal RJ, Viergever MA (1989) Nonlinear and active two-dimensional cochlear models: Time-domain solution. J Acoust Soc Am 85:803–812

Duifhuis H (1988) Cochlear macromechanics. In: Edelman GM, Gall WE, Cowan WM (eds) Auditory Function. Neurobiological basis of hearing, A Neurosciences Institute Publication, Wiley, New York, chap 6, pp 189–211

Flanagan JL (1962) Computational model for basilar-membrane displacement. J Acoust Soc Am 34:1370–1376

Fletcher H (1951) On the dynamics of the cochlea. J Acoust Soc Am 23:637–645

Gold T (1948) Hearing. II. The physical basis of the action of the cochlea. Proc Royal Soc London, Series B, Biol Sc 135(881):492–498

Greenwood DD (1961) Critical bandwidth and the frequency coordinates of the basilar membrane. J Acoust Soc Am 33:1344–1356

Greenwood DD (1990) A cochlear frequency-position function for several species—29 years later. J Acoust Soc Am 87:2592–2605

Hall JL (1974) Two-tone distortion products in a nonlinear model of the basilar membrane. J Acoust Soc Am 56:1818–1823

von Helmholtz HLF (1863) Die Lehre von den Tonempfindungen, 1st edn. Vieweg und Sohn, Braunschweig, english edition: On the Sensations of Tone, transl. by A.J. Ellis (1885) of 4th German edition (1877), publ. by Dover in 1954

Holmes MH, Cole JD (1984) Cochlea mechanics: Analysis for a pure tone. J Acoust Soc Am 76:767–778

Hubbard AE, Geisler CD (1972) A hybrid-computer model of the cochlear partition. J Acoust Soc Am 51:1895–1903

Kemp DT (1978) Stimulated acoustic emissions from within the human auditory system. J Acoust Soc Am 64:1386–1391

Kemp DT (1979) Evidence of mechanical nonlinearity anf frequency selective wave amplification in the cochlea. Arch Otorhinolaryngol 224:37–45

Lamb H (1895) Hydrodynamics, 2nd edn. Cambridge University Press, Cambridge

Lesser MB, Berkley DA (1972) Fluid mechanics of the cochlea. Part 1. J Fluid Mech 51:497–512

Lien MD (1973) A mathematical model of the mechanics of the cochlea. PhD thesis, Washington Univ. School of Medicin, St. Louis

Lighthill J (1981) Energy flow in the cochlea. J Fluid Mech 106:149–213

Lynch TJ, Nedzelnitsky V, Peake WT (1982) Input impedance of the cochlea in cat. J Acoust Soc Am 72:108–130

Patterson RD, Robinson K, Holdworth J, McKeown D, Zhang C, Allerhand M (1991) Complex sounds and auditory images. In: Cazals Y, Demany L, Horner K (eds) Auditory physiology and perception, Pergamon Press, Oxford, pp 429–443

Peterson LC, Bogert BP (1950) A dynamical theory of the cochlea. J Acoust Soc Am 22:369–381

Puria S, Allen JB (1991) A parametric study of cochlear imput impedance. J Acoust Soc Am 89:287–309

Ranke OF (1950) Theory of operation of the cochlea: A contribution to the hydrodynamics of the cochlea. J Acoust Soc Am 22:772–777

Rayleigh JWS (1896) The Theory of Sound, vol II, 2nd edn. MACMILLAN, London, [Dover edition: 1945]

Siebert WM (1974) Ranke revisited – a simple short-wave cochlea model. J Acoust Soc Am 56:595–600

Steele CR, Taber LA (1979a) Comparison of WKB and finite difference calculations for a two-dimensional cochlear model. J Acoust Soc Am 65:1001–1006

Steele CR, Taber LA (1979b) Comparison of WKB calculations and experimental results for three-dimensional cochlear models. J Acoust Soc Am 65:1007–1018

Viergever MA (1977) A two-dimensional model of the cochlea II. The heuristic approach and numerical results. J Eng Math 11:11–28

Viergever MA (1980) Mechanics of the inner ear. PhD thesis, Delft University of Technology, Netherlands

Zweig G, Lipes R, Pierce JR (1976) The cochlear compromise. J Acoust Soc Am 59:975–982

Zwislocki J (1948) Theorie der Schneckenmechanik: qualitative und quantitative Analyse. Acta Oto-Laryngol Suppl 72:1–72

Zwislocki J (1950) Theory of the acoustical action of the cochlea. J Acoust Soc Am 22:778–784

Zwislocki J (1953) Review of recent mathematical theories of cochlear dynamics. J Acoust Soc Am 25:743–751

Part II
Anatomy and Function of the Nonlinear Cochlea

The second part of the book starts with a report of the experimental and theoretical developments around 1980. It focuses on nonlinear phenomena, in particular on oto-acoustic emissions. The interest in CT generation, and even more so in two-tone inhibition or suppression, greatly diminished, although CTs interest soon came back through DPOAEs.

The second chapter presents the core of this book: it proposes and develops a time-domain model of the cochlea that can deal with the nonlinear problems as elementary parts of an intact cochlea. The setup provides a working, and still expandable, framework. Some of the first tests are mentioned in this part. Further evaluation is postponed to Part III.

Chapter 4
Nonlinear Auditory Phenomena (I) Knowledge Around 1980

Abstract This chapter describes characteristics of auditory phenomena that appear to be attributable to a properly described nonlinear cochlea. In view of a reliable quantification, the discussion is here largely limited to relatively recent experimental data.

4.1 Aural Combination Tones

Aural CTs have been of interest to both musicians and scientists for a long time. The scientist have presented detailed quantitative information and have developed several hypotheses and theories, on the one hand based on mathematics and on the other on anatomical and physiological knowledge. The following sections will primarily address psychoacoustical, neurophysiological, and mathematical developments. Distortion product otoacoustic emissions (DPOAEs) are addressed in Sect. 4.4.

4.1.1 Psychoacoustics of Combination Tones

The term *combination tone* (CT) was introduced by Vieth in 1805 to denote a third, lower tone that was perceived when two other tones were presented.

> Weil es bequem ist, für ein einzelnes Phänomen ein einzelnes Wort zu haben, so sey es mir erlaubt, diesen Ton, der aus der Combination der Schwingungen zweier anderer Töne entsteht, *Combinationston* zu nennen.[1]

[1]Because it is useful to have a single term for a single phenomenon, I would like to be permitted to designate this tone, which arises from the combination of the vibrations of two other tones, as *combination tone*.

H. Duifhuis, *Cochlear Mechanics: Introduction to a Time Domain Analysis of the Nonlinear Cochlea*, DOI 10.1007/978-1-4419-6117-4_4,

Vieth reflects on the discussion amongst physicists and musicians, in particular in the English journals of that time, and reports contradictory interpretations from T. Young and J. Gough. Basically, Young proposes that the CT, because it sounds very similar to the presented tones, is generated acoustically—proposing a similarity with surface waves, which is disputed by Vieth—whereas Gough claims that it is "mental and imaginary". At the time, the frequencies of the CTs were marked by musical notes, which are often not sufficiently accurate to characterize the CT as odd ($2f_1 - f_2$ family) or even (the difference tone $f_2 - f_1$ and its overtones). Discussions about the source remained overshadowed by the imprecision of the description of the phenomenon for a long time. Helmholtz (1863), e.g., reports only the even order difference tones ("differential tones" in Ellis' 1877 translation) referring to those as the first class, and he claims to have discovered himself a second class of summation tones. He proposes that the source would be in the middle ear: eardrum and/or ossicle joints, and that CT generation occurs only at high levels when the small amplitude linear approximation of middle ear operation is no longer valid.

Accurate systematic studies of frequencies and strengths of the CTs were made almost a century later, e.g., by Zwicker (1955) and Plomp (1965). The latter also gave an extensive historical introduction. Starting with an analysis of available data, Plomp arranges these in diagrams which are related to the groups of odd- and even-order intermodulation products.

Combination Tone Terminology

The term *combination tone* originally just referred to an audible low pitched third tone that becomes audible when a second tone is presented simultaneously to another tone.

From this point on, we will consider the primary tones to be purely sinusoidal with frequencies f_1 and f_2, where $f_1 < f_2$. The tones have similar amplitudes.

The audible CTs have frequencies $m\,f_1 - n\,f_2$, where m and n are integers and $m\,f_1 > n\,f_2$. The CTs belong to the even family if $m + n$ (and therefore also $m - n$) is even, and to the odd family otherwise.

In signal analysis, the family of components with frequencies $m\,f_1 \pm n\,f_2$ is called the family of intermodulation products, which arises with nonlinear time-invariant systems (also called: systems with instantaneous nonlinearity). Any such nonlinear function can be written as the sum of an even component and an odd component (see Fig. 4.1).

It can be shown that an even nonlinearity only generates even order intermodulation products, and an odd nonlinearity generates only odd-order intermodulation products. Formally the overtones or harmonics also belong to the family, but in those cases either m or n is equal to 0.

If the nonlinearity can be approximated by a power-law nonlinearity, $y = x^{\nu}$ where ν can be any real number, then the amplitudes of the intermodulation products can be related analytically to the amplitudes of the f_1 and

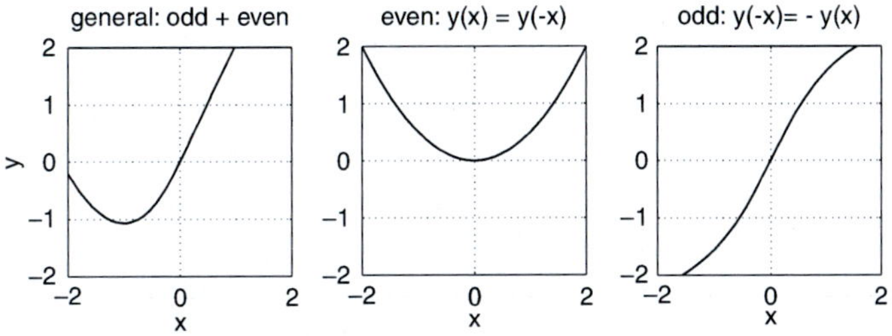

Fig. 4.1 Schematic nonlinear functions. *Left*: a general arbitrary function can always be decomposed in odd and even parts. *Middle*: even function, output symmetric around vertical axis. *Right*: odd function, output point symmetric with respect to origin

f_2 components. This is straightforward for quadratic or cubic nonlinearities ($\nu = 2$ or 3), but it can also be done for noninteger exponents. Of particular interest for hearing is the case where $\nu < 1$. For that case the common Taylor-expansion is rather useless (because of its very slow convergence), but the related Bernstein-expansion is useful. The odd function shown in Fig. 4.1 is of the type "compressive", which means that the slope decreases with increasing x. Similarly, the even function shown in the middle panel is of the type "expansive". But these relations were purely coincidental: both odd and even functions can be either expansive or compressive.

The term *CDT* or *cubic difference tone* for $2f_1 - f_2$ (used, e.g., by Zwicker 1955, 1979 and Plomp 1965 any in many studies thereafter) is an unfortunate misnomer, and should be avoided. It suggests, a.o., that $2f_1 - f_2$ is (must be) generated by a cubic nonlinearity ($\nu = 3$). The truth is that:

- Practically any odd nonlinearity generates $2f_1 - f_2$, and other members of the odd-order family.
- The auditory nonlinearity is *not* expanding whereas a cubic exponent is!

In addition, the pure cubic nonlinearity does not generate intermodulation products beyond $2f_1 \pm f_2$, $2f_2 \pm f_1$, $3f_1$ and $3f_2$. So it misses distortion products that are perceived, and overestimates others. A similar concern applies to the use of the term *quadratic* for the difference tone, where even order would be the appropriate alternative.

In line with the earliest reports (before 1800), only the CTs with frequencies below the values of the primaries are perceptible. We show that the other high-frequency CTs are also generated but not effectively fed back into the cochlea; they remain masked below the excitation profile of the primaries.

Plomp's diagrams provide numerical relations between the primary tone frequencies and the CTs, covering intermodulation frequencies by straight lines. All experimental points were on or close to the $mf_1 - nf_2$ lines. Most CTs occur at m, n combinations for which the CT is lower than f_1. Plomp also measured the primary level thresholds that were required to elicit audible CTs, at the primary frequencies 800 and 1,000, and 800 and 1,400. He concluded, a.o., that intra-subject variability is large, that the predominant CTs occurred at $f_2 - f_1$, $2f_1 - f_2$, and $3f_1 - 2f_2$), and that CT thresholds were significantly lower for the smaller frequency interval between the primaries. Plomp considered this to disprove the Helmholtz hypothesis that CTs were generated in the middle ear.

Goldstein (1967) obtained estimates of the CT strength by two different techniques: the loudness balance method, where the level of a presented single tone is adjusted to match its loudness to that of the CT. Secondly, he proposed the cancellation method, where the CT-percept is suppressed by adding a third tone at the CT frequency and where phase and amplitude of the third tone are adjusted to minimize the loudness level of the CT. The last method may seem cleaner at first sight, but it should be noted that it involves three-tone interaction, which complicates its analysis. The basic result, however, was very important: for equal primary levels $L_1 = L_2$, the $2f_1 - f_2$ CT was detectable down to threshold, and its level changed linearly with the primary levels. This ruled out the low-level linearity, and the cubic nonlinearity, because for a cubic nonlinearity the distortion product amplitude should grow as $A_1^2 A_2$ (see Table 4.1), and therefore the level growth should follow $\Delta L_{CT} = 2\,\Delta L_1 + \Delta L_2$.

The results were confirmed in an extensive study by Smoorenburg (1972a), who varied levels and frequencies of the primaries systematically. An example of his results is shown in Fig. 4.2. In a similar experiment, Zwicker (1979) measured four to five points on the tops of such curves, over a range from 30 to 70 dB. For $f_1 = 1{,}620$ and $f_2 = 1{,}728$ Hz, i.e., with a smaller ratio $f_2/f_1 = 1.067$, he finds somewhat larger L_{CT} levels. Above 40 dB, the peaks grow practically linearly with level. The decaying slopes for $L_2 > L_1$ tend to be steeper than -1 dB/dB (estimate is based on two points per line), the increasing slopes cannot be estimated reliably. Zwicker concludes:

> Nonsimultaneous (Smoorenburg) as well as simultaneous (Zwicker) measurements on the time and duration-dependence of the cubic difference tone have indicated that the nonlinear characteristic involved works instantaneously, similar to the behavior of masking.

We will address the masking behavior in Sect. 4.2.1.

The point raised here is: are we dealing with an instantaneous nonlinearity, or with a feedback system? A feedback system could be linear and still compress the dynamic range, but it would not generate "instantaneous-like" distortion products. On the other hand, if a feedback is extremely fast, it remains to be seen if it can be distinguished from an instantaneous nonlinearity. This issue is of theoretical interest, and was raised explicitly by Smoorenburg (op. cit.). On the basis of his data, he allows for a feedback within 20 ms. [Nowadays a 20 ms feedback would be well discernible from a 2 ms or faster feedback.] Others apparently just assumed an

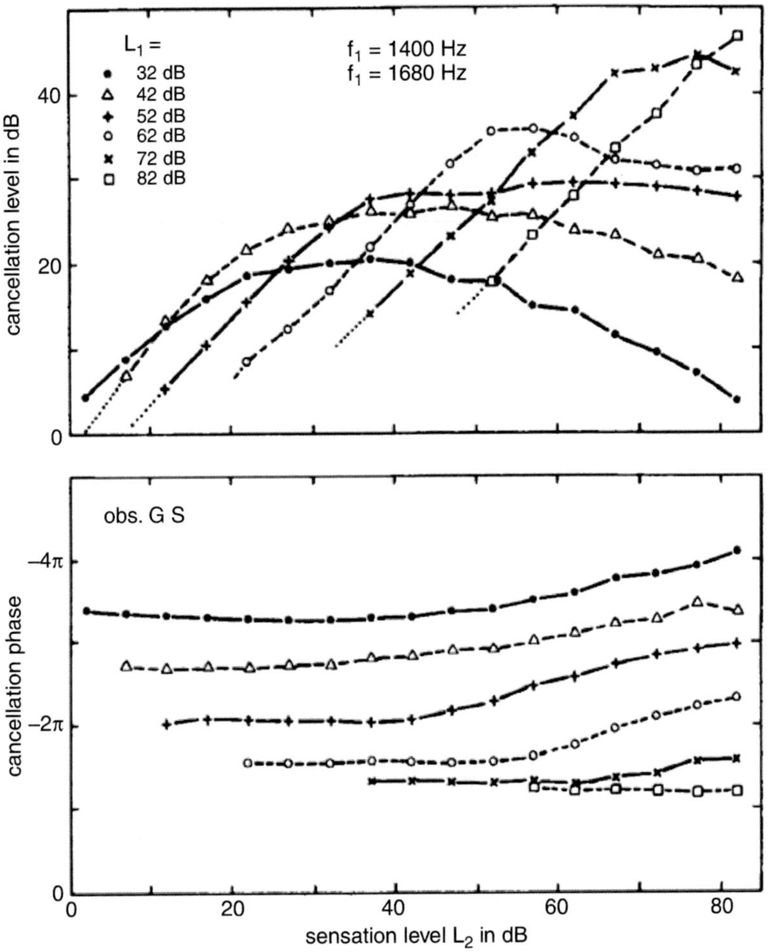

Fig. 4.2 Cancellation measurements of $2f_1 - f_2$ levels (*upper panel*) and phase (*lower*). Responses are measured as a function of L_2, with L_1 as a parameter. *Dotted line* extensions of data lines are extrapolations to the points where the CT was just audible. Note how close all rising slopes approach slope of 1 dB/dB; beyond the points where $L_1 = L_2$, which occur close to the maxima, further increase of L_2 tends to produce a decreasing slope, between 0 and −1. The phases are close to constant up to the point of equal levels, and to grow slightly for higher levels. Reprinted with permission from GF Smoorenburg, Journal of the Acoustical Society of America, Vol. 52, Page 615 (1972). Copyright 1972, Acoustical Society of America

instantaneous nonlinearity. Goldstein (op. cit.) used the term *essential*, and before that the primary question was not so much the timing, but the size and range of occurrence of the level effects: distortion was generally assumed to take place at (very) high levels, similar to what was known from many instruments. Zwicker and Plomp found that the occurrence extended to much lower levels. In this context, Smoorenburg proposed

Assume that the nonlinearity can be described by a νth-law device described by the odd transfer function $f(x) = x^\nu$ if $x > 0$ and $f(x) = -|x|^\nu$ if $x < 0$; $\nu > 0$, $\nu \neq 1$. A canceled CDT means that the Fourier component $2f_1 - f_2$ at the output of the νth-law device equals zero.

and continues with the implication that this type of nonlinearity compresses the response amplitude but does not change the relative shape with level, as long as all components change by the same amount (amplitude ratios constant). He concludes the paper with the remark that both generation of combination tones and suppression effects (see Sects. 4.1.4 and 4.3, in particular the reference to Engebretson and Eldredge 1968) might be explained by a compressive nonlinearity as the νth-law device with $0 < \nu < 1$.

4.1.2 *Neurophysiology of Combination Tones*

The neurophysiology of CTs started with the analysis of whole nerve responses around 1940 (see footnote 13 on page 11), the progress of which continued for several decades (e.g., Dallos 1973, Chap. 6),[2] until significant progress was made with the recording and analysis of single nerve responses.

A crucial neurophysiological experiment was set up by Goldstein and Kiang (1968). Action potential responses (spikes) in auditory nerve fibers were recorded from cats. The stimulus contained f_1 and f_2, which could be complemented with a third tone that was used for cancellation. Most stimuli were sufficiently of low-frequency to provide phase locking spike responses. Hence, responses could be synchronized to the primaries f_1 and f_2 and to the CTs, as $2f_1 - f_2$ and $2f_2 - f_1$. Stimulus frequencies were adjusted to a fiber's CF, such that $2f_1 - f_2$ was close to f_{CF}. For relatively close primary frequencies ($f_2/f_1 \approx 1$), the response would provide synchronization to the primaries (f_1 and f_2) and to $2f_1 - f_2$ over a large range of amplitudes, and possibly to $2f_2 - f_1$ over a much smaller range. For a somewhat larger difference between the primaries, several cases were reported here the response to the primaries was lacking completely, and yet a response to $2f_1 - f_2$ was distinctly present. Their Fig. 5, from a fiber with a low spontaneous activity, is particularly convincing, because it shows no response to f_1 or f_2, which had a frequency ratio 3:4, and the response synchronized to the fundamental or difference tone at $1 \times f_0$ [$f_2 - f_1 = (4-3) \times f_0$, but in this case also $3f_1 - 2f_2 = (9-8) \times f_0$] shows synchronization to $2 \times f_0$ [$2f_1 - f_2 = (6-4) \times f_0$, and in this case also $2f_2 - 2f_1 = (8-6) \times f_0$]!

[2] The occurrence of subharmonics and fractional harmonics in cochlear potentials is an additional nonlinear phenomenon (see e.g. Dallos 1973, Chap. 6 Sect. III for a review) which is not explainable in terms of passive NL processes. However, these products can be generated by active NL processes, which are discussed in Chap. 5.

The response synchronized at the CT could be suppressed by adding a cancellation tone, just like in the psychophysical experiments.

Goldstein and Kiang concluded, a.o., that

> Neighboring spectral components f_1 and f_2 in the acoustical stimulus interact to produce both physiological and psychophysical responses that are equivalent to responses produced by a spectral component at the combination frequency $2f_1 - f_2$, which is absent from the original stimulus spectrum.
>
> The equivalences found in the physiological experiments reported above are:
>
> (a) synchrony between the neural responses and individual cycles of $2f_1 - f_2$
> (b) the stimulus tones can be chosen so that the fiber responds only to the CT ($2f_1 - f_2$) when the stimulus tones (f_1 and f_2) are outside the fiber's tuning curve, although the fiber responds to all three tones when they are all within the tuning curve
> (c) synchrony with $2f_1 - f_2$ can be canceled without increasing the average response rate of the fiber.

They also conclude that the cochlea must contain a nonlinear mechanism that accounts for $2f_1 - f_2$ generation, even though the nature of the nonlinearity remained a puzzle, except for the following clear constraints: the nonlinearity should be predominantly odd order, approximately level normalized, and frequency dependent.

The results were generally supported by later studies, e.g., in a study of AVCN responses by Smoorenburg et al. (1976) although Kim et al. (1974) report a faster (than "linear") decrease of CT responses at low levels, which leads them to the conclusion that the AN response to a CT does not show an essential nonlinearity.

4.1.3 Mechanics of Combination Tones

Although the mechanical studies of BM responses after 1970 show nonlinear behavior, it took approximately two decades before reliable and repeatable CT recordings could be reported (Nuttall et al. 1990; Robles et al. 1990, 1991; Ruggero et al. 1992). As suggested by Ruggero et al., this might have been due to

> a combination of inadequate vibration-measurement technologies and the poor physiological state of the experimental cochleae. During the last year mechanical CTs have finally been measured, using laser interferometry, ...

Ruggero is cautious about the state of their own data, because the effect of primary frequency separation was virtually insignificant, in contrast with physiological and psychophysical data.

In most other experimental studies of CMs, the measurements focused on other aspects of linear behavior, such as level dependence, as in Rhode's (1971) classical measurements.

4.1.4 Mathematics of Combination Tones

Since the Taylor expansion is a common, and mostly very useful tool for the analysis of arbitrary functions, it should be no surprise that it also has been used to describe auditory nonlinearity. The series provides integer-power terms, so that the expansion for $y = y(x)$ has the form:

$$y(x) = \sum_{n=0}^{\infty} a_n\, x^n \qquad \text{where } n\text{-values are integer.} \tag{4.1}$$

Unfortunately, all terms with an exponent >1 are of the expansive type, although the nonlinearity to be matched very often is compressive. The imperfect match implies that the series converges slowly (requires a large number n-terms), and over a very restricted range only (limits $x_{\max}$). Compressive functions can be expanded more efficiently with compressive basis functions. A good example is the νth-law device, proposed by Smoorenburg (see Sect. 4.1). More information about the properties are given in Sect. 9.1, which gives the responses of the νth-law device to single tone and two-tone stimuli. Some amplitude factors obtained for the two-tone stimulus are presented in Table 4.1. Note that the sum of the powers found in the amplitude terms is always equal to the value of the power of the power-law device. This is also the case for quadratic and cubic nonlinearities. This factor is corrected by the modified Bennett function which plays the most important role in the range where the amplitudes of the tones are comparable, and the nonlinear interaction becomes prominent.

For comparison, Table 4.1 also gives the response amplitudes for quadratic and cubic nonlinearity, which can be derived directly for the stimulus $x(t) = A_1 \cos(\omega_1 t + \phi_1) + A_2 \cos(\omega_2 t + \phi_2)$:

$$y_{\mathrm{q}} = A_1^2 \cos^2(\omega_1 t + \phi_1) + 2A_1A_2 \cos(\omega_1 t + \phi_1) \cos(\omega_2 t + \phi_2) + \\ + A_2^2 \cos^2(\omega_2 t + \phi_2), \tag{4.2}$$

$$= \frac{1}{2} A_1^2 \{1 + \cos 2(\omega_1 t + \phi_1)\} + A_1\, A_2 \cos((\omega_1 + \omega_2)t + \phi_1 + \phi_2) \\ + A_1\, A_2 \cos((\omega_1 - \omega_2)t + \phi_1 - \phi_2) + \frac{1}{2} A_2^2 \{1 + \cos 2(\omega_2 t + \phi_2)\}, \tag{4.3}$$

$$y_{\mathrm{c}} = A_1^3 \cos^3(\omega_1 t + \phi_1) + 3A_1^2A_2 \cos^2(\omega_1 t + \phi_1) \cos(\omega_2 t + \phi_2) + \\ + 3A_1A_2^2 \cos(\omega_1 t + \phi_1) \cos^2(\omega_2 t + \phi_2) + A_2^3 \cos^3(\omega_2 t + \phi_2), \tag{4.4}$$

$$= \frac{1}{4}\, A_1^3 \cos\{3(\omega_1 t + \phi_1)\} + \left\{\frac{3}{4}\, A_1^3 + \frac{3}{2}\, A_1\, A_2^2\right\} \cos(\omega_1 t + \phi_1) + \\ + \frac{3}{4}\, A_1^2 A_2 \{\cos\{(2\omega_1 + \omega_2)t + 2\phi_1 + \phi_2)\} +$$

Table 4.1 Amplitudes for lower order intermodulation components in two-tone responses of quadratic, cubic, and νth = law devices

Components	Quadratic	Cubic	Odd νth-law device Case $A_1 > A_2$	Case $A_2 > A_1$
0 (DC)	$\frac{1}{2}(A_1^2 + A_2^2)$	0	0	0
f_1	0	$\frac{3}{4}A_1^3 + \frac{3}{2}A_1A_2^2$	$2A_1^{\nu}\mathbf{B}_{10}$	$2A_1A_2^{\nu-1}\mathbf{B}_{01}$
f_2	0	$\frac{3}{4}A_2^3 + \frac{3}{2}A_1^2A_2$	$2A_1^{\nu-1}A_2\mathbf{B}_{01}$	$2A_2^{\nu}\mathbf{B}_{10}$
$f_2 - f_1$	A_1A_2	0	0	0
$2f_1$	$\frac{1}{2}A_1^2$	0	0	0
$2f_2$	$\frac{1}{2}A_2^2$	0	0	0
$2f_1 - f_2$	0	$\frac{3}{2}A_1^2A_2$	$2A_1^{\nu-1}A_2\mathbf{B}_{21}$	$2A_1^2A_2^{\nu-2}\mathbf{B}_{12}$
$2f_2 - f_1$	0	$\frac{3}{2}A_2^2A_1$	$2A_1^{\nu-2}A_2^2\mathbf{B}_{12}$	$2A_1A_2^{\nu-1}\mathbf{B}_{21}$
$3f_1$	0	$\frac{1}{4}A_1^3$	$2A_1^{\nu}\mathbf{B}_{30}$	$2A_1^3A_2^{\nu-3}\mathbf{B}_{03}$
$3f_2$	0	$\frac{1}{4}A_2^3$	$2A_1^{\nu-3}A_2^3\mathbf{B}_{03}$	$2A_2^{\nu}\mathbf{B}_{30}$
$3f_1 - 2f_2$	0	0	$2A_1^{\nu-2}A_2^2\mathbf{B}_{32}$	$2A_1^3A_2^{\nu-3}\mathbf{B}_{23}$
$3f_2 - 2f_1$	0	0	$2A_1^{\nu-3}A_2^3\mathbf{B}_{23}$	$2A_1^2A_2^{\nu-2}\mathbf{B}_{32}$

Note that the sums of the powers of the amplitudes or amplitude products correspond to the overall device power

$$
\begin{aligned}
&+\cos\{(2\omega_1-\omega_2)t+2\phi_1-\phi_2)\}\}+\\
&+\frac{3}{4}A_1A_2^2\{\cos\{(2\omega_2+\omega_1)t+2\phi_2+\phi_1)\}+\\
&+\cos\{(2\omega_2-\omega_1)t+2\phi_2-\phi_1)\}\}+\\
&+\left\{\frac{3}{2}A_1^2A_2+\frac{3}{4}A_2^3\right\}\cos(\omega_2t+\phi_2)+\frac{1}{4}A_2^3\cos\{3(\omega_2t+\phi_2)\}. \qquad (4.5)
\end{aligned}
$$

It is noted in Sect. 9.1 that the sign of the Bennett function depends on the indices and value of the power ν. For instance, for $2f_1 - f_2$ the sign flips if ν changes from above 1 to below 1, or from compressive to expansive. This is relevant for the study of some observed phase conversions in the data. Some of the low-level CT amplitude data (e.g., Smoorenburg 1972a; Zwicker 1979) indicate that at the lowest levels, below 30 dB SPL, the growth of the CTs follows an underlying expansive nonlinearity. This issue will come back at several points in this study.

The primary strength of the general power-law nonlinearity is that over a significant dynamic range the characteristics are approximated closely by a single νth-law term.

4.2 Two-Tone Level effects

4.2.1 Masking

Some of the first accurate quantitative tone-on-tone masking experiments, where a pure tone masker M masks a probe P, making it undetectable, were carried out by Wegel and Lane (1924). This was half a century after Mayer (1876a,b) had reported a distinct asymmetry in tone-on-tone masking: a low-frequency could mask a high-frequency tone, but a high-frequency tone was unable to mask the low-frequency probe tone. In Wegel and Lane's systematic level study (half a century later, better controllable equipment), the masker level required to mask a certain fixed probe tone was studied for probes between 200 and 3,500 Hz, and maskers between 150 and 4,500 Hz, depending on the probe frequency. Two of their datasets are presented in Fig. 4.3. As long as the masker frequency is above the probe[3] frequency, the growth of masking follows a slope <1; for greater masker frequencies $f_M > f_P$ increases to values above 1. Note that the scales are logarithmic amplitude scales: a factor of 10 corresponds to a 20 dB level difference.

Basically, later studies confirmed Wegel and Lane's results. Better control of signal quality over a large range of levels, in combination with techniques to avoid difficulties of different detection criteria (cf., e.g., beating vs. "masking") only changed some minor details. The probably best known second study is that by Egan and Hake (1950). They presented predominantly masking patterns (termed "masking audiograms") for fixed maskers (their Figs. 1, 2, 5–8) at 410 Hz, and two masking growth figures, similar to Fig. 4.3, but only for masker frequencies above the probe frequency. Maskers were either pure tones or narrow bands of noise (3 dB bandwidth: 90 Hz). For $f_P \rightarrow f_M$ the growth of masking tends to be linear, although the linear range is possibly smaller for the tone masker (40–80 dB) than for the noise-band (30–90 dB). More remote maskers show a steeper slope. The "masking audiograms" show the effect in a different way: around the noise-band masker the threshold growth linearly with masker level, for higher frequencies we see "upward spread of masking." This term denotes the effect that a tone or a narrow band of noise easily mask a signal that is higher in frequency: the effect reported by Mayer (see above). All data support that this hf-expansion increases with level.

[3]In several older studies, the probe has been indicated as signal, and its frequency was denoted by f_S.

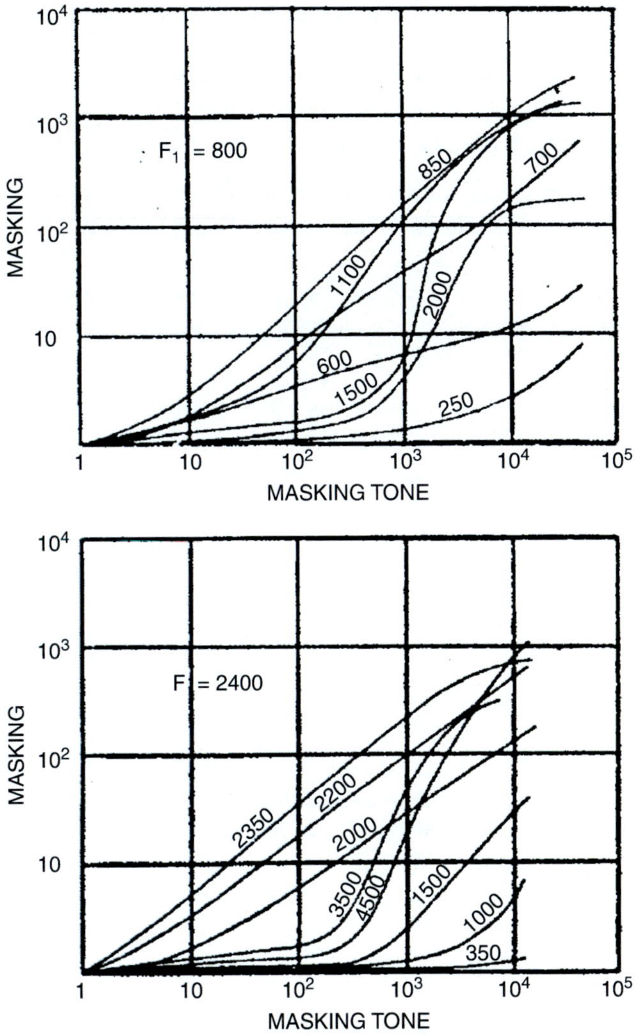

Fig. 4.3 Two from the nine tone-on-tone masking graphs presented by Wegel and Lane in 1924 (their Fig. 3). The figures present the amount of masking that is produced at a fixed probe frequency by maskers of different frequencies, as a function of masker level. Intersection points with a horizontal line are equivalent to tuning "curves." The curves are "corrected" representations of the raw data, some of which are presented in their Fig. 2. Reprinted figure with permission from RL Wegel and CE Lane, Physical Review 23, p.266, 1924. Copyright 1924 by the American Physical Society

Small (1959) presented masker profiles for fixed probes, or masking tuning curves, which puts steep and shallow masking slopes at different sides, but the effect is the same. Asymmetry in masking, and an increment with level that is approximately linear if $f_P \approx f_M$, growth more slowly for $f_P > f_M$, and faster if

$f_P < f_M$, thereby broadening the masking tuning curve particularly towards lower frequencies. Small's results were confirmed, e.g., by Vogten (1978a,b) who also showed Iso-L_M curves, which look like horizontally flipped masked threshold (e.g., his Fig. 7 in 1978b).

The important point from these data is that:

- All data confirm the asymmetry of masking (the upward spread of masking) at least at medium ($\approx$45 dB) or higher levels; at very low levels the asymmetry is not obvious.
- The amount of masking is approx. linear if probe and masker frequencies are similar (and beating is avoided), but it deviates strongly from linear behavior if probe and masker frequencies are different.

The conviction that the cause of this effect is most likely of cochlear origin is supported by the finding that after hearing impairment of known or suspected cochlear origin, the growth of masking changes considerably (see, e.g., Steinberg and Gardner 1937; Smits and Duifhuis 1982).

4.2.2 *Two-Tone Suppression*

Auditory two-tone suppression was first demonstrated in neurophysiological experiments (Sect. 4.2.2.1). Originally these investigators referred to the phenomenon as "two-tone inhibition," but it did not take too long to ascertain that the phenomenon was not comparable to known neural inhibition. Instead, it primarily reflected a nonlinear cochlear process, which was also observable in other cochlear responses, and was affected by cochlear damage.

4.2.2.1 Neurophysiology

During the 1960s, Nomoto et al. (1964) and Sachs and Kiang (1968) demonstrated the first two-tone suppression data from auditory nerve fibers. Originally the term "(lateral) inhibition" was used, and it remained to be used in the field until about 1975. Nomoto et al. measured inhibition effects in auditory nerve fibers of the monkey. Sachs and Kiang confirmed the basic findings in the cat. They demonstrated that the spike rate response to a tone could be reduced by the simultaneous presentation of a second tone. Currently, the first tone is called the suppressee, the second the suppressor. A tone can be active as a suppressor at frequencies and levels where its own rate response is negligible. It was clearly demonstrated that response reduction pertained during the interval that the suppressor was present, and as long as the suppressor was in a certain frequency range around the suppressee, or probe tone. Over the next decades many additional studies have been added.

A schematic picture of the two observed suppression areas is given in Fig. 4.4, after their Fig. 11. The point at CTCF (Constant Tone at CF) indicates frequency

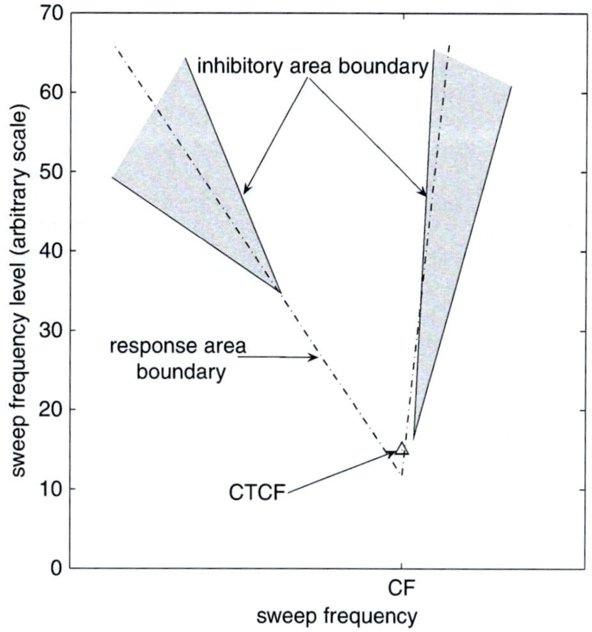

Fig. 4.4 Schematic two-tone suppression (inhibition) after Sachs and Kiang. The *dot-dash line* indicates the tuning curve. Point CTCF characterizes the parameters of the suppressed probe tone, and the *gray areas* indicate the frequency-level areas within which a suppressor produces more than a preset reduction of the rate response. Narrower subareas would indicate subfield with more suppression. Similar results have been observed for all auditory nerve fibers, irrespective of the tuning frequencies

and level of the suppressed tone (the suppressee), and the response area boundary represents a standard tuning curve at CF. Generally, two separate inhibitory or suppression areas can be found, partly outside but overlapping the response area. These areas show asymmetry: at the steeper, hf-slope of the tuning curve the suppression area remains rather narrow, and can extend down to (or even below) the CTCF level; at the lf-side the slopes of the suppression area are shallower, especially the lower boundary, and more remote from the CTCF point. The extension of the lower suppression area over the frequency axis is greater, and over the level axis shorter than at the hf side.

The high-frequency branch of the suppression areas was also observed in the squirrel monkey by Rhode (1977). His measurements required relatively high SPLs, and showed optimal suppression for $f_{\text{suppressor}}/f_{\text{CF}} \approx 1.14$.

Later studies by (e.g. Geisler and Nuttall 1997) suggest a distinction to be made between phasic and tonic suppression, in order to explain the different behavior between high-frequency and low-frequency (re CF) suppression. We have not yet fully accepted or checked that explanation, and consider it a topic for further study.

4.2.2.2 Psychophysics

A few years later, Houtgast (1972) successfully presented a psychophysical counterpart of two-tone suppression, at that time also under the label "lateral inhibition." He developed the paradigm *pulsation threshold* to successfully measure the suppression of the perceived strength or loudness. In short: a tone that acts as suppressee

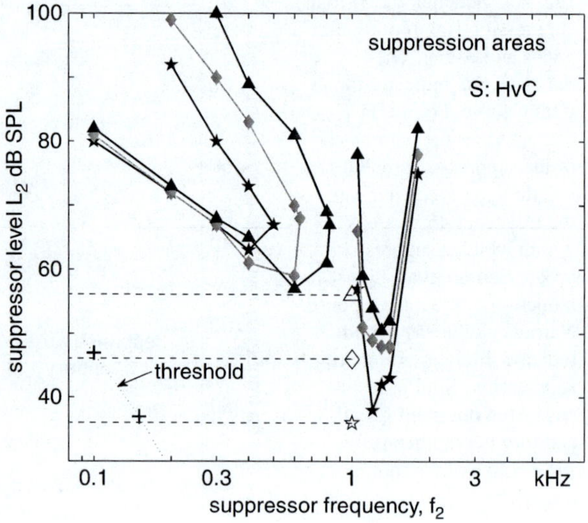

Fig. 4.5 Psychophysical two-tone suppression areas measured around a 1 kHz suppressee. The points are reconstructed 3 dB suppression boundaries. Note that the low-frequency suppression area expands if the suppressee level increases. At the high-frequency side the areas seem to follow the probe level, at least partly (after Duifhuis 1980)

(at f_1) is suppressed (inhibited, partially masked) by a simultaneous suppressor (at f_2), for a duration of the order of 100 ms (in this experiment: 125 ms). This two-tone complex was presented alternately with a probe tone at the suppressee frequency, so that pitches of the two were equal. Phase discontinuities at the transitions were avoided by using a single carrier. With a probe tone of the same duration, he presented stimuli at a repetition rate of about 4 Hz. The idea was that if the suppressor inhibits the response to the suppressee, then the amplitude would go up and down at 125 ms intervals, generating a perceived loudness modulation with a period of 250 ms. If the modulation depth would be more than about 1 dB, it would be detectable as a 4 Hz amplitude modulation. Adjustment of the level of the comparison tone then allows the observer to minimize this amplitude modulation. At the point where the modulation is no longer audible, the observer has reached the pulsation threshold level.

In addition, Houtgast also used a more standard forward masking technique.

Houtgast measured suppression for a fixed suppressor, e.g., at 1 kHz and 60 dB, for variable suppressees at 40 dB. He found significant suppression for a suppressor frequency above the suppressee, and no significant effect if the suppressee was above the suppressor. This behavior is in agreement with the prediction from a horizontal cross-section of Fig. 4.4 below the lf suppression area.

Duifhuis (1980) presented individual two-tone suppression data from three subjects. His subjects measured vertical cross-sections of Fig. 4.4 for suppressees at 1 kHz, at several levels, and for a number of different suppressor frequencies, both below and above 1 kHz. For S:HvC, a suppression area plot was drawn for three "CTCF" levels (Fig. 4.5). The general shape is very similar to the auditory nerve data from Sachs and Kiang. It was also shown that the suppression effect reduced, or even disappeared, when the experiment was repeated against a sufficiently strong white noise background.

4.3 Theoretical Steps

Several ideas about the mechanism(s) underlying two-tone suppression and CT generation have been developed during the seventies. They were inspired by the successful experimental results on two-tone suppression, the revived interest in CTs (in particular $2f_1 - f_2$) and in nonlinear growth in masking. One of the major problems with interpreting the results was, and still is, that although the knowledge of linear system analysis and signal detection theory was at a more than adequate level, the knowledge of nonlinear systems was and is not.

We will discuss some of the steps that were set during this decade. They were useful in sharpening the tools. But awareness that nonlinear analysis had to restart from the beginning did not really catch on. Solutions were sought for in small modifications of proven linear tools.

We start with the introduction of a study by Pfeiffer (1970) who concluded that

> the "inhibition" is easily explained by signal suppression inherent to bandpass nonlinearities.

The paper introduced the BPNL (bandpass nonlinear) model into cochlear analysis. Actually, it was a follow-up on a proposed analog model presented by Engebretson and Eldredge (1968) to investigate cochlear microphonic responses to single tones and two tones. The BPNL acronym was introduced somewhat later. The model contains a single cube-root nonlinearity $f(u) = k\,u^{1/3}$, sandwiched between two bandpass filters (Fig. 4.6). The cochlea could then be modeled as a bank of parallel BPNL systems. The NL element could be identical in all BPNLs of the bank. The type of interpretation that uses two filters, led to—or supported—the introduction of the term *second filter*. BPNL elements have in common with a linear filterbank that they operate unidirectionally.[4] A number of later studies has used (M)BPNL filter-banks (with the additional M for multi) or similar networks (e.g., Duifhuis 1976; Goldstein 1990).

Kim et al. (1973) presented a nonlinear model that looks somewhat like a larger section from a cochlea model. A CP-point is modeled by ten linked nonlinear

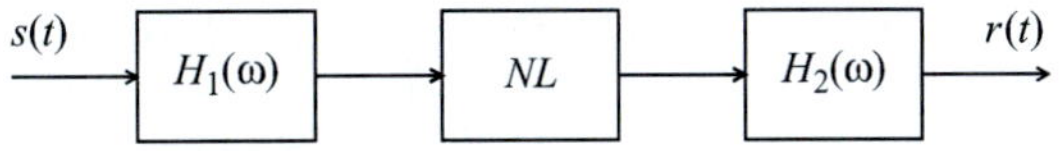

Fig. 4.6 The BPNL structure: a time-invariant nonlinearity NL is placed between two bandpass filters, $H_1(\omega)$ and $H_2(\omega)$. The first limits the input frequency band, the second reduced harmonic distortion (overtones), but allows intermodulation within the passband

[4]Actually, in a linear system, any feedback from the system to a source is usually limited to proper use of the input impedance. This works fine for stationary responses. For nonlinear systems, this simplicity, unfortunately, breaks down completely, and unidirectional coupling is clearly incorrect, because it does not allow distortion products to couple back into the system. The point will be addressed in more detail in Sect. 5.1.3.2.

oscillators, but with unidirectional coupling, and without fluid mass inertia involved. This puts it in the category of nonlinear filterbank models, where each line is a 20th order system. Nonlinearity was introduced in the damping term in each 2nd order section. Assuming a simple parabolic damping profile leads to an output/input power-law relation with a power of 1/3, equal to the value proposed by Pfeiffer. This leads to odd-order nonlinear behavior, with odd-order CTs. Kim et al. emphasize nine different nonlinear phenomena that had been reported either from basilar membrane data, cochlear microphonics or single nerve data, or in combinations. The open points in their matrix have been filled up later, and some items have been modified by newer data which showed that significant nonlinearity was observable everywhere. Their major argument to explore the parabolic damping nonlinearity was that this would provide nonlinear compression, decreasing Q with level, and generation of odd-order CTs, as well as the other points from their list.[5] The correlation hypothesis was not shared by everybody, others addressed different phenomena separately.

Sachs and Abbas (1974, 1976), e.g., focused on the auditory nerve response and assumed a strong effect of saturation (including a power $\approx$1.8) in addition to a nonlinear term that would apply at levels above 73 dB. However, accumulating data from related experiments, including more recent studies of CMs, favor the interpretation that the effect is primarily of cochlear origin.

Hall (1974) made other combinations. He started from Peterson and Bogert's transmission line model, and knowing later developments at Bell Labs (e.g., Flanagan 1962, and Schroeder 1973) analyzed the properties of a 175 element transmission line model. He started to model CT generation, and using similar arguments as Kim et al., assumed that a major nonlinear contribution should be attributable to the damping. Goldstein and Kiang's data showed that the CTs must be present within the cochlea. Hall compared both odd-order nonlinearity (with a parabolic damping profile, similar to Kim et al.) and a combined version with an additional linearly increasing damping term, which provides an even nonlinear behavior. His basic conclusion was that both even-order and odd-order DPs are probably generated within the cochlea. Later he used the same approach to analyze two-tone suppression (Hall 1977).

Duifhuis (1976) also started with a (M)BPNL network where the nonlinearity was implemented as a νth-law device, but with a power of about 0.6 rather than the cube-root value proposed above. This value was also claimed to fit Houtgast's psychophysical two-tone suppression data (Sect. 4.2.2.2).

[5] In this chapter, we reduced the number of topics that appear in the literature as relevant and related. Our primary argument is the conviction that through renewed insight some of the different relevant studies reflect images, or sides, of the same phenomenon rather than different phenomena.

4.3.1 Summarizing the Seventies

Theoretical studies converged at the notion that the cochlear nonlinearity is compressive. The nonlinearity is apparent before the neural transduction process: some of the nonlinear distortion products propagate along the cochlear partition. This propagation is an unavoidable, natural spin-off of a transmission line model, i.e., a model in which consecutive elements are coupled by the cochlear fluid. Filterbank models, on the other hand, are traditionally conceived to be constructed with unidirectionally coupled elements. Thereby they generate a number of problems, some of which are reduced by the introduction of a second filter in every band.

Although the discussion was shaken up by the additional otoacoustic emission data, the development did not yet make the choice between the two options sufficiently obvious to lead to a general choice. In a number of cases, the choice for a filterbank model was primarily based on practical arguments for technical applications. Unfortunately, this may have slowed down comprehension of the underlying biophysics.

Next we have to go into the otoacoustic emissions in some detail, and analyze where new elements have to be inserted in the cochlea models.

4.4 Auditory Emissions

Toward the end of the seventies Kemp presented the first evoked auditory emission data (Kemp 1978), followed by reports of recordings of spontaneous auditory emissions (Kemp 1979a,b). Kim et al. (1980) and Wit and Ritsma (1980) were among the first to confirm Kemp's results, with which they had been confronted at the 1978–Workshop in Münster, organized by Hoke and De Boer (1979). They also presented possible theoretical interpretations, somewhat different from Kemp's first ideas. Kemp focused on reflections from impedance transitions or discontinuities within the cochlea, and did not directly imply nonlinear behavior. He also recognized the active properties of the emissions, which he related to the ideas presented by Gold in 1948 (see also footnote 10 in Chap. 3). Both Kim et al. and Wit et al. suggested a cochlear mechanism with local negative damping, i.e., with a net local power supply rather than power dissipation.[6] In the first proposal activity is limited to a specific area, basalward of the site of maximum excitation. In the second proposal it is stabilized by the specific local damping profile. The papers were presented at the 5th International Symposium on Hearing (Noordwijkerhout) where Johannesma, one of the participants, suggested that a Van der Pol-oscillator would be a good model for spontaneous emissions. It has negative damping for small

[6]Nowadays this property is usually denoted as cochlear activity, or active behavior.

amplitudes (or velocities) only, and stabilizes rather quickly to a positive damping value with the quadratic damping increase (with increasing velocity) (Johannesma 1980). This oscillator produces a stable limit cycle oscillation. In both cases, some local activity is supposed to be present. Note that the damping nonlinearity is of the same type as the compressive nonlinearity mentioned above. A parabolically increasing damping can be shown to generate an input–output compression with exponent 1/3, as had been proposed by Pfeiffer for the BPNL model (Sect. 4.3).

Cochlear emissions have become a popular research topic, which has successfully provided clinical applications as it provides noninvasive tool for direct measurement of certain properties of the cochlea. It provides valuable information in cases that a behavioral response can hardly be obtained.

The growing interest has produced a wealth of scientific papers, including several review papers, and, so far, about a dozen specific textbooks, often focusing on the clinical relevance. An elaborate review has been given by Probst et al. (1991), who distinguish four basic subcategories. Although their introduction as well as some other interpretative remarks deserve a serious discussion, the overview is very valuable, and the discussion of prominent subcategories of emissions in the following sections largely matches their selection. And although their unlimited support for application of otoacoustic emission measurements as noninvasive techniques for the external measurement of inner ear properties deserves some credit, the conclusion should be reduced to realistic constraints. Otoacoustic emission measurements usually do not require invasive tools such as needles or knives. However, experimental data acquisition or observation is fundamentally impossible without interfering with the source. Complete noninvasiveness simply does not exist; best experimental practice aims at minimizing the invasive interference. In the case of acoustic emissions: auditory fields cannot be measured without the use of a specific microphone, and placing a microphone in a sound field does affect the sound field. This applies to spontaneous as well as evoked emissions. Microphones can be used for sound detection as well as for sound generation. Their size is never completely negligible in the sound field, so that their presence affects the field, especially at high frequencies.

An additional problem with evoked emissions is that the applied stimulus affects the physiological "set point" of the system. In practice: critical testing of probe effects remains mandatory, and claims of 1 dB or better experimental accuracy remain suspicious.

4.4.1 Spontaneous Otoacoustic Emissions: SOAE

From a theoretical point of view, the spontaneous otoacoustic emissions (SOAEs) appear to provide the strongest evidence that cochlear power sources are actively involved in sound processing. However, it took a relatively long time to unequivocally distinguish SOAEs from (or alternatively: relate SOAEs to) phenomena that so far had been considered a subgroup of tinnitus. Around 1970 this subgroup was

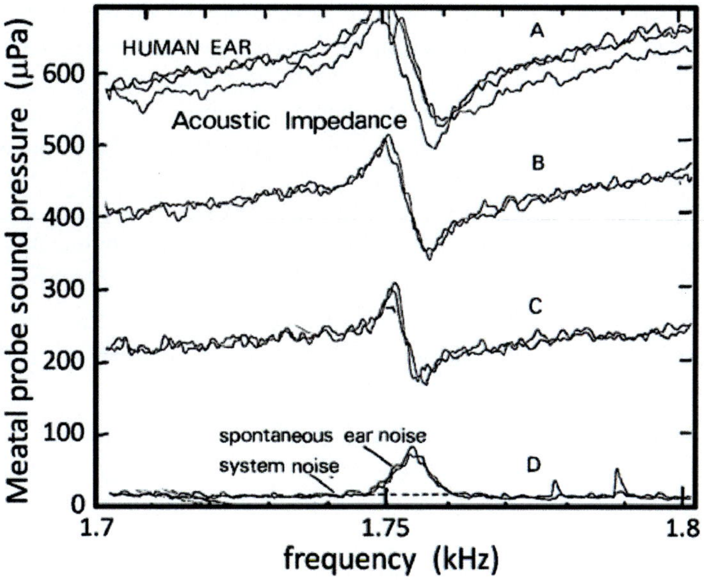

Fig. 4.7 Sound pressure in the acoustic meatus is displayed as a function of frequency for three values of constant volume velocity sound input. Note the linear amplitude scale; 200 μPa is at 20 dB SPL, 400 μPa is at 26 dB SPL, etc. At the time, the focus was on the change in response at about 1.75 kHz, at the same point where auditory sensitivity was maximum, and where a spontaneous emission was measured, at the time denoted as "spontaneous ear noise." Currently, it would be termed a SOAE at about 10 dB SPL. Reprinted with permission from DT Kemp, Archives of Otolaryngology, 224: 37–45, 1979, Fig. 3. Copyright ©(1979) American Medical Association. All rights reserved

specified as *objective tinnitus* by Glanville et al. (1971). The formal distinction between *objective* and general tinnitus is that general tinnitus has been defined as perceived by the subject, whereas detectable OAE, or objective tinnitus usually is not. Note that the categorization did not imply an explanation of the phenomenon. The term was used

> again, for want of a better term that is still easily recognizable.

The authors concluded that:

> The condition is believed to be harmless and due to venous abnormalities at the base of the skull,

in other words, the phenomenon remained unexplained.

The older clinical reports concerned rather special cases with relatively high-level emissions (Glanville et al. report examples up to 38 dB SPL in a 4-year old subject). The development of special equipment for the measurement of evoked emissions led to an increasing number of reports of SOAEs, at first, however, primarily as byproducts (e.g., Kemp 1979a; Wilson 1980a; Wit and Ritsma 1983), and originally not labeled with the term SOAE. For example, Kemp clearly shows a spontaneous emission at ≈1.755 kHz in his Fig. 3, reproduced here in Fig. 4.7, which he describes as "spontaneous ear noise." Zurek (1981) explicitly discusses "Spontaneous narrowband acoustic signals emitted by human ears," starting with a review of related clinical reports from 1962 to 1975. Most of these reports did not speculate about the possible emission source, but in one case a relation had been proposed to streaming of blood over a fiber under tension. Zurek introduced the term *otoacoustic emission* (OAE), which gradually replaced the older terminology.

Although in 1980 the cochlear origin of (S)OAEs was generally accepted, debates about the details remained. Wilson (1980b), e.g., commenting Wit and Ritsma (1980) and Rutten (1980b), states that (in particular in relation to evoked emissions):

> An active local mechanical feedback onto BM displacement appears to be (i) unlikely, (ii) unsuitable as a sharpening mechanism, and (iii) unnecessary.

Of course these points were substantiated with arguments, but also disputed by Rutten and Kemp on the subsequent papers, where Kemp advocates to keep an open mind, at least "until the latter process[7] is finally understood."

Around the same time Bialek and Wit (1984) endeavored to proof that the spontaneous emission signal was a (noisy) oscillation rather than narrowband filtered noise. Theoretically, the amplitude distributions of the two signals are fundamentally different. Filtered noise has a normal (Gaussian) distribution with average 0 and a single maximum at Zero-amplitude, whereas a (noisy) oscillation also will have the average at 0, but it will have two peaks close to $\pm$ the oscillation amplitude, but convolved with a noise distribution if additional noise is present.

The method has two practical limits. First, if the noise amplitude is large compared to the oscillation, then the double peak will disappear, and the distribution of the oscillation will not be detectable. Secondly, if the duration of the signal is not sufficiently long, the narrowband filtered noise will also show the double peak distribution. This is the consequence of the $f \longleftrightarrow T$ properties of narrowband noise: the accuracy of an estimate of the amplitude distribution is proportional to the sample duration, and a short sample can look very similar to a slightly disturbed sinusoid.

The distinction is relevant for the role of nonlinearity in SOAEs. Filtered noise can be generated by a linear process, whereas an active oscillator—even though its response may appear to be linear—contains at least one essential nonlinear element.[8]

Although Bialek and Wit's conclusion was not immediately accepted by the auditory community, the additional "circumstantial" evidence that the intact cochlea has nonlinear properties anyway, has convinced a substantial fraction. Their conclusion was corroborated by Talmadge et al. (1991) who reviewed several alternatives, and addressed the issue directly.

---

[7]This refers to the transduction process(es) at the hair cell(s) level.

[8]This is one of the minimum requirements for a limit-cycle oscillator.

Currently, the alternative hypothesis of coherent reflection at cochlear irregularities, promoted by Shera and Zweig (1995) and very much in line with Kemp's original ideas, seems to be favored. The basic principles will be presented with an alternative implementation in Sect. 5.3.5.5.

4.4.2 Click Evoked Otoacoustic Emissions:[9] CEOAE

As mentioned above, Kemp's first studies of emissions regarded evoked emissions, at the time termed "evoked cochlear mechanical response" or ECMR (Kemp 1978). The response in the earcanal to a click appeared to consist of more than just the expected click response shape. It showed prolonged damped oscillations, originating from imperfect impedance matching at the ear drum, and possibly from other points in the middle ear, and from the cochlea input at the stapes–oval window interface. However, based on estimates of the impedances involved, the conclusion that the responding oscillations lasting over some tens of milliseconds could not originate from the middle ear was inevitable. Kemp also established that the emissions had nonlinear properties, and were inhibited by "damage to the sensory system such as occurs in cochlea deafness" all in support of the conclusion that these emissions had a cochlear origin, and he proposed that the phenomenon might be a byproduct of the transduction process.

The travel-time of the response became an important issue, in connection with the concept of the cochlear traveling wave. The emission was seen as a retrograde traveling wave. However, here we have to be cautious. First, the forward traveling wave does not travel as a wave on a string or a drum, but it originates as a resultant effect of propagation through the cochlear fluid, with the fluid inertia, in tune to the cochlear partition impedance (cf Fig. 3.7). The emitted wave builds up with the responses to the incoming wave, and it will be generated over the entire range where its dissipation is significant. This means that the net result will be the vector summation, or integration, of all local contributions. Accurate prediction of the net result of such a summation is nontrivial, in particular because of the dependence on the precise local phase responses. This can be complicated further by unforeseen nonlinear effects; unforeseen because our systems analysis expertise is usually limited to linear systems.

The earcanal responses to a click are subject dependent (e.g., Kemp 1978, Fig. 2 A–D). The total response appears to reflect one or more oscillations with different carrier frequencies. The high-frequency components appear first, low-frequency parts appear later.

[9]An alternative term covering much of the same group of emissions is: transiently evoked otoacoustic emissions, or TEOAEs.

This phenomenon has received considerable attention in subsequent research of evoked emissions, but its analysis and interpretation deserve further analysis and a careful reassessment. The primary points are that:

1. The experimental definition of the delay is indirect, or it is valid for linear systems.
2. The experimental definition of a response delay requires specification of a threshold (a criterion) against the background noise; this implies a fundamental low-level uncertainty.
3. Spectral analysis of signals with a varying carrier frequency—which applies to most CEOAEs—requires more than common caution; amplitude spectra imply averaging about some time window, and short time windows provide poor spectral resolution, and instantaneous nonlinear effects tend to be smoothed.

For example, Rutten (1980a) refers the response to the click onset, and although a rather narrow-band response can be recognized from the waveform (see his Fig. 1), the precise onset of that response is not, and no threshold criterion was specified. Qualitatively, there is no doubt that slower oscillatory parts of the earcanal responses occurred later with respect to the click onset, but accurate quantitative estimates appeared virtually impossible, and this uncertainty is stressed by uncertainty about the precise estimates of the carrier frequencies (cf. Kemp's Fig. 2 cited above).

This point will be addressed in the following section.

4.4.3 Stimulus Frequency Evoked Otoacoustic Emissions: SFOAE

Since both objective tinnitus and CEOAEs showed band-limited responses, the step to investigate the responses to narrow-band stimuli was a logical next one. Traces E – G in Fig. 2 of Kemp (1978) present some examples, which clearly show that short tonal stimuli can also evoke emissions in the earcanal, although the strength of these emissions, the stimulus frequency evoked otoacoustic emissions (SFOAE) varies with frequency.

One of the first quantitative techniques used by Kemp to estimate the delays of the responses was the use of the group delay, τ_{gr}. Although already mentioned in Kemp (1979b), the technique was used extensively in Kemp and Chum (1980), and even more for DPOAEs (next section) than SFOAEs. The technique is described in some detail in Sect. 8.2.1, where a note of sincere caution is added, because the validity of the use of this linear frequency domain technique has not been validated thoroughly for application to nonlinear systems. Kemp and Chum were well aware of this type of problem, which motivated them not to compare the results of the (continuous) tone evoked emissions with the very short click evoked emissions. The authors developed a "vector subtraction technique which exploits the nonlinear properties of the emission generator." They assumed that at 80 dB SPL, the stimulus level was sufficiently high to justify negligence of the returned emission, so that the measure earcanal pressure would be a proper estimate of the stimulus contribution only. Scaling down to the lower levels employed for the emission measurements then provides the *stimulus only* reference, and the emission amplitude and phase follow from the vector subtraction of this signal from the *stimulus + response* measurement.

The method relies on the observation that evoked emissions are relatively strong for low-level stimuli.

The delay obtained in a frequency band around 1 kHz was around 10 ms (their Fig. 5). This was the result of a spectral group delay analysis (8.40), thereby implicitly assuming the validity of the linear approximation of this technique.

4.4.4 Distortion Product Otoacoustic Emissions

In hindsight, the phenomenon of DPOAE could—or even should—have been the least surprising of the different emissions. Given that distortion products, which may be perceived as CTs, are generated in any intact cochlea, and that they are observable in mechanical responses within the cochlea, side-effects could have been expected. The acoustic coupling from auricle and earcanal to the cochlear partition is not unidirectional, and the set of differential equations that describes the transmission of sound into the inner ear also prescribes how much of the distortion generated within the cochlea remains there to be dissipated internally and what fraction leaves the cochlea as acoustic radiation loss.

DPOAEs require a nonlinear element in the auditory system, but they do not require additional power supply. Compressive nonlinearities, proposed as the best predictors of physiological auditory nonlinearity (e.g. Engebretson and Eldredge 1968) or perceptible auditory distortion products (e.g., Smoorenburg 1972b) lead to increased power loss rather than increased activation. Relevant details about aural CTs have been discussed extensively in Sect. 4.1, and details about the prediction of CT and DPOAE generation will be presented in Chap. 6.

An important but largely neglected aspect of CT generation is that the distortion product is generated over a cochlear range with rapidly changing phases, which are transmitted to a DP-phase profile that has a significant impact on the total integrated response. Assumptions that one or even two sites characterize the major effect are overly simplistic.

The relevance of DPOAE-data is that they provide fairly direct information about the nonlinear properties of the cochlea. A linearized cochlea does not generate any DPOAEs. This consideration has promoted the availability of specialized equipment for its measurement, which enables assessment of normal cochlear nonlinearity without requiring a subjective response.

4.5 Linearization Through Outer Hair Cell Loss

An important finding is that cochlear hearing loss can reduce, or even eliminate, otoacoustic emissions and CTs. Hearing loss can be correlated with a linearization of the cochlea, thereby limiting the input dynamic range to the output dynamic range. This type of hearing loss is accompanied by the threshold increase.

Of particular relevance to these effects are physiological and psychophysical data concerning response growth functions, masking growth functions, loudness growth functions (not discussed), suppression data in relation to cochlear hearing loss, as in noise trauma and other cases of hair cell loss.

Linearization of growth of masking functions in ears with presumed cochlear hearing loss was reported in several psychophysical studies (e.g. Smits and Duifhuis 1982; Nelson and Schroder 1997; Oxenham and Plack 1997).

Smoorenburg (1972b) was one of the first to compare CT generation in subjects with normal hearing and one subject with a specific high-frequency hearing loss, and he observed that CTs where significantly reduced if one of the primary frequencies was within the loss region.

The result from DPOAEs in relation to hearing loss is more complicated. Studies appear to agree that DPOAE thresholds increase with increasing loss, but at sufficiently high primary levels DPs are still elicited (e.g., Probst et al. 1991; Moulin et al. 1991).

References

Bialek W, Wit HP (1984) Quantum limits to oscillator stability: theory and experiments on acoustic emissions from the human ear. Phys Letters 104A:173–178

Dallos P (1973) The Auditory Periphery. Academic, New York

Duifhuis H (1976) Cochlear nonlinearity and seond filter: Possible mechanism and implications. J Acoust Soc Am 59:408–423

Duifhuis H (1980) Level effects in psychophysical two-tone suppression. J Acoust Soc Am 67:914–927

Egan JP, Hake HW (1950) On the masking pattern of a simple auditory stimulus. J Acoust Soc Am 22:622–630

Engebretson AM, Eldredge DH (1968) Model for the nonlinear charateristics of cochlear potentials. J Acoust Soc Am 44:548–554

Flanagan JL (1962) Computational model for basilar-membrane displacement. J Acoust Soc Am 34:1370–1376

Geisler CD, Nuttall A (1997) Two-tone suppression of basilar membrane vibrations in the base of the guinea pig cochlea using "low-side" suppressors. J Acoust Soc Am 102:430–440

Glanville JD, Coles RRA, Sullivan BM (1971) A family with high-tonal tinnitus. J Laryngol Otol 85:1–10

Gold T (1948) Hearing. II. The physical basis of the action of the cochlea. Proc Royal Soc London, Series B, Biol Sc 135(881):492–498

Goldstein JL (1967) Auditory nonlinearity. J Acoust Soc Am 41:676–689

Goldstein JL (1990) Modeling rapid waveform compression on the basilar membrane as multiple-bandpass-nonlinearity filtering. Hear Res 49:39–69

Goldstein JL, Kiang NYS (1968) Neural correlates of the aural combination tone 2f1-f2. Proc IEEE 56:981–992

Hall JL (1974) Two-tone distortion products in a nonlinear model of the basilar membrane. J Acoust Soc Am 56:1818–1823

Hall JL (1977) Two-tone suppression in a nonlinear model of the basilar membrane. J Acoust Soc Am 61:802–810

von Helmholtz HLF (1863) Die Lehre von den Tonempfindungen, 1st edn. Vieweg und Sohn, Braunschweig, english edition: On the Sensations of Tone, transl. by A.J. Ellis (1885) of 4th German edition (1877), publ. by Dover in 1954

Hoke M, de Boer E, eds. (1979) Models of the Auditory System and Related Signal Processing Techniques. Scandinavian Audiology (Suppl. 9), Proc. from the workshop held at Münster, Germany, Sept. 1978

Houtgast T (1972) Psychophysical evidence for lateral inhibition in hearing. J Acoust Soc Am 51:1885–1894

Johannesma PIM (1980) Narrow band filters and active resonators. Comments on papers by D. T. Kemp & R. A. Chum, and H. P. Wit & R. J. Ritsma. In: van den Brink G, Bilsen FA (eds) Psychophysical, physiological, and behavioural studies in hearing, Delft University Press, Delft, pp 62–63

Kemp DT (1978) Stimulated acoustic emissions from within the human auditory system. J Acoust Soc Am 64:1386–1391

Kemp DT (1979a) Evidence of mechanical nonlinearity anf frequency selective wave amplification in the cochlea. Arch Otorhinolaryngol 224:37–45

Kemp DT (1979b) The evoked cochlear mechanical response and the auditory microstructure – evidence for a new element in cochlear mechanics. In: Hoke M, de Boer E (eds) Models of the Auditory System and Related Signal Processing Techniques, Stockholm, pp 35–47

Kemp DT, Chum RA (1980) Observation of the generator mechanism of stimulus frequency acoustic emission – two-tone suppression. In: van den Brink G, Bilsen FA (eds) Psychophysical, Physiological and Behavioural Studies in Hearing, Delft University Press, Delft, Netherlands, International Symposium on Hearing, pp 34–41/2

Kim DO, Molnar CE, Pfeiffer RR (1973) A system of nonlinear differential equations modeling basilar membrane motion. J Acoust Soc Am 54:1517–1529

Kim DO, Littlefield WM, Pfeiffer RR, Molnar CE (1974) Combination tone $2f_1 - f_2$ in responses of single auditory nerve fibers: evidence against essential nonlinearity. J Acoust Soc Am 55:467, (Abstract NN5)

Kim DO, Molnar CE, Matthews JW (1980) Cochlear mechanics: Nonlinear behavior in two-tone responses as reflected in cochlear-nerve-fiber responses and ear-canal sound pressure. J Acoust Soc Am 67:1701–1721

Mayer AM (1876a) Mayer's recent acoustical researches. Nature 14:318–320

Mayer AM (1876b) LXI. Researches in acoustics.–No. VIII. Phil Mag S 5 2:500–507

Moulin A, Bera JC, Collet L (1991) Distortion product otoacaoustic emissions and sensorineural hearing loss. Audiology 33:305–326

Nelson DA, Schroder AC (1997) Linearized response growth inferred from growth-of-masking slopes in ears with cochlear hearing loss. J Acoust Soc Am 101:2186–2201

Nomoto M, Suga N, Katsuki Y (1964) Discharge patterns and inhibition of primary auditory nerve fibers in the monkey. J Neurophysiol 27:768–787

Nuttall AL, Dolan DF, Avinash G (1990) Measurements of basilar membrane tuning and distortion with laser doppler velocimetry. In: Dallos P, Geisler CD, Matthews JW, Ruggero MA, Steele CR (eds) The Mechanics and Biophysics of Hearing, Springer, Berlin, pp 288–295

Oxenham AJ, Plack CJ (1997) A behavioral measure of basilar-membrane nonlinearity in listeners with normal and impaired hearing. J Acoust Soc Am 101:3666–3675

Pfeiffer RR (1970) A model for two-tone inhibition of single cochlear nerve fibers. J Acoust Soc Am 48:1373–1378

Plomp R (1965) Detectability threshold for combination tones. J Acoust Soc Am 37:1373–1378

Probst R, Lonsbury-Martin BL, Martin GK (1991) A review of otoacoustic emissions. J Acoust Soc Am 89:2027–2067

Rhode WS (1971) Observation of vibration of the basilar membrane in squirrel monkeys using the Mössbauer technique. J Acoust Soc Am 49:1218–1231

Rhode WS (1977) Some observations on two-tone interaction measured with the Mössbauer effect. In: Evans EF, WIlson JP (eds) Psychophysics and Physiology in Hearing, London, pp 27–38 (41)

Robles L, Ruggero MA, Rich NC (1990) Two-tone distortion products in the basilar membrane of the chinchilla cochlea. In: Dallos P, Geisler CD, Matthews JW, Ruggero MA, Steele CR (eds) The Mechanics and Biophysics of Hearing, Springer, Berlin, pp 304–311, (disc: 312–313)

Robles L, Ruggero MA, Rich NC (1991) Two-tone distortion in the basilar membrane of the cochlea. Nature 349:413–414

Ruggero MA, Robles L, Rich NC, Recio A, Brown AM, Evans EF (1992) Basilar membrane responses to two-tone and broadband stimuli [and discussion]. Phil Trans R Soc Lond B 336(1278):307–315

Rutten WLC (1980a) Evoked acoustic emissions from within normal and abnormal human ears: comparison with audiometric and electrocochleagraphic findings. Hear Res 2:263–271

Rutten WLC (1980b) Latencies of stimulated acoustic emissions in normal human ears. In: van den Brink G, Bilsen FA (eds) Psychophysical, Physiological, and Behavioural Studies in Hearing, Delft University Press, Delft, pp 68–71, comment on Wit and Ritsma (1980) and Rutten (1980)

Sachs MB, Abbas PJ (1974) Rate versus level functions for auditory-nerve fibers in cats: tone-burst stimuli. J Acoust Soc Am 56:1835–1847

Sachs MB, Abbas PJ (1976) Phenomenological model for two-tone suppression. J Acoust Soc Am 60:1157–1163

Sachs MB, Kiang NYS (1968) Two-tone inhibition in auditory-nerve fibers. J Acoust Soc Am 43:1120–1128

Schroeder MR (1973) An integrable model of the basila membrane. J Acoust Soc Am 53:429–434

Shera CA, Zweig G (1995) The origin of periodicity in the spectrum of evoked otoacoustic emissions. J Acoust Soc Am 98:2018–2047

Small AM (1959) Pure-tone masking. J Acoust Soc Am 31:1619–1625

Smits JTS, Duifhuis H (1982) Masking and partial masking in listeners with a high-frequency hearing loss. Audiology 21:310–324

Smoorenburg GF (1972a) Audibility region of combination tones. J Acoust Soc Am 52:603–614

Smoorenburg GF (1972b) Combination tones and their origin. J Acoust Soc Am 52:615–632

Smoorenburg GF, Gibson MM, Kitzes LM, Rose JE, Hind JE (1976) Correlates of combination tones observed in the response of neurons in the anteroventral cochlear nucleus of the cat. J Acoust Soc Am 59:945–962

Steinberg JC, Gardner MB (1937) The dependence of hearing impairement on sound intensity. J Acoust Soc Am 9:11–23

Talmadge CL, Tubis A, Wit HP, Long GR (1991) Are spontaneous otoacoustic emissions generated by self-sustained cochlear oscillators? J Acoust Soc Am 89:2391–2399

Vieth D (1805) Ueber Combinationstöne, in Beziehung auf einige Streitschriften über sie zweier englischen Physiker, Th. Young und Jo. Gough. Annal d Physik 21(3):265–314

Vogten LLM (1978a) Low-level pure-tone masking: A comparison of "tuning curves" obtained with simultaneous and forward masking. J Acoust Soc Am 63:1520–1527

Vogten LLM (1978b) Simultaneous pure-tone masking: the dependence of masking asymmetries on intensity. J Acoust Soc Am 63:1509–1519

Wegel RL, Lane CE (1924) The auditory masking of one pure tone by another and its probable relation to the dynamics of the inner ear. Phys Rev 23:266–285

Wilson JP (1980a) The combination tone, $2f_1 - f_2$, in psychophysics and ear-canal recordings. In: van den Brink G, Bilsen FA (eds) Psychophysical, Physiological, and Behavioural Studies in Hearing, Delft University Press, Delft, pp 43–50(2)

Wilson JP (1980b) Recording of the Kemp echo and tinnitus from the ear canal without averaging. J Physiol 298:8P–9P

Wit HP, Ritsma RJ (1980) On the mechanism of the evoked cochlear mechanical response. In: van den Brink G, Bilsen FA (eds) Psychophysical, Physiological, and Behavioural Studies in Hearing, Delft University Press, Delft, pp 53–60

Wit HP, Ritsma RJ (1983) Two aspects of cochlear acoustic emissions: response latency and minimum. In: de Boer E, Viergever MA (eds) Mechanics of Hearing, Martinus Nijhoff and Delft University Press, Delft, Netherlands, pp 101–107

Zurek PM (1981) Spontaneous narrowband acoustic signals emitted by human ears. J Acoust Soc Am 69:514–523

Zwicker E (1955) Der Ungewöhnliche Amplitudengang der Nichtlinearen Verzerrungen des Ohres. Acustica 5:67–74

Zwicker E (1979) A model describing nonlinearities in hearing by active processes with saturation at 40 dB. Biol Cybernetics 35:243–250

Chapter 5
Modeling the Nonlinear Cochlea

Abstract This chapter presents an introduction to nonlinear cochlear modeling. It starts with some general definitions, goes to the basic DEs, and discusses tools to solve these. The art of minimizing errors by not using signal analysis tools that are only valid in linear systems is advocated. Starting with a circuit example, we move on to the cochlea.

5.1 Nonlinear Systems

We start this chapter with the fundamental definition of nonlinearity of a functions and (or) systems. This definition contains two essential components. Then we explore tools for the analysis of simple nonlinear systems. Additional tools are presented in Chap. 9.

5.1.1 Definition of NL-Functions and Systems

Definition 5.1. Let f define an operation on an input variable x, and let y be the result of that operation, also termed the response or output variable, or:

$$y = f(x) \tag{5.1}$$

then the function f is defined to be linear if and only if:

1. For any constant (real or complex) factor α applied to the input x, the result y scales by exactly the same constant: $f(\alpha\, x) \equiv \alpha\, y$.
2. For any combination of two different inputs x_1 and x_2, the response to the sum of the inputs is identical to the sum of the responses to the separate inputs: $f(x_1 + x_2) \equiv f(x_1) + f(x_2)$ and f is nonlinear otherwise.

H. Duifhuis, *Cochlear Mechanics: Introduction to a Time Domain Analysis of the Nonlinear Cochlea*, DOI 10.1007/978-1-4419-6117-4_5,

The input variable can be multidimensional as well as time-dependent, as in: $\mathbf{x} = \mathbf{x}(t)$. If so, then the operation of the function f applies to all the elements of $\mathbf{x}(t)$, and propagates to all elements of the response $\mathbf{y}(t)$.

A well-known nonlinear example of a simple NL-function is the quadratic function, which fails on both criteria. A less obvious case is a linear half-wave rectifier, which meets criterion 1 but fails on criterion 2.

Input–output graphs usually provide a straightforward and fast tool for a linearity check (criterion 1): any linear relation will show up as a straight line of slope 1 on a dB–dB plot (at least if one has taken the effort to use equal scales). Of course, it also would show up as a linear relation on linear scales, but for practical reasons those are hardly used for acoustic signals. Any straight line that deviates from slope 1 dB/dB reflects a power-law relation $y = x^{\nu}$ with a power ν that differs from 1.

The definition for linear and nonlinear systems is similar to the definition for functions: systems can be considered a specific class of functions. More in particular, in linear systems, the functional relation between input and output of a system is usually expressed in terms of signal processing in the time domain (convolution with the systems impulse response) or in the frequency domain (multiplication of the input spectrum by the transfer function of the system), as:

$$y(t) = h(t) * x(t), \tag{5.2}$$

$$Y(\omega) = H(\omega)\, X(\omega), \tag{5.3}$$

where

$$H(\omega) = FT\{h(t)\}, \tag{5.4}$$

[see Sect. 8.5, and (8.55)].

Representing the response of the system by its transfer function does imply (and therefore requires!) linearity: if $H(\omega)$ is a proper transfer function, i.e., independent of the input variables, then by definition the response scales linearly with the input, as follows from (5.3), and the second condition for linearity is met because of commutativity of addition and multiplication:

$$H(\omega)\,(X_1(\omega) + X_2(\omega)) = H(\omega)\, X_1(\omega) + H(\omega)\, X_2(\omega). \tag{5.5}$$

5.1.2 *Breakdown of the Linear Approach*

For a linear system, i.e., a system that is described by a linear function, the classical signal analysis approach is very efficient. The analysis is performed either in the time domain or in the frequency domain.[1]

[1]Instead of the complex frequency domain, the related Laplace domain can be used. Nowadays its use is somewhat less common in (applied) physics and mathematics, but it can be particularly advantageous if causality issues are important, as in electrical engineering problems.

Two points that require caution are:

1. In practice, the methods of signal analysis involve averaging over a chosen time window, the shape and duration of which have to be specified.
2. The system transfer function is usually treated unidirectionally, incorrectly suggesting that changes in properties further on have no effect on preceding steps. For stationary stimuli such effects tend to be hidden: embedded in net amplitude and phase adjustments.

A linear system cannot generate signal components that were not present in the stimulus spectrum, but any nonlinear system will produce harmonic distortion products in response to simple tonal stimuli, and more complex stimuli produce more complex distortion product spectra.

There is, however, also a class of broadband stimuli which is hardly affected by time-invariant nonlinearities. One specific example is a click or pulse shaped signal. This shape is not modified by a stationary nonlinearity, only the input–output amplitude ratio may be affected. Another example is noise. The response to white noise will remain white noise, although its amplitude distribution can be modified, an effect which may show up in the r.m.s.-amplitude of the noise. These properties of broadband signals have been addressed in detail in a some classical signal analysis textbooks, such as Middleton (1960).

5.1.3 Relevant Fundamentals

If we have to reconsider our general signal analysis tools for applications to nonlinear systems, where do we start?

As long as we have to do with time-independent (time-invariant, instantaneous) nonlinearities, the time domain approach remains appropriate. The frequency domain analysis becomes immediately suspect, and should be used only as a supporting tool that helps in evaluating a time domain solution. Time domain analysis also remains applicable to systems in which parameters adapt over time, as in regular feedback or feed-forward systems.

The natural way to start the time domain analysis is to setup all relevant system equations completely in the time domain, using differentiation and integration wherever appropriate.[2]

In practice, it is usually impossible to analyze the set of system equations analytically, which leaves methods of numerical time domain analysis or of linear approximation and perturbation techniques. These options are worked out in more detail in the next subsections.

---

[2]Of course, the same considerations apply to the place–wavenumber relations that are used to characterize traveling waves.

But first we discuss another relevant issue, applicable to both linear and nonlinear systems: the benefit of the use of conservation laws.

5.1.3.1 Conservation Laws

Since the introduction of Newtonian mechanics, conservation laws have played an important part in what is now called "Classical Dynamics." They also have been used setting up the system equations for fluid mechanics. Conservation of energy translates to the balance of sound-energy flux in CMs. The *net input* sound-energy flux can be determined in the earcanal, e.g., at the eardrum, and all of that is dissipated in the ear. The *net* input corrects for the sound emitted by the ear as well as the additional energy that would be supplied by a cochlear amplifier.

The sound-energy flux carried by an acoustic traveling wave is defined as the product of the sound intensity and surface area perpendicular to the propagation direction, as was mentioned before (Sect. 3.2). The parameter values for the propagation in air (earcanal) differ from those inside the cochlea, which involve fluid and CP-properties. One of those relevant variables is the sound pressure. Sound is carried by relatively small and fast variations of the total atmospheric pressure.[3] That justifies why the analysis of the physics of sound in most textbooks immediately assumes linearity.

Although in many cases the analysis of the energy balance and the power balance are very similar, there are also differences. Sound power is defined as the amount of energy radiated by a source per unit of time (ANSI 2004, Definition 3.63). A *simple sound source* is a source that radiates uniformly in all directions.[4] Technically this can not be realized by an acoustic point source, but beyond a realistic radius it is equivalent to a spherical source (cf. Sect. 8.2). Many available sources do not radiate in all directions. For those that do, computation of the sound energy radiated uniformly from a simple source is straightforward: the transmission area at distance r is simply $4\pi r^2$. This can be extended to plane waves by considering the spherical wave at sufficiently large r, where effectively it has become independent of the distance r. Ideal plane waves also have limitation in a physical environment: the extension of the wave is limited, and edge effects modify the net energy content. The idealized plane wave and spherical wave primarily have considerable theoretical relevance.

Except for the specialists, scientists usually do not worry about the precise sound source properties, such as: are we dealing with a pressure source or an intensity source, what are the source limitations, what is the internal acoustic impedance, and what are the acoustic properties of the environment into which the sound is emitted? These consideration are rather similar to the electrical analog of a source connected

[3]To be verified by the reader. Compare sound pressure (Pa) in the range from 0 to 120 dB SPL to a realistic atmospheric pressure.

[4]Formally: under free-field conditions.

to a load. What are the source properties, and what are the properties of the load? Together they determine how much power and energy the source can dissipate into the load. For a linear system any incorrect initial estimate can be corrected by a simple complex factor (amplitude and phase correction).

The situation becomes more complicated if the load (electrical or acoustical) is nonlinear, because in that case the load becomes a new source—at least in the classical terms—viz. it generates distortion products that were not present in the stimulus. Since these internal second sources meet finite "impedances" in the cochlear structure, some of this internally generated distortion will come back toward the external source.

Even if auditory stimuli are presented to the ear in a controlled way, as by earplugs, earphones, or loudspeakers, the acoustic input impedance is not trivial, especially not at high frequencies (>4 kHz), and neither is the output impedance. These properties have been studied quite extensively over recent decades[5] and lf-transmission through outer and middle ear is largely understood for the linear case. All current studies of outer and middle ear do use frequency domain analysis at some stage, tacitly assuming linear behavior.

In the selected time domain approach, the underlying characteristics of the frequency domain representation, or of the equivalent network, are transformed to time domain equations. These provide the relevant coupling elements between cochlea and external sources and loads. We start the analysis of the cochlea with a simplistic outer- and middle ear approximation that allows acoustic input and output to the environment or to a plugged earcanal.

Back to the energy flow. Let us consider the sound intensity first. If we analyze the motion of an average elementary air-particle,[6] then the sound intensity it can produce per second I is

$$I(\mathbf{x}, \mathbf{r}, t) = \frac{1}{T} \int_{t-T}^{t} p(\mathbf{x}, \tau)\mathbf{u}(\mathbf{x}, \tau)\, \mathrm{d}\tau. \tag{5.6}$$

The transmitted sound-energy flux J (influx to the ear) becomes $J = I \times A$, where A is the relevant surface area.[7] The parameter T indicates a time interval which for harmonic (periodic) signals is an integral number of cycles, and which is not well defined for nonperiodic signals. The parameter $\mathbf{r}$ denotes the propagation direction of the flux, which is equal to the direction of velocity vector $\mathbf{u}$, at least as long as it is not modified by a gradient in the pressure profile. I is expressed in W/m^2. Integrated over the meatus area, it provides the sound-energy influx (J) per second.

[5] Supported by a tri-annual series of symposia on Middle-Ear Mechanics in Research and Otology.

[6] Note that fluid mechanics is one of the areas of physics where the use of global properties and global parameters is successful. This means that the minimum physical scale on which the analysis applies is above the molecular size of the relevant molecules.

[7] Tentatively, we use $A = A_{\mathrm{TM}}$, the tympanic membrane area.

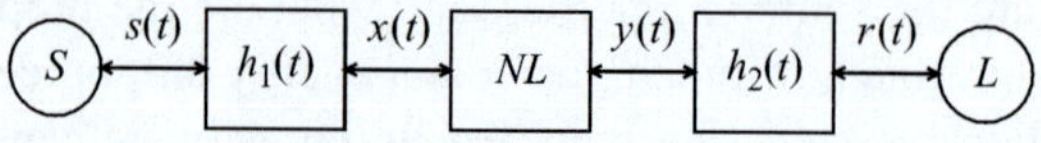

Fig. 5.1 The time domain version of a simple BPNL structure: a nonlinear element NL is placed between two linear elements with impulse responses, $h_1(t)$ and $h_2(t)$. The first has the properties of a band-pass filter, the second is a simple low-pass element which reduces distortion product harmonics (overtones). Stimulus source properties and load properties also play a role (Elaborated in the text)

For a harmonic sound pressure source $p(t) = p_0 \cos(\omega t)$, which in a linear system generates a particle velocity $u(t) = u_0 \cos(\omega t + \varphi)$, the transmitted intensity is $I = \frac{1}{2} p_0 u_0 \cos(\varphi)$. For the plane wave, pressure and velocity are in phase and the $\cos(\varphi)$-term vanishes. For the spherical wave the phase relation changes at short distances. In general, the temporal fine-structure relations are important for the full analysis.

Following the sound-energy flux into the auditory system is useful and can be done accurately up to the point where unknown additional power sources may appear. The measurement techniques for both velocity and pressure have improved substantially over recent years. For instance, optical techniques such as laser interferometry have been developed both for direct velocity measurements of BM or CP structure, and indirect velocity measurements of condensor microphone membranes. An example of application of the techniques can be found in the experiments by Olson in the base of the gerbil cochlea (Olson 1998, 1999, 2001, 2004). Measurements were made over a wide range of stimulus frequencies, but—as in other mammals—the explored base area is specialized for the high-frequency part of that spectrum. Her addition of fluid mechanics is useful for the interpretation of the results, but was limited to frequency domain details.

5.1.3.2 Time Domain Analysis

To elucidate the relevance and the relative ease of time domain analysis, we present a simple example. A BPNL structure as presented in Fig. 4.6 is analyzed in the time domain. The setup is adjusted to account for the consequences of the occurrence of a nonlinear element, and coupling between blocks is now bidirectional Fig. 5.1.

For the analysis, we consider an electrical analog. The system parameters are specified in more detail than might be expected to be required for the precise differential equation setup: we cannot combine impedances in the usual (linear) way. We will treat the elements successively, but note that the equations all have to be solved simultaneously (or at least as parallel snapshots).

Assume that the source is an ideal oscillator that can be switched on with an onset ramp of adjustable duration. This source is delivering a voltage of $10\,V_{\mathrm{r.m.s.}}$ when

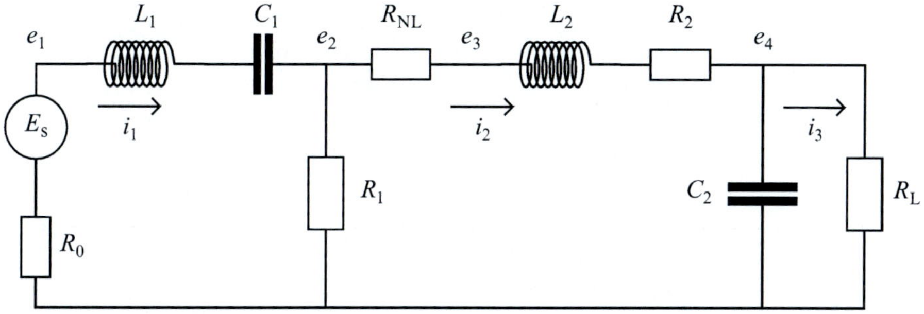

Fig. 5.2 A simple circuit with a single nonlinear element, R_{NL}. As in Fig. 5.1, the nonlinear resistance is placed between two linear elements: an *LCR*-band-pass, with index 1, and an *LRC*-low-pass, with index 2. The source is an Thevenin equivalent circuit of an AC voltage-source with internal resistance R_0

linked properly to an audio-network. This means that it delivers $E_s = 10\,\mathrm{V_{r.m.s.}}$ over $600\,\Omega$. The frequency is set to 1 kHz. The idcal internal oscillator is assumed to have a stable voltage amplitude.[8]

The current $i_1(t)$ leaving the source is determined by both the internal source properties and the total network load. Assume that the former is characterized by a simple low resistance R_0, which will be significantly less than $600\,\Omega$, and let the remaining system be specified by a second-order band-pass, a single NL-resistive element, a second-order low-pass, and a load resistance. The system is drawn in Fig. 5.2 The system is described completely by the parameters $e_1(t)\cdots e_4(t)$ and $i_1(t)\cdots i_3(t)$, specified by the following seven equations, where the stimulus signal $e_s(t) = w(t)\,E_s\sqrt{2}\cos(2\,\pi\,1000)$:

$$e_1(t) = e_s(t) - i_1 R_0, \tag{5.7}$$

$$e_1(t) - e_2(t) = \frac{1}{C_1}\int i_1\,\mathrm{d}t + L_1\frac{\partial i_1}{\partial t}, \tag{5.8}$$

$$e_2 = (i_1 - i_2)R_1, \tag{5.9}$$

$$e_2(t) - e_3(t) = i_2 R_{NL}(1 + \varepsilon i_2^2) = i_2 R_{NL}^*, \tag{5.10}$$

$$e_3(t) - e_4(t) = L_2\frac{\partial i_2}{\partial t} + i_2 R_2, \tag{5.11}$$

$$e_4(t) = i_3 R_L, \tag{5.12}$$

$$e_4(t) = \frac{1}{C_2}\int (i_2 - i_3)\,\mathrm{d}t. \tag{5.13}$$

[8] Note that this is only approximately realizable in practice.

The onset-window $w(t)$ defines a smooth onset envelope of the stimulus (10 ms in this example). These seven equations (5.7)–(5.13) can be reduced to three in $i(t)$ (step 1), but containing different time specifications: the time derivative of $i(t)$ and its integral (step 2).

Step 1 involves three straightforward additions of these equations:

1. −(5.7)+(5.8)+(5.9) ⇒ (5.14)
2. −(5.9)+(5.10)+(5.11)+(5.12) ⇒ (5.15)
3. −(5.12)+(5.13) ⇒ (5.16)

Step 2 involves the rewriting of $\int i_n \, \mathrm{d}t = y_n(t)$, yielding the combined result (dropping the time dependence notation):

$$\frac{1}{L_1}e_s + \frac{R_1}{L_1}i_2 - \frac{1}{L_1C_1}y_1 - \frac{R_0+R_1}{L_1}i_1 = {i_1}', \tag{5.14}$$

$$-y_1 + \frac{R_1+R_2+R^*_{\mathrm{NL}}}{R_1}y_2 + \frac{L_2}{R_1}i_2 + \frac{R_L}{R_1}y_3 = 0, \tag{5.15}$$

$$-y_2 + y_3 + R_LC_2i_3 = 0. \tag{5.16}$$

In (5.10), R_{NL} is a constant parameter, and ε the scaling and criterion factor for the NL-effect: if $\varepsilon i_2^2 \ll 1$, then the nonlinear effect is relatively small, and if $\varepsilon i_2^2 \gg 1$ then the nonlinear character dominates. Note that after step 2, the remaining differential equations are (5.14–5.16) and $\frac{\partial y_n}{\partial t} = i_n$. These equations are solved using the Runge–Kutta 4 method. Given the solutions for the currents, the voltages follow straightforwardly.

An example of solutions is given in Fig. 5.3. They show that the nonlinearity is already present in $e_1(t)$ (although almost 90 dB down) and $i_1(t)$ (at about 50 dB down). It is pronounced more clearly in the amplitude spectra than in the time waveforms (not shown), even though the effect is weak for relatively small εi_2^2. The spectra are computed from a response section obtained after the onset-duration. The low-pass character of the second section leads to some reduction of the higher harmonics.

The results are critically dependent on the value of ε with respect to i_2^2, as indicated above. For the given parameters, the amplitude of $i_2(t)$ is of the order of 2 mA. For the given value of ε, this creates a situation where the nonlinearity begins to become important.

The most obvious nonlinear effect is the occurrence of odd overtones in the spectra, i.e., components at 3, 5, 7... times the stimulus frequency. No significant even-order products appear. This is due to the fact that the introduced nonlinearity is of the odd type (cf. right-hand panel of Fig. 4.1), which is caused by the introduced nonlinear damping property: $R^*_{\mathrm{NL}} = R_{\mathrm{NL}}(1 + \varepsilon i_2^2)$. It should also be clear that the

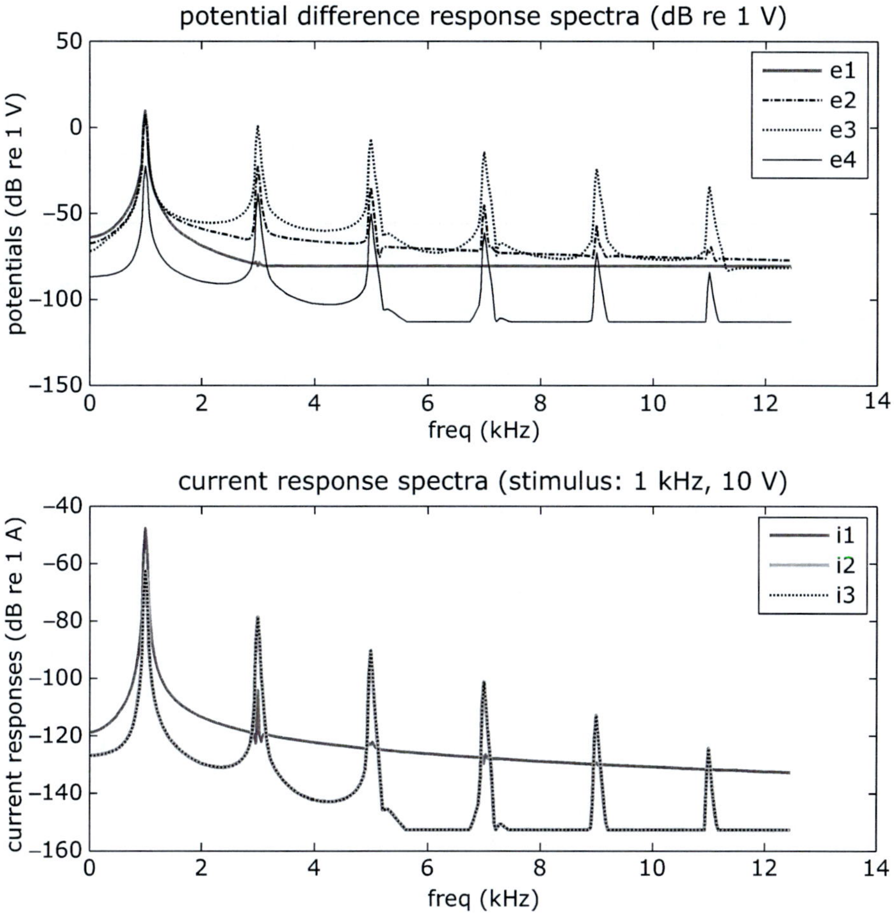

Fig. 5.3 Spectra of current and voltage responses of the simple nonlinear network in Fig. 5.2. Dynamic ranges of the dB-scales are limited to 90 dB for each parameter. The results in this figure were obtained for the following parameter settings: $R_0 = 10$, $R_1 = 600$, $R_2 = 60$, $R_{\mathrm{NL}} = 400$, $R_L = 100\ \Omega$, $C_1 = 3/(4\,\pi^3\,10^6)$ F, $C_2 = 0.5\,C_1$, $L_1 = 3/\pi$ H, $L_2 = 0.5L_1$, and $\varepsilon = 10^6$ A^{-2}

sharpness of the peaks in the spectrum is limited by shape and duration of the time window that was used to obtain the plot.[9] Increasing ε by a factor of 100 gives a much more pronounced nonlinear effect, with significantly stronger overtones.

---

[9]In this example, the window was a 3,200 point Kaiser–Bessel window (parameter: $4.0/\pi$ or 4.0, depending on software definition[10]) at 10 μs steps, or a 320 ms duration.

[10]This difference by a factor of π may be confusing. I follow the definition by Harris (1978, Sect. I), where the argument of the zero-order modified Bessel function I_0 is $\pi\alpha$. In that case, the first side lobe is 45 dB down for $\alpha = 2$, and more than 70 dB down for $\alpha = 3$. Packages like Matlab tend to apply the parameter $p = \pi\alpha$, in which case a value 4.0 represents a sharp peak at the cost of a not very selective sideband suppression.

You might want to check this, either by writing your own analysis program or by using the provided test routine. Note that numerical test routines are sensitive to sufficiently high time sampling rate.

If the pass-band ranges of the band-pass element and the low-pass element overlap, then the nonlinear intermodulation effects can be more pronounced than the harmonics. Substituting the one-tone stimulus by a two-tone stimulus demonstrates this phenomenon. It is a consequence of the reduced frequency separation of the resulting components: for an odd-order nonlinearity the distortion products $2f_1 - f_2$ and $2f_2 - f_1$ are much closer to f_1 than $3f_1$ or $3f_2$ are. A test of the simple network response to a two-tone stimulus provided results where the intermodulation products were smaller than the overtones. For power-law devices, the relation between the strengths of overtones/harmonics and intermodulation products can be obtained analytically (see Sect. 9.1). The power-law approximation for the current simple example is $\nu = 1/3$.

If the right-hand part of the network in Fig. 5.2 would be tuned to higher frequencies, then the response at the load resistance would reflect the harmonics more strongly. We will not discuss this option further, because it does not appear to be the one that mimics the cochlea most closely, where more apical elements are found to be tuned to lower frequencies.

The proof of stability of the results of a numerical method such as RK4 partly relies on analytical studies.[11] An important other validating argument relies on the correctness of the assumed hypotheses, i.e., the set of DEs to be solved, and on practical expertise. An interesting, and rather simple, procedure that can be used for an evaluation is to rewrite the system equation(s) in a set of first order DEs, and assume solutions (for each DE) with terms $e^{\lambda_i t}$. Then a region of convergence for values of λ_i can be found, which depends on the numerical method that has been used. The regions characterize the areas within which the products $\lambda\, dt$ give stable solutions. The time step dt involved in this product is the step between two successive numerical sampling points: $dt = 1/f_s$. The stability area for the most common RK4 method is presented in Fig. 5.4. Note that it has a reasonable behavior along the imaginary axis, even with small extensions into the positive half plane. The absolute value of this product, $|\lambda\, dt|$ apparently must not be greater than order 1, and occurs as a new sampling criterion for vibrations. Stable points along the (negative) real axis represent an exponential decay. To operate reliably within the audio frequency range, we propose to use sampling frequencies above 400 kHz, for linear as well as for nonlinear models of auditory processing. In general, the RK4 method is suited to provide stable oscillatory solutions over a limited range along the imaginary axis in the (complex) frequency plain, whereas most other algorithms do not even reach this axis and are stable for solutions with a negative real component only.

[11] It also depends on the exact specification of the Runge–Kutta 4 method. Different specifications are used. We present additional discussion later in this Chapter, in particular in Sect. 5.3.4.

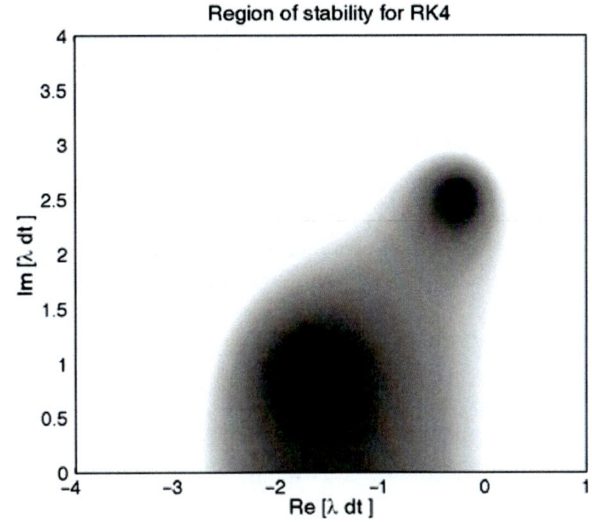

Fig. 5.4 Region of stability of the RK4 method. Parameter is $\lambda_i dt$. Deviation from stable points is minimum at the dark peaks. The values of λ are the roots of the DE-matrix, dt is the sampling interval: $dt = 1/f_s$. The area is symmetric for positive and negative imaginary values, only the upper half is presented (see, e.g., Butcher and Wanner 1996, Fig. 12)

5.1.3.3 Perturbation Theory Approach: Options and Limitations

The alternative to the numerical analysis in the time domain is the perturbation theory approach. As long as it can be proved that the nonlinear effects are relatively small byproducts, the behavior can be approximated by a linear set of equations, which in first approximation can be solved either in frequency or in time domain. Next, a number of iteration steps adjusts for the relatively small nonlinear deviations.

In some cases where the deviation from linear is larger, the perturbation method still works reasonably well. The approaches vary from advanced mathematical analysis to rather classical applied approaches. The first category includes analytical approaches used, e.g., by Guckenheimer and Holmes (1983). In Chap. 2, they treat the van der Pol oscillator, and study the stability conditions. Remember that the van der Pol oscillator was introduced as a possible cochlear partition element in the late seventies (cf Sect. 4.4). We will present some common, well-known, results and work out the application in CMs later in this chapter, and deal with the (applied) properties of the nonlinear oscillators in more detail in Sect. 9.2.2.

The application of the van der Pol-oscillator (VDP) will be treated later in this chapter, but first we present some of its properties. Precise definitions are given in Sect. 9.2.1, within a context of related definitions.

Basically the oscillator is described by an ordinary differential equation (ODE) similar to one that describes a damped mass–spring system, but, the damping term is an even nonlinear term, which has a negative value for small amplitudes, which would make the oscillator unstable. At high amplitudes, the damping is positive again and monotonically increasing (at least nondecreasing). With mass, damping and stiffness terms scaled to unity, it takes the following form:

$$\ddot{x} + \varepsilon\phi(x)\dot{x} + x = s(t) \tag{5.17}$$

The term $s(t)$ represents a stimulus variable (can be 0), ε is a crucial VDP parameter which determines major characteristics of the response, and $\phi(x)$ is the nonlinear term. It looks similar to R_{NL} used in the previous example, but note that for the van der Pol it is negative for $x = 0$. The transition from the classical response (positive linear damping term) to the oscillatory response (negative linear damping term) marks a Hopf-bifurcation (see Definition 9.19).

The original value used by van der Pol was $\phi(x) = -1 + x^2$. The van der Pol oscillators converge to periodic solutions when undriven, which solutions are called *limit cycles*. If ε is sufficiently small ($\varepsilon \ll 1$), then the limit cycle is very close to a sinusoid, with a frequency close to that of the undamped oscillator: in this case $\omega \approx 1$. If on the other hand ε is >1, then the limit cycles change shape dramatically and the oscillators become relaxation oscillators.

Some examples of responses are presented in Fig. 5.5. For three values of ε, we show the evolving response for the same initial conditions (close to 0 amplitude and velocity). The following points are of particular relevance:

- ε appears to have a significant effect on the stimulus onset duration.
- For low ε the oscillator frequency is $\omega \approx 1$ so that the period is $\approx 2\pi$, for high values, however, the period becomes much longer.
- The response for large ε, the relaxation oscillator response, is markedly deformed from a circle, and the period duration T depends on ε.

More examples of the analysis of nonlinear oscillators are presented in Chap. 9. The focus is on the van der Pol- (and/or Rayleigh-) oscillators.

A somewhat similar approach, although applied to a different but related set of DEs, is used by Eguíliuz et al. (2000), Magnasco (2003) and Kern and Stoop (2003). Starting with a harmonic approximation or "Ansatz" the solution is computed for relatively small perturbations. Their approaches, which also focus on modeling the cochlear partition response properly, is discussed in more detail in Sect. 7.3.

A different example is the use of the WKB or WKBJ method for the analysis of a 2D fluid mechanics model of the cochlea, which was introduced in Sect. 3.3.1. In that case, the difficulty of the wave equation was not the existence of nonlinear terms, but the fact that the equation had a nonconstant parameter. The solution

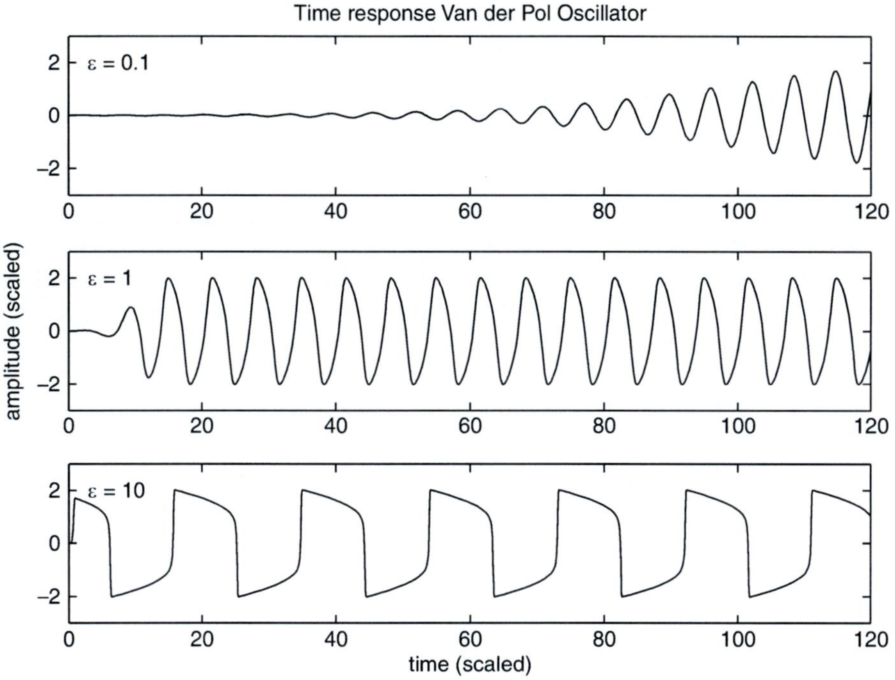

Fig. 5.5 Plot of the development of the solution of the Van der Pol equation for three values of ε. For the lowest value, $\varepsilon = 0.1$, the solution begins to approach a sinusoid, with an angular frequency $\omega_0 = 1$ or a period of 2π. For $\varepsilon = 10$, the period has increased significantly and the wave shape has altered too

changes from long-wave to short-wave approximation over a relatively short range. The approximation is reasonably accurate from base to slightly over the point of maximum response, as explained by Zweig et al. (1976) and de Boer (1980).

Less convincing applications have been presented as well: regular science education generally teaches that a Taylor expansion is an always converging all-purpose expansion. However, although convergence in the expansion does occur, there is no independent criterion for its efficiency. It may require many steps before a proper stable solution is approached, and limitation to a few steps can lead to disastrous errors. This type of approach starts with a linear approximation of the solution, using standard LTI tools, and then tries to improve the solution by using iteration steps. At each step, the number of terms in the expansion is increased, up to the point where the remaining error is deemed acceptable.

Many current studies of auditory nonlinearity use the last method and claim that it does not make much difference, at least not for fitting the data. Unfortunately, these claims are not always justified.

5.2 Elements of a Model Cochlea

5.2.1 *Parallel Oscillator Models*

As mentioned in Sect. 1.3, von Helmholtz has been a dominant promoter of cochlea models where the BM tuning was completely attributed to mechanical properties of the radial fibers within the membrane. Neither the fluid coupling[12] nor an interactive coupling (either longitudinal within the BM, or more complex coupling involving the organ of Corti) played a crucial role—in fact those were believed to play at best a secondary part.

The mechanical properties of a string at relatively small deflections was well understood at the time, and the profile of the membrane width, from narrow at the base up to broad near the helicotrema—even though counteracting the gradient of the cochlear duct—could provide a reasonable frequency map.

This is also the type of model discussed by Gold (1948), who realized that radial membrane fibers must be subject to damping, either within the membrane structure or within the cochlear fluid, and from estimating the fluid damping he concluded that cochlear tuning could not be sufficiently sharp.[13]

Anatomical studies showed that auditory systems in nonmammals often do not have a BM equivalent—at least not with respect to all its properties. These systems still have tuning, and also may have some frequency mapping.

Already bird ears have some rather different inner ears properties, but they also have a frequency map. Moreover, the hair cells change more globally from "inner" to "outer" than the "binary" change in mammals. But the fact that some birds can mimic human speech implies that those systems can pick up—and recognize—at least some relevant acoustic parameters that characterize speech properties.

Many other studies involve lizards and frogs. In lizards the origin of the frequency map is not always completely understood, but hair bundle length may be a candidate. In frogs there usually are two different populations of hair cells in different physical environments (tectorial structure present or absent; also differences in size of hair cells and structural type of hair cells).

Within the mammalian cochlea, some of these properties may also play a role. From base to apex:

1. Hair cell length increases, in particular the OHC-body length
2. Hair bundle length increases
3. OC cross-sectional area increases
4. TM cross-sectional area increases
5. Cuticular plate angle (re BM) increases

[12] But the fluid was supposed to transmit the sound wave practically instantaneously, to provide simultaneous input to all fibers.

[13] Note the discussions in footnote 10 in Chap. 3 and in Sect. 4.4. Gold postulated additional active behavior to reduce the net damping.

Both CM and auditory nerve data obtained since the seventies leave no room for a global interpretation in terms of parallel second-order filters or harmonic oscillators. This has led to an alternative of higher-order data matching, where the underlying physics no longer was the important boundary condition. A rather successful example is the gamma-tone filter bank mentioned before. Almost exclusively, the filterbank models use linear system analysis techniques, and unidirectional coupling of elements in the system. For this reason, the success for applications to nonlinear behavior will remain limited.

5.3 Coupling the Elements

5.3.1 *The System Equations*

The system equations for the linear cochlea have been introduced in Chap. 3. They have been setup such that the inclusion of nonlinear elements is a straightforward step. We summarize the resulting set of equations and discuss the introduction of nonlinear elements.

We start the analysis using a globalized cochlear partition model, which does not cover the 3-D interactions within the OC in detail. This uses a longitudinal grid of the order of 50 to 100 μm, or averages over 5 to 10 cross-structures.

It describes the movement of the cochlear partition in terms of local mass, damping, and stiffness, and these are supposed to represent a first approximation of the "integrated" effect over the 10 μm sections.

In Chap. 3 where we dealt with a linear analysis (of an assumed linear system), the impedance concept was introduced. This concept is invalid for the nonlinear system. So starting from (3.9) we have to stick strictly to the time domain formulation, but up to that point, no fundamental differences have occurred.[14] Hence, the force balance equation at the partition can no longer use an impedance definition, but it will be related directly to the mechanical properties of a section, combining (3.9) and (3.10) and dropping the $wZ_{CP}(x)$ term.

The full 3D analysis requires detailed consideration of fluid behavior in the radial (y-axis) and vertical (z-axis) directions (Fig. 3.2), as well as the interactions at the cochlear partition. We will simplify the fluid motion by reducing it to laminar flow approximations. Where appropriate, averages across a dimension are denoted by a bar over the component axis. Ignoring details along the y-axis by integration or averaging across the width and assuming a first-order vibration mode, we obtain

$$\begin{aligned} p_v(x,\bar{y},0,t) - p_t(x,\bar{y},0,t) &= 2\,p_v(x,\bar{y},0,t) \\ &= -\left\{ m(x)\,\frac{\partial w}{\partial t} + d(x)\,w + s(x)\int w\,\mathrm{d}t \right\}, \end{aligned} \tag{5.18}$$

[14]Except, of course, for the point that the linear approach tends to consider unidirectional coupling.

where $w(x)$ is the average across partition velocity at x. As in (3.9) and (3.10) it follows the force balance law. If the fluid viscosity is still negligible, then the same relation applies here:

$$w(x,\bar{y},0,t)\,b = \frac{\partial U_v(x,t)}{\partial x}. \tag{5.19}$$

Again Newton's law couples the longitudinal pressure gradient to the longitudinal acceleration (the time derivative of the volume velocity), or

$$b\,h\,\frac{\partial p_v(x,\bar{y},\bar{z},t)}{\partial x} = -\rho\,\frac{\partial U_v(x,t)}{\partial t}. \tag{5.20}$$

An important difference between the pressure terms in (5.18) and (5.20) is that the first averages across y at $z = 0$, and the second averages across y and z. We introduce the factor $\alpha(x)$, which translates this effect into a 1D parameter, for the laminar flow assumption:

$$\alpha(x) = \frac{\text{average pressure over partition } (z=0) \text{ at } x}{\text{average pressure over duct at } x} \tag{5.21}$$

$$= \frac{p_v(x,\bar{y},0,t)}{p_v(x,\bar{y},\bar{z},t)}. \tag{5.22}$$

Putting this together, one obtains

$$\frac{\partial^2}{\partial x^2}\left[\frac{1}{\alpha(x)}\left\{m(x)\,\frac{\partial^2 w}{\partial t^2} + d(x)\,\frac{\partial w}{\partial t} + s(x)\,w\right\}\right] - 2\frac{\rho}{h}\,\frac{\partial^2 w}{\partial t^2} = 0. \tag{5.23}$$

It might be intuitively clear that the dependence of α on x is related to the direction of the fluid flow shown in Fig. 3.7. In the basal part of the cochlea, the fluid flow in response to a narrow-band stimulus will remain prominently parallel and laminar, and follow the long-wave approximation.[15] When approaching the point of resonance the flow will bend towards the partition, still without significant circular motion other than required for the adjustment of the direction of propagation. Here, we approach the short-wave area, but without reaching it: the input power is dissipated before arriving at the point of resonance (see: Sect. 3.6.1).

It is useful to reconsider the physical meaning of the constituting terms in (5.23) at this point. For themoment, the dependence on α is neglected, which implies that

[15]However, one should note that this does not apply to the response at the extreme base, where the stapedial volume influx has to find its way into the cochlea. The actual stapes entrance is to the side of the duct, implying a different spatial response for the locally mapping high-frequencies than for the low-frequencies that map apically.

we use the long-wave approximation for this discussion. Rewriting (5.23), but using subscripts to denote differentiation yields:

$$[m(x)\, w_{tt} + d(x)\, w_{\mathrm{t}} + s(x)\, w]_{xx} - \frac{2\rho}{h}\, w_{tt} = 0. \tag{5.24}$$

The bracket term represents the across partition behavior. The full term is equal to the time derivative of transpartition pressure $p_v(x,t) - p_t(x,t)$, denoted as p_i in the equivalent circuit (Figs. 3.6 and 5.8). If one considers the pressure—or force per unit area—to be a proper equivalent for the electrical potential e (V), then the volume velocity U is a similar equivalent for the electrical current i (A). Note that the product $e.i$ represents (electrical) power P (W), as does the product of the acoustic quantities[16] $p.U$. Therefore, the equivalent electrical network is very useful for the analysis of the acoustic quantities. This would also apply to a mechanical $F.v$ (force–velocity) combination.

Henceforth we will frequently use the function φ for the time derivative of the pressure, defined as

$$\varphi(x,t) = \frac{\mathrm{d}}{\mathrm{d}t}\,(p_v(x,t) - p_t(x,t)) = \frac{\mathrm{d}p_i}{\mathrm{d}t}, \tag{5.25}$$

which allows us to rewrite (5.24) as

$$\varphi_{xx} - \frac{2\rho}{h}\, w_{tt} = 0. \tag{5.26}$$

The volume velocity U appeared naturally for the fluid motion in the scalae. It also compares in a simple way to the transpartition flow U, which represents the product of the local area ($\mathrm{d}x \times b$) and the (average) membrane velocity w_{t} which is also written as $v(x,t) = w_{\mathrm{t}}(x,t)$. In general, the canal flow will differ from the transpartition flow. We will indicate the former by superscript indices and the latter by subscripts (cf Fig. 5.8).

Details of the structure of the cochlea have effects on the precise profile of the resulting cochlear frequency map, and also on the local properties of points along the partition. Several studies have been made on the global profile of the cochlea; some opposites are listed here:

1.	Rectangular cochlear duct	–	Tapered duct
2.	Straight cochlea	–	Rolled-up cochlea
3.	Simple rectangular cross-section	–	More realistic rounded or (semi-)circular cross-section
4.	Inviscid lymph	–	Viscous lymph

[16]Note that the product $p.v$ represents acoustic intensity I (W/m^2).

Although some of these points have been addressed for over about a century, so far they have been solved partly at best. Several lucid biophysical studies have addressed one or more of the points, such as the studies by von Békésy, Zwislocki (1965, 1974), Dallos (1973, Chap. 4), Inselberg (1978), and Puria and Allen (1991). A complicating factor remains that all earlier studies focus on transfer function behavior, using a linear system analysis approach, and insignificant attention is paid to bidirectional information flow and power flow.

Ad 1: Towards the apex tapering of the duct can affect the transpartition pressure difference, and hence the resulting local cross-partition volume flow, because the tapering causes an additional gradient in the longitudinal direction. If the fluid is loss-less, the associated power must be dissipated in the organ of Corti at the sensory transduction mechanism. The factor is similar to adjusting overall cochlear partition parameters (mass, damping, and stiffness) that would retain the resonance frequency and increase dissipation or decrease the Q_{3dB}. Tapering has been brought into view already by Peterson and Bogert (1950), and by Zwislocki (1965), where the latter proposes the following approximation (see also Dallos 1973):

$$S_v = S_t \approx 2.5\mathrm{e}^{-0.05x} \qquad \mathrm{mm}^2, \tag{5.27}$$

where parameter and constants are in mm-measures. The maximum factor is approximately $\mathrm{e}^{-0.05\times 35} = \mathrm{e}^{-1.75} = 0.174$. Dallos analyzed a linear cochlea model around 1970, trying to match available data, and using both partial differential equations (PDEs) and system analysis. He noticed that the function of tapering would be more obvious if the cochlear perilymph and endolymph would have a nonnegligible viscosity. The associate boundary layers would dissipate more acoustic energy, thereby preventing possible reflections from the apical end, where proportionally the boundary layer would be largest.

An other approach to the tapering of the cochlea is presented in studies that consider the BM-traveling wave as a shallow water wave. Originally, those waves were introduced for surface waves on the water–air interface, and the interface properties play a fundamental role in the results. But instead of surface tension, the cochlear partition—coupling to the other scala—provides an impedance of a completely different order of magnitude, and hence the coupling is incomparable with the normal surface wave. Therefore, the approximation assumptions in the original (Lamb 1895; Rayleigh 1894, 1896) analysis have to be adjusted, and the use of common textbook formulas for these waves may be invalid.

Ad 2: The coiled cochlea was examined by Viergever (1978), where he concluded (tentatively, but not updated as far as I know) that

> the mechanical properties of the cochlea are only affected slightly by the spiral form, even near the apex where the spiral is very tightly wound.

A similar conclusion was obtained in a local thesis study where a 3D finite element method with flexible boundary conditions was used to estimate this effect by Berk

Hess.[17] But a more recent study by Cai et al. (2005) claims a significant effect of the curvature on the shear gain at the apical part of cochlear partition. This gain relates to hair bundle stimulation to cochlear partition/BM deflection. It relies more on the change in shape of the CP across cochlear length than on a possible role of viscosity and boundary layers. The paper also lists additional references, which present conclusions more in line with those of Viergever.

Ad 3: Several investigators have examined the possible effect of the canal shape, other than the tapering. A number of studies approximate the scala by semicircularly or circularly shaped ducts. Some examples are Dallos (1973), Steele and Zais (1985), and Kohllöffel (1990).

Here too, there is no evidence that the effect is very strong. In the basal area, the power flow through the duct is parallel anyway, and near the point of resonance the focus is on the partition and not on the wall.

Following experimental results, both psychophysical, BM, and AN-data, it appears obvious that nonlinearity is probably dominant in the mechanical damping term, it may also be present weakly in the stiffness term, and it is probably absent in the mass term. The statement that a nonlinearity is present at a damping term limits the possible sites to damping elements in the BM–OC structure, such as viscous elements in fluid boundary layers, and power transition at the sensory elements. Obviously, the loss of acoustic energy is not intended to warm up the system, which suggest that natural optimization probably focuses to sensory transmission. But changes in stiffness of the structure, either BM or OC (e.g., hair bundle stiffness), also will have some effects. These points will be discussed later in more detail.

Ad 4: The study of the strength of the effect of fluid viscosity has gone through several phases. Of course perilymph and endolymph do have a finite viscosity, but so has water. The question is: does this lead to boundary layers that become significant with respect to the sizes of the cochlear ducts (S_v and S_t), or of the organ of Corti or cochlear partition.

For several decades, the common opinion has been that the boundary layer is sufficiently thin to exclude a strong friction effect, except probably at the apex, in a tapered cochlea. Currently, there is a revived interest in the general relevance of viscosity as determining the cochlear damping parameter.

My personal view disagrees with the latter: the ear supposedly has evolved as a very sensitive auditory detector. Any detector has to pick up some power or energy from the quantity it wants to detect. In other words, it is a fundamental property of a detector to transform some incoming energy for further processing. If this detector has to be very sensitive, it should loose as little power as possible heating up the system, which is what a significant viscous component would do. In addition, a spring or structural stiffness cannot be used as a detector, because it does not absorb and transmit energy, it only can temporarily process onset and offset effects. Hence, in terms of global parameters the detector is unlikely to be stiffness dominated.

[17] B. Hess. *A mathematical model of the inner ear of a bat.* Master's thesis in Applied Mathematics, Dept. Mathematics, University of Groningen, Dec. 1995.

5.3.2 *The Middle Ear As an Efficient Coupler*

The middle ear is an essential interface between the acoustic environment and the cochlea. If the auditory system were linear, we could state that the middle ear has to provide a proper match between the acoustic impedance of air and the input impedance of the cochlea. As noted before, a nonlinear cochlear model cannot operate with a frequency domain property as impedance, so we have to convert the properties established so far from the regular signal analysis description to the underlying time domain equations.

We have planned to work with two different classes of interfaces: ME-0, a very simplistic version, just matching the order of magnitude of power transmission coupling within the mid-frequency range (around 2 kHz), and the more elaborate version ME-1 after O'Connor and Puria (2008). In the latter case, we use their detailed network and write down the corresponding differential equations rather than using the common linear system analysis approach. An open end with this model is that it starts with an earphone driver entry, thereby assuming a practically closed auditory canal, excluding the possibly of an open coupling through the auricle. We propose to consider the two different options, to be labeled as *plugged* or *open* canal with the application of a simple approximation: treating plugged as a closed pipe and open as a flanged pipe (e.g., Kinsler et al. 1982, Chap. 9). So far we have only been able to check the simple interface ME-0 in any detail.

Several other middle ear models have been proposed over the recent decades, but most of those either have not been matched to human data or are rather similar.

Middle ear coupler-studies tend to use the acoustic pressure–volume velocity pair of parameters as the relevant combination. Acoustic parameters, such as the acoustic impedances: acoustic mass, acoustic damping and acoustic stiffness, fit within this category. However, one should realize that the transmitted *acoustic power* depends on the pressure–particle velocity pairs (Sect. 3.6.1, 3.45), and that the acoustic parameters involve the selection of global averages.

5.3.2.1 The Zero-Order Approximation, ME-0

In first approximation, the middle ear realizes efficient acoustic power transmission from the open environment into the cochlea. In linear approximation, this means that the specific acoustic impedance of air should match the specific input impedance of the cochlea. The latter, by the way, is not determined by the compressibility of the cochlear fluid, but by dynamical fluid mass displacement properties. But the former does depend on the compressibility in air, which depends on the average density (see Table 8.1) as well as on several other physical parameters.

For the cochlea, the power influx is the relevant quantity. It is determined by the average pressure–volume velocity product. The ratio of these two quantities is the acoustic impedance rather than the specific acoustic impedance. Thus, already in the earcanal a proper match has to be made from the free-field condition to the

Table 5.1 Parameter values for the simplest middle ear model, ME. It provides a bandpass filter, centered (on a logarithmic scale) around 2 kHz, with cut-offs at 0.7 and 5.7 kHz, and relates the acoustic properties of air at the auditory entrance to those at the entrance of the cochlea, assuming $Z_a = Z_{sa}/A_{\text{TM}}$. Independent variables: Z_{sa}, n_{t}, $Q_{\text{3dB,me}}$, A_{TM}, and A_{fp}

	Canal open		Canal plugged	
Z_{sa}	415	Ns/m^3 = Pa s/m	$2 \times Z_{sa}$	Ns/m^3
Z_a	6.92×10^6	Ns/m^5 = Pa s/m^3	14.0×10^6	Ns/m^5
n_{t}	30	–	Same	
m_{me}	198×10^3	kg/m^4	401×10^3	kg/m^4
s_{me}	3.13×10^{13}	N/m^5	6.33×10^{13}	Nm
$Q_{\text{3dB,me}}$	0.4	–	Same	
A_{TM}	60×10^{-6}	m^2	Same	
A_{fp}	3.16×10^{-6}	m^2	Same	

earcanal condition. In other words, we have to estimate the input power flow at the auricle for a given external sound pressure or intensity. In this case, the acoustic intensity I in W/m^2 is the most directly related physical quantity. The power input is the product of I and a reasonable estimate of the canal cross-sectional area $A_{\text{ec}} \approx 50$ mm^2. Note that the size of this area is close to the tympanic membrane surface area (see Table 5.1). If necessary, an amplification factor 2 covers the correction for the difference between open canal and flanged canal.[18]

The simplest approximation of the middle ear frequency characteristic is a bandpass filter, centered around $f_{\text{c,me}} = 2$ kHz, and with a low-frequency cut off at $f_{\text{l,me}} = 0.7$ kHz, and a high-frequency cut-off at $f_{\text{h,me}} = 5.7$ kHz. This can be represented by a second order system, which is described in terms of DE elements. From bandwidth, center frequency, and resistance parameter R_{me}, we can derive an estimate of the capacitance C_{me} and inductance L_{me}:

$$Q_{3dB} = \frac{2\pi\, f_{\text{c,me}}\, m_{\text{me}}}{d_{\text{me}}} \qquad \Rightarrow \qquad Q_{3dB} = \frac{2\pi\, f_{\text{c,me}}\, L_{\text{me}}}{R_{\text{me}}} \tag{5.28}$$

where we show the relation between the physical parameters and the equivalent electrical network representations. Moreover,

$$f_{\text{c,me}} = \frac{1}{2\pi}\sqrt{\frac{s_{\text{me}}}{m_{\text{me}}}}. \tag{5.29}$$

To couple the external ear parameters to the cochlea parameters, we still have to account for the cochlear transformer, which, going inward, increases the force and

[18] Kinsler et al. (1982, Chap. 9) give for the flanged pipe transmission coefficient $T_\pi \doteq 2\,(ka)^2$ (9.16b), for wavenumber $k = 2\pi/\lambda$ and pipe radius $a \ll \lambda$, or $ka \ll 1$. The similarly derived transmission coefficient for the unflanged open pipe is $T_\pi \doteq (ka)^2$ (19.7b).

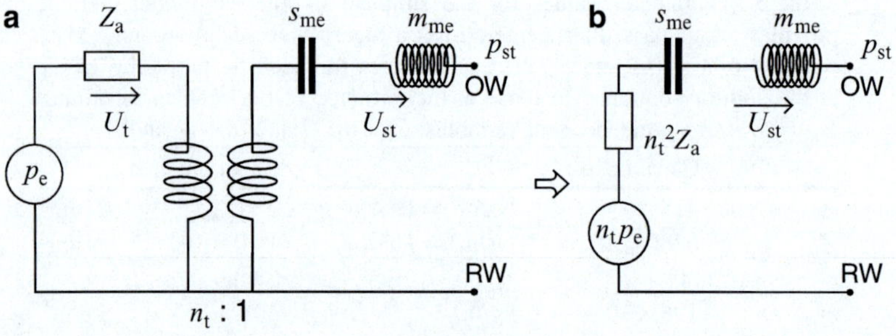

Fig. 5.6 A very simple middle ear model that optimizes acoustic flow into the cochlea. It contains the transformer ratio, a very crude band-pass filter, and uses the tympanic membrane area as an estimate of the factor between the specific acoustic impedance of air and the acoustic impedance of the area at the auricles[20]

pressure, and decreases the responding movement. Here this is characterized by the factor $n_t = 30$, which is an approximation of current data and models (Table 5.1).

The simple network representing this middle ear model is given in Fig. 5.6a. An equivalent version where the transformer is omitted is shown in Fig. 5.6b. Note that the input pressure amplitude now also has to be scaled by n_t, the acoustic impedance is scaled by n_t^2, but the global middle ear mass and stiffness parameters remain unaltered.

For the closed earcanal, we use the simple approximation that the impedance values are increased by a factor of 2, as indicated above. The parameters are listed in Table 5.1. Tympanic membrane area and oval window / footplate area are also listed, but functionally they are fully incorporated in n_t.

Actually, besides $f_{c,me}$ the independent variables are Z_{sa}, A_{TM}, n_t, and $Q_{3dB,me}$. Given this set,[21] the other parameters follow from (5.28) and (5.29).

The transformer specifies the "volume velocity"–"pressure input" pair, and for a steady- state signal it provides the parameters required to determine the power flow. Local particle velocity inside the cochlea in the stapedial area can be obtained from the volume velocity by multiplication of the latter with the stapes area A_{st}.

5.3.2.2 O'Connor and Puria's 2008 Approximation, ME-1

The recent version of O'Connor and Puria's middle ear model goes back to earlier versions of Puria and Allen (1991, 1998), which—like many other models—had been developed for the cat middle ear. These studies, however, did start at the auricle entrance (pressure ($p_e c$ and volume velocity $U_e c$ at the earcanal entrance), whereas O'Connor and Puria begin their analysis at the tympanic membrane. In other words,

[20] See Sect. 8.3 for the definitions.

[21] Or any alternative set of independent variables.

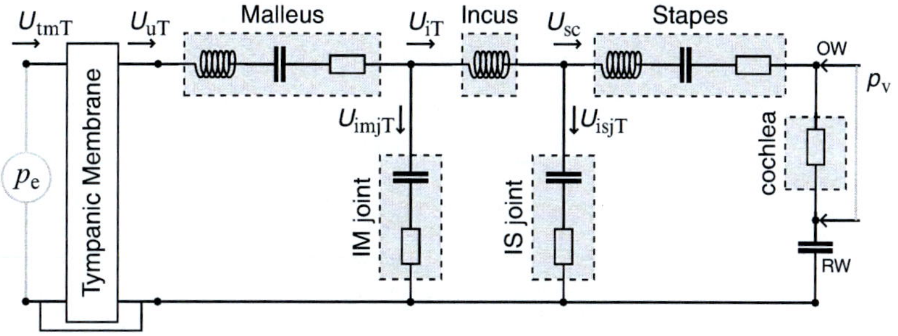

Fig. 5.7 A slightly modified version of the alternative middle ear model proposed by O'Connor and Puria (2008) where the transformers have been "removed" such that all variable are represented by their acoustic equivalents as seen from the *right* of all three transformers. Variables with an attached "T" in their subscripts are redefined to follow this transformation. The redefinitions are listed in Table 5.2, in combination with the original values

at the front end it needs an additional section to account for this difference. This is discussed briefly in connection with their Fig. 5.1, but details are consciously left open.

Similar to our proposal in the previous section, they also propose a version of the model where the transformer factors are incorporated in the network. Again this means that physical parameters are scaled. Figure 5.7 gives a representation of their result. It is a close copy of their Fig. 2b, with the modification that we start the scaling from the OW outward, as in Fig. 5.6b. The parameters are briefly indicated in the figure caption, and a detailed list is given in Table 5.2, grouped by the blocks that are indicated in the figure. The tympanic membrane is providing a delay term of about 50 μs and a damping term $Z_{0\rm tm}$, which is about $10 \times Z_a$ (cf Table 5.1), the term that would complete the network with an external acoustic impedance.

5.3.3 One-Dimensional Approximation

Now that we have a first approximation to the coupling properties, we can setup the 1D equations and discuss the solution. The approach will be numerical, based on the advise of our colleague H. W. Hoogstraten (Appl. Math., Univ. Groningen). After some introductory discussions, he started with an analytical approach and provided a first numerical solution in 1984, which at the time took considerable computation power.[22] At the time, proper scaling and rewriting of the ODEs representing the transmission line network were essential to meet stability criteria. Presently the scaling requirements are less tight because parameters can be represented in much more detail, but the other conditions still apply. The original program was limited

[22]Of the order of 12 h on a lab computer for 20 ms stimulus time.

Table 5.2 Parameter values middle ear model ME-1, with the formulas that apply to the mechanical section has been transformed to acoustic equivalent

Parameter		Data ($n = 16$)		Scaling	
Name	Units	Mean	Sem	Formula	Units
TM:					
T_{tm}	s	5.15×10^{-5}	0.80×10^{-5}		
$Z_{0\,\text{tm}}$	kg/s m^4	13.8×10^7	3.6×10^7	$\frac{Z_{0\,\text{tm}} A_{\text{tm}}^2 N_{LR}^2}{A_{\text{fp}}^2}$	kg/s m^4
A_{tm}	m^2	6.0×10^{-5}	0		
Malleus:					
M_{m} (m_{m})	kg	4.52×10^6	0.89×10^6	$m_{\text{m}} N_{LR}^2/A_{\text{fp}}^2$	kg/m^4
K_{m} (s_{m})	kg/s^2	530	71	$s_{\text{m}} N_{LR}^2/A_{\text{fp}}^2$	kg/m^4 s^2
R_{m} (d_{m})	kg/s	0.118	0.015	$d_{\text{m}} N_{LR}^2/A_{\text{fp}}^2$	kg/m^4 s
N_{LR} (n)	–	1.3	0		
Im-joint:					
K_{imj} (s_{imj})	kg/s^2	1.52×10^3	0.11×10^3	$s_{\text{imj}}/A_{\text{fp}}^2$	kg/m^4 s^2
R_{imj} (d_{imj})	kg/s	2.66×10^{-2}	0.43×10^{-2}	$d_{\text{im}}/A_{\text{fp}}^2$	kg/m^4 s
Incus:					
M_{i} (m_{i})	kg	4.60×10^{-6}	0.63×10^{-7}	m_i/A_{fp}^2	kg/m^4
Is-joint:					
K_{isj} (s_{isj})	kg/s^2	10.6×10^3	1.0×10^3	$s_{\text{isj}}/A_{\text{fp}}^2$	kg/m^4 s^2
R_{isj} (d_{isj})	kg/s	1.64×10^{-2}	0.40×10^{-2}	$d_{\text{isj}}/A_{\text{fp}}^2$	kg/m^4 s
Cochlea:					
A_{fp}	m^2	$3.14 \times s10^{-6}$	0		
M_{s} (m_{s})	kg	$3.55 \times s10^5$			
$K_s c$ (s_{sc})	kg/s^2	1.21×10^{14}	0.11×10^{14}		
$R_s c$ (d_{sc})	kg/s	2.83×10^{10}	0.19×10^{10}		

The mechanical data are given with the standard error of the mean (sem). The right-hand columns provide the correction formulas and resulting units

to one text-page in F77. With some interfaces, and an upgrade to modern Fortran, it expanded considerably. Both are available through the URL. The core of the original program by Hoogstraten is also presented in a gray-box at the end of this chapter. We have to admit that in 1983 we fell in the trap that we did not use a middle ear interface, and drove the cochlea at the stapes.

The transmission line network is almost identical to that shown in Fig. 3.6, the important difference is that implementation of NL elements in the system means that the elements in the cross-partition equivalents can now be nonlinear. We treat this problem in detail for the case of NL-damping, introducing a Van der Pol type of damping term for d_{a_i}, or Van der Pol oscillators for the CP-sections. The mass terms will always be treated as linear, but for the stiffness we will allow variations at a later stage.

The transmission line network is a numerical approximation of the cochlear structure. The constituting equations, as long as they were linear, could represent a smooth, continuous system. It may be questionable if that is also true for the real cochlea, which does have a grid structure along the length axis: the number of cells that setup the organ of Corti is of the order of 3200. This is the quotient of the BM-length up to the helicotrema, and the width of a pillar cell or a hair cell. In other words, a discrete representation along that parameter may be more natural than a continuous representation. Moreover, it set a limit to the maximum number of points that might be useful for the analysis.

For the cochlear fluid, however, the discretization provides a limitation. Therefore, we will start from (5.23) earlier in this chapter. We assume tentatively that $\alpha(x) = 1$, and use the subscript notation for differentiation. Finally, we follow the common assumption that the acoustic mass, or the mass per unit area, $m(x) = m$, a constant. Application of these conditions modifies (5.24) slightly to

$$(m\, w_{tt} + d(x)\, w_{\mathrm{t}} + s(x)\, w)_{xx} - \frac{2\rho}{h}\, w_{tt} = 0. \tag{5.30}$$

Note that details of damping and stiffness functions are still unspecified in this formula.

The first step that is set to solve this PDE properly is to transform it into two coupled ODEs. Next the spatial differential equation is transformed to a difference equation. The remaining ODEs in time are treated separately. The second-order equations are rewritten as sets of first-order equations, which are solved with the Runge–Kutta–method (RK4).

5.3.3.1 Separating Space and Time Differentiation

In Sect. 5.3.1 the full bracket term in (5.30) was defined as $\varphi = \varphi(x,t)$. Here we introduce an additional short-hand for the sum of the damping and stiffness terms[23] as $g = g(x,t)$, or:

$$\varphi = m\, w_{tt} + d(x)\, w_{\mathrm{t}} + s(x)\, w, \tag{5.31a}$$

$$g = d(x)\, w_{\mathrm{t}} + s(x)\, w. \tag{5.31b}$$

The difference gives a new specification of the first term mw_{tt}:

$$mw_{tt} = \varphi - g. \tag{5.32}$$

With these definitions (5.30) can be rewritten as

$$\varphi_{xx} - \kappa\varphi = -\kappa g, \tag{5.33}$$

[23]Actually, the function g introduced here represents the standard cross-partition force (or pressure) term. The dimensions of g and φ are $[\mathrm{M\,L^{-1}T^{-2}}]$, and the units: Pa.

where the constant κ combines the remaining parameters: $\kappa = \frac{2\rho}{hm}$. This equation is an ordinary nonhomogeneous second-order spatial differential equation. At a single point in time, the spatial part of (5.33) can be solved, and this solution yields all required local differential equations in time. Introduction of the partition velocity $v = w_t$ enables the rewriting of (5.31) into a set of two first-order equations:

$$w_t = v, \tag{5.34a}$$

$$v_t = (\varphi - g)/m, \tag{5.34b}$$

where

$$g = d(x)\, v + s(x)\, w \tag{5.35}$$

and has to be updated, the function specifying the CP-properties which also must be updated at each time step.

Equation (5.33) is solved using Gauss-elimination, and (5.34a,b) with RK4.

5.3.3.2 The Spatial Difference Equation

We go into more detail of this part of the procedure because the boundary conditions for the spatial equation are important.

Let the discrete version of the transmission line have n regular sections, from point 1 as the first point at the base, and up to point $i = n$ representing the last partition point before the apex. In other words, the apex is beyond point $i = n$ but directly connected in parallel, and point $i = 0$ represents the connection at the cochlear entrance.

As will be shown below, the spatial differentiation can be represented as a square matrix product on the vector φ. In other words, the differential equation is transformed into a $(n+1)\times(n+1)$ matrix equation, where the boundary conditions involve the points at 0 and at n. This equation is of the general shape

$$D\{\varphi\} - \kappa\varphi = -\kappa\mathbf{g}, \tag{5.36}$$

except at the boundaries.

The standard transform of the second-order spatial derivative to the difference equation involves three adjacent terms:

$$\varphi_{xx}(i) = \frac{1}{\Delta^2}\left(\varphi(i-1) - 2\varphi(i) + \varphi(i+1)\right). \tag{5.37}$$

Δ is the spatial discretization step size, which equals the relevant length along the BM divided by the number of sections n. The second-order spatial difference is symmetric around the point of interest, and therefore relatively insensitive to the spatial discretization. On the edges, however, only two points are available, and more subtlety is useful. The three adjacent elements that represent the differentiation generate a tridiagonal matrix, in which the κ term can be incorporated. Next we multiply all terms in (5.33) by Δ^2, set $\kappa \times \Delta^2 = \beta$, and replace $-2 - \beta$ by K. Leaving the boundary conditions (A, B, Y, Z) open for the moment, we obtain:

$$\begin{pmatrix} A & B & 0 & & \cdots & & & 0 & 0 \\ 1 & K & 1 & 0 & \cdots & & & & 0 \\ 0 & 1 & K & 1 & & & & & \\ & 0 & 1 & K & & & & & \\ \vdots & \vdots & & & \ddots & & & \vdots & \vdots \\ & & & & & K & 1 & 0 & \\ & & & & & 1 & K & 1 & 0 \\ 0 & & & & \cdots & 0 & 1 & K & 1 \\ 0 & 0 & & & \cdots & & 0 & Y & Z \end{pmatrix} \varphi = -\beta \mathbf{g} + \begin{pmatrix} s \\ 0 \\ 0 \\ 0 \\ 0 \\ 0 \\ 0 \\ 0 \\ apex \end{pmatrix}, \tag{5.38}$$

where φ and $\mathbf{g}$ are $n + 1$-dimensional vectors.

The equation at the cochlea base (top line in the matrix, index $n = 0$) must involve the stapedial boundary condition, including a potential stimulus input. This means that it is connected to the middle ear interface. This is represented in line $n = 0$:

$$A.\varphi(0) + B.\varphi(1) = -\beta_0 g(0) + s(t). \tag{5.39}$$

To solve this line—and ultimately the full matrix—we must find the proper values for A and B, and the stimulus term. The coupling terms $g(i)$, which determine the fluid coupling, also have to be specified, so here we start analyzing $\mathbf{g}(0)$.

If

$$\varphi_{xx}(0) = \frac{1}{\Delta^2}(\varphi(-1) - 2\varphi(0) + \varphi(1)) \tag{5.40}$$

and the first derivative / differential at $x = 0$ is

$$\varphi_x(0) = \frac{1}{\Delta}(\varphi(0) - \varphi(-1))$$

then (5.40) can be rewritten as

$$\varphi_{xx}(0) = \frac{1}{\Delta^2}(-\varphi(0) + \varphi(1)) - \frac{\varphi_x(0)}{\Delta}, \tag{5.41}$$

which leaves us with the question of finding $\varphi_x(0)$. This is the point where the stimulus comes into play, through the middle ear interface. The physical reason for this is that the quantity $\varphi(i)$ represents the time derivative of the pressure at point i, or

$$\varphi(i) = \frac{\mathrm{d}p(i)}{\mathrm{d}t} \qquad \text{and} \qquad \varphi(0) = \frac{\mathrm{d}p(0)}{\mathrm{d}t},$$

which couples directly to the OW-pressure stimulus.

Additional adjustments must be made in the second matrix line (index $n = 1$) if one takes the spatial differences between the footplate—BM coupling, and the coupling between elements 1 and 2 along the CP into account. The former is basically independent of the spatial sampling discretization along the BM (simply distance between OW and BM), whereas the second is (BM-length / number of points). This leads to the correction factor γ_0 in most of the terms. Because of the tridiagonal matrix form, the second line still couples directly to the stimulus $s(t)$.

For the simple middle ear (ME-0) we can now specify the relevant coupling mass in the input branch and with the results of this analysis we can rewrite the line at $n = 0$ as (see Figs. 5.6 and 5.8 for more details):

$$-(1 + \gamma_0\kappa)\varphi(0) + \varphi(1) = -\gamma_0\kappa g(0) - \gamma_0\kappa n_{\mathrm{me}}\varphi_e(t), \tag{5.42}$$

where

$$\kappa = \frac{\rho h}{m} \qquad \text{and} \qquad \gamma_0 = \frac{m\,\Delta_0\,b}{m_{\mathrm{me}}\,h^2}$$

with $m = m(x)$, the partition mass in kg/m^2, and m_{me} the acoustic middle ear mass in kg/m^4. Hence, the product $\gamma_0\kappa$ is dimensionless. The solutions for unknowns A and B are: $A = K - 1$ and $B = 1$. The remaining terms in row 0 involve $g(0)$ and $n_{\mathrm{me}}p_e(t)$. The first involves the fluid mass coupling from oval window to CP characterized by Δ_0, which generally differs from the spatial discretization step Δ.

The general matrix row, i.e., from $i = 1$ to n, remains as discussed above,

$$\varphi(i-1) + K\varphi(i) + \varphi(i+1) = -\beta g(i), \tag{5.43}$$

where local BM properties are contained in $g(i)$. We also can specify the underlying physical parameters of K and $g(i)$.

The acoustic fluid mass in a section i of x-length Δ over a scala cross-section is

$$m_a^i = \frac{\rho\Delta}{bh} \qquad \mathrm{kg/m^4}$$

per section. If we consider the motion in the scala vestibuli and scala tympani together, as combined drivers of the CP this provides an additional factor 2:

$$m_a^i = \frac{2\rho\,\Delta}{b\,h}. \tag{5.44}$$

We obtain the acoustic mass of a BM section in a similar way. Starting from the given mass per surface area $m(x)$, the acoustic mass of a CP-section is:

$$m_{a_i} = \frac{m(x)}{b\,\Delta}. \tag{5.45}$$

Finally, the conditions at the apex must be specified. A historical issue is that many cochlea models have assumed an apical shortcut, and got away with it.

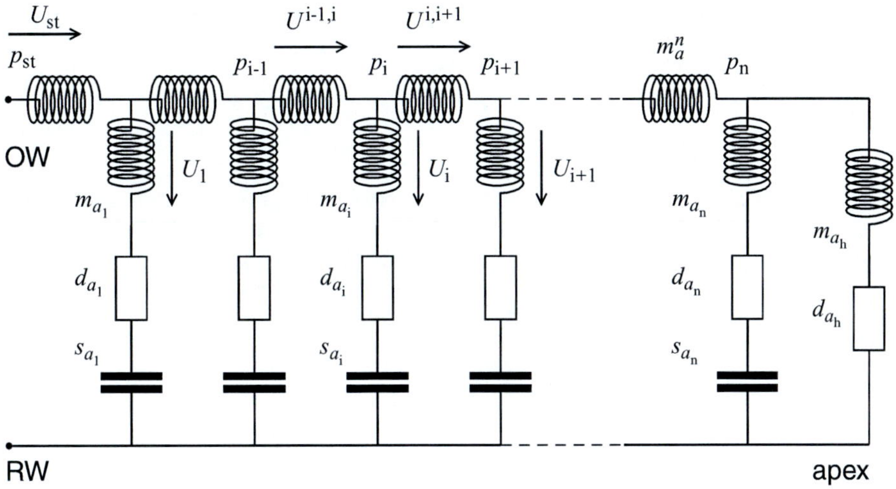

Fig. 5.8 The same network representation of a 1D cochlea model as a transmission line as in Fig. 3.6, with a slightly different—but equivalent—representation of the helicotrema. Moreover, the elements of the cochlear partition are now considered to be potentially nonlinear second order networks elements. The fluid mass coupling in the scalae is represented by the (mostly unlabeled, except for m_a^n) inductances in the horizontal branch. If the spatial discretization is uniform, then the elements are equal, and for each element m_a^i the mass is $\rho \Delta x A_{sc}$, and the acoustic mass is $\rho\Delta/A_{sc}$. Over the same Δx the CP-mass is $m(x)\Delta x b$, and the acoustic CP-mass for that section is $m_{a_i} = m(x)/(\Delta.b(x)$. Note that the (fluid mass)/(partition mass) ratio is independent of Δx. The difference with Fig. 3.6 is that index n now refers to the most apical CP-section, and that all acoustic fluid mass involved in the helicotrema flow is combined in m_{a_h}

This goes back to the underlying assumption that cochlear response amplitudes to narrow-band signals in the mid-frequency range have decayed by many orders of magnitude once they reach the helicotrema. For stimuli in the low-frequency range, however, this is not true, and interferences and reflections occur.

Insertion of apical boundary conditions involves the interaction with BM-points $n-1$ and n. We follow the network as depicted in Fig. 5.8, and leave it to the reader to proof that it is completely equivalent[24] to Fig. 3.6. Line n in the matrix defines the most important apical conditions: the unknowns Y, Z and the variable $apex$, which appear in the equation that represents the final row

$$Y\varphi(n-1) + Z\varphi(n) = -\beta_n g(n) + apex \tag{5.46}$$

have to be resolved. The analysis is based on the right-hand part of Fig. 5.8. The volume velocity through section n is U_n, and through the helicotrema U_h. The sum

[24] i.e., the one can be transformed into the other.

of these passes through the scala section from $n-1$ to n. Hence, the force balance in x-direction is:

$$p_{n-1} - p_n = m_a^n \frac{\partial U^{n-1,n}}{\partial t} = m_a^n \frac{\partial}{\partial t}(U_n + U_\mathrm{h}). \quad (5.47)$$

For the apical point $i = n$, we have two branches: branch n representing the partition and branch h representing the helicotrema coupling. The first appears to follow the regular partition conditions, the second misses the spring term. However, the remaining mass and damping term are quite different.[25] We obtain:

$$p_n = m_{a_n} \frac{\partial U_n}{\partial t} + d_{a_n} U_n + s_{a_n} \int U_n \mathrm{d}t \quad \text{and} \quad p_n = m_{a_h} \frac{\partial U_\mathrm{h}}{\partial t} + d_{a_h} U_\mathrm{h}. \quad (5.48)$$

Taking the time-derivative of (5.47) and replacing $\partial p_n/\partial t$ by φ_n, and $\partial U_n/\partial t$ and $\partial U_\mathrm{h}/\partial t$ by V_n and V_h gives:

$$\varphi_{n-1} - \varphi_n = m_a^n \frac{\partial}{\partial t}(V_n + V_\mathrm{h}). \quad (5.49)$$

The time derivatives of the two parts of (5.48) provide substitutions for the right-hand side of this equation:

$$\frac{\partial V_n}{\partial t} = \frac{1}{m_{a_n}}[\varphi(n) - g(n)] \quad \text{and} \quad \frac{\partial V_\mathrm{h}}{\partial t} = \frac{1}{m_{a_h}}\left[\varphi(n) - d_{a_h} V_\mathrm{h}\right] \quad (5.50)$$

thereby finally providing

$$-\frac{m_a^n}{m_{a_n}}\varphi_n = -m_a^n \frac{\partial V_n}{\partial t} - \frac{m_a^n}{m_{a_n}} g(n), \quad (5.51)$$

$$-\frac{m_a^n}{m_{a_h}}\varphi_n = -m_a^n \frac{\partial V_\mathrm{h}}{\partial t} - \frac{m_a^n}{m_{a_h}} d_{a_h} V_\mathrm{h} \quad (5.52)$$

or

$$\varphi_{n-1} - [1 + \frac{m_a^n}{m_{a_n}} + \frac{m_a^n}{m_{a_h}}]\varphi_n = -\frac{m_a^n}{m_{a_n}} g(n) - \frac{m_a^n}{m_{a_h}} d_{a_h} V_\mathrm{h}. \quad (5.53)$$

Note that the general expression for the ratio of fluid mass and partition mass is:

$$m_a^i = \frac{2\rho\Delta}{b\,h} \quad \text{and} \quad m_{a_i} = \frac{m(x)}{b\,\Delta}$$

---

[25]The mass term m_{a_h} is estimated from a "semicircular pipe", and the damping from the viscous loss through the same. The estimates were refined by optimization of the resulting termination impedance which minimizes apical reflections. This approach is a compromise between the use of data and the assumption of optimum processing.

or

$$\frac{m_a^i}{m_{a_i}} = \frac{2\,\rho\Delta^2}{m(x)\,h}. \tag{5.54}$$

In our model, the connection at the apex is specified to minimize reflections at the helicotrema because we tacitly assume that this is a logical biological optimization. We approximate it by an estimate of the characteristic impedance of the transmission line at the apical end (obviously a linear approximation). The required parameters involve a mass term and a resistance, which match the properties of fluid moving back and forth through a tube. The properties at point n determine the amount of the canal fluid mass through the helicotrema. More precisely, the requirements for minimum reflection match the properties of the apical fluid mass and fluid flow damping.

Reflections at the end of a transmission line are suppressed by application of the matching termination impedance. A common example is the 50 Ω resistance for common coax-cables. For those transmission lines, the cable properties are independent of x. The properties of the cochlea do change over the length, but near the apex the gradient decreases. Hence, we will just consider the properties at position $i = n$, and compare that to the helicotrema properties.

The characteristic acoustic impedance $Z_{a,0}$ is given by

$$Z_{a,0} = \sqrt{\frac{Z_{x,n}}{Y_{x,n}}}, \tag{5.55}$$

where $Z_{x,n}$ is the longitudinal impedance per unit length, and $Y_{x,n}$ the transversal admittance over the same step. From the analysis in this section it should be clear that in the linear systems analysis approximation

$$Z_{x,n} = \frac{2\mathrm{i}\,\omega\,\rho}{h\,b} \qquad \text{and} \qquad Y_{x,n} = \frac{b}{\mathrm{i}\,\omega\,m_{a_n} + d_{a_n} + s_{a_n}/\mathrm{i}\,\omega}, \tag{5.56}$$

so that;

$$Z_{a,0} = \sqrt{\frac{2\,\rho\,m_{a_n}}{h\,b^2}\left(\frac{s_{a_n}}{m_{a_n}} - \omega^2 + \mathrm{i}\,\omega\,\frac{d_{a_n}}{m_{a_n}}\right)}. \tag{5.57}$$

The network specification of the apical termination impedance is

$$Z_{a_h} = \mathrm{i}\,\omega\,m_{a_h} + d_{a_h}, \tag{5.58}$$

and it turns out to be possible to set $Z_{a_h} = Z_{a,0}$, even though we have three equations and at first glance only two free parameters. Squaring (5.57) and (5.58) and equalizing the results for all powers of ω gives a frequency independent match for m_{a_h} and d_{a_h}. The "missing" parameter is "hidden" in the local Q-value in branch $i = n$. The result implies that the value $Q_{3dB} = 0.5$ is a necessary ingredient for optimum termination.

The verifications that the involved fluid mass and "tube" resistance are of the proper orders of magnitude are left to the interested reader.

At this point, we have solved the matrix elements Y and Z, the apical version of β and the term $apex$:

$$Y = 1, \tag{5.59}$$

$$Z = -\left[1 + \frac{m_a^n}{m_{a_n}} + \frac{m_a^n}{m_{a_h}}\right], \tag{5.60}$$

$$\beta_n = \frac{m_a^n}{m_{a_n}}, \tag{5.61}$$

$$apex = -\frac{m_a^n}{m_{a_h}} d_{a_h} V_{\mathrm{h}}. \tag{5.62}$$

Note that in fact β_n is independent of position, because the ratio of the acoustic masses is (5.54), but it does depend on the discretization grid size. The factor m_a^n/m_{a_h}, however, generally is orders of magnitude smaller that g, because the termination mass term is much greater than m_a^n or m_{a_n}. The most right-hand term in (5.53) also contains the component d_{a_h}, which is of a similar order of magnitude as m_{a_h}. Therefore, the fluid motion through the helicotrema is not simply negligible, and will need proper treatment.

5.3.4 *Solving the Matrix Equation and the Differential Equations*

Remember that the matrix equation represents the second order spatial difference equation, which is coupled to second order differential equations in time. For one step in time we have to solve the matrix equation, and with the new parameters we estimate the next step in time for all points x.

The matrix equation of the type (5.38) is solved by a straightforward Gaussian elimination technique. For the solution of a tridiagonal matrix it needs two runs. In principle it does not matter where the procedure starts, in practice accuracy is optimized by limiting propagation of results of computations based on small numbers as much as possible.

Details of the Gauss Elimination Procedure

The last three lines of matrix (5.38) were

$$\begin{pmatrix} 0 & & & & 1 & K & 1 & 0 \\ 0 & 0 & & \cdots & 0 & 1 & K & 1 \\ 0 & 0 & 0 & \cdots & & 0 & 1 & Z \end{pmatrix} \varphi = -\beta \mathbf{g} + \begin{pmatrix} 0 \\ 0 \\ A_p \end{pmatrix}$$

or, written out in detail:

$$\begin{aligned} \varphi_{n-3} + K\varphi_{n-2} + \varphi_{n-1} &= -\beta g(n-2), \\ \varphi_{n-2} + K\varphi_{n-1} + \varphi_n &= -\beta g(n-1), \\ \varphi_{n-1} + Z\varphi_n &= -\beta g(n) + A_p. \end{aligned}$$

Now set $-\beta g(n) + A_p \Rightarrow k_n$, and divide the last row by Z. This gives

$$\frac{1}{Z}\varphi_{n-1} + \varphi_n = \frac{1}{Z}k_n \tag{5.63}$$

$$\text{or:} \quad \varphi_{n-1} - b_n\varphi_n = k_n \qquad \text{with} \qquad b_n = -Z. \tag{5.64}$$

After subtraction of (5.63) from row $n-1$, the φ_n-term disappears:

$$\varphi_{n-2} + (K - \frac{1}{Z})\varphi_{n-1} = -\beta g(n-1) - \frac{k_n}{Z} = k_{n-1} \tag{5.65}$$

or, rewritten in shorthand:

$$\varphi_{n-2} - b_{n-1}\varphi_{n-1} = k_{n-1}. \tag{5.66}$$

The complete results so far can be written as

$$\begin{pmatrix} -b_0 & 0 & 0 & \cdots & & 0 & 0 \\ 1 & -b_1 & 0 & \cdots & & & 0 \\ 0 & 1 & -b_2 & & & & \\ & 0 & 1 & & & & \\ \vdots & \vdots & & \ddots & & \vdots & \vdots \\ & & & & 0 & 0 & \\ 0 & & & & -b_{n-2} & 0 & 0 \\ 0 & 0 & & \cdots & 1 & -b_{n-1} & 0 \\ 0 & 0 & 0 & \cdots & 0 & 1 & -b_n \end{pmatrix} \varphi = \begin{pmatrix} k_0 \\ k_1 \\ k_2 \\ k_3 \\ \\ k_{n-3} \\ k_{n-2} \\ k_{n-1} \\ k_n \end{pmatrix} \tag{5.67}$$

with properly computed values for the b's and k's. The step to the final solution for the vector φ continues with the downward steps:

$$\begin{aligned}\varphi_0 &= -k_0/b_0\\ \varphi_1 &= -k_1/b_1 + \varphi_0/b_1\\ \ldots &= \ldots\\ \varphi_n &= -k_n/b_n + \varphi_{n-1}/b_n.\end{aligned} \tag{5.68}$$

Solving the time steps, or the NL cochlea, ODEs involve the application of a discrete approximation of the analytic solution for which we use the RK4 method. In Sect. 5.1.3.2 this method was introduced as one of efficient and stable methods for obtaining stable oscillating solutions. The efficiency was studied by Diependaal et al. (1987) in a comparison with the Heun method and a modified Sielecki method and the conclusions summarized in their TABLE II, which shows that for equal computation time the Runge–Kutta method is much more accurate than the alternatives. The stability properties of the RK4 method have been indicated in Fig. 5.4, where points outside the grayed area are unstable, and the highest stability is obtained at the two darkest areas, and their conjugates. The criterion for stability is different from the Nyquist criterion, which applies to linear systems.

The RK4 method that is commonly used for the solution of second-order differential equations begins with the step to transform all second-order differential equations in sets of coupled first-order equations. This has been established in (5.34a,b), where both deflection w and velocity v of the cochlear partition were used.

The program lines shown below are involved with one RK4 time step, including the update of the matrix and Gauss elimination. The variable V represents v and Y represents w. The method uses four sub-steps per step. Estimates for the next values of Y use the latest value of V: $\mathrm{d}Y = V\mathrm{d}t$, but estimates for the next values of V have to be computed separately at all points i, and is renewed in the $(n+1) \times 4$ matrix $\mathbf{M}1234(0:n, 1:4)$. An example of the full program text is available on the URL.

```
do RK4step=1,4
  select case (RK4step)
  case (1)
      F0=Stimulus(0)
      Ytmp = Y
      Vtmp = V
  case (2)
      F0=Stimulus(1)
      Ytmp = Y + Vtmp * half_dt
      Vtmp = V + M1234(0:n,1) / 2.d0
  case (3)
      Ytmp = Y + Vtmp * half_dt
      Vtmp = V + M1234(0:n,2) / 2.d0
  case (4)
```

```
        F0=Stimulus(2)
        Ytmp = Y + Vtmp *dt
        Vtmp = V + M1234(0:n,3)
    end select

    CALL Calculate_g

    U_helicotrema = -stapesArea(parameterSet) * Vtmp(0) &
           - SUM (Vtmp(1:n)) * dx * bm_width
    k(n) = -Asq * g(n) - phi * U_helicotrema
    DO i = n-1, 1, -1
        k(i) = -Asq * g(i) + k(i+1) * b(i+1)
    ENDDO
    k(0) = -Asq0 * (p0x * F0 + g(0) + r_Xtr0 * Ytmp(0)) &
           + k(1) * b(1)
    q(0) = -k(0) * b(0)
    DO i = 1, n
        q(i) = (q(i-1) - k(i)) * b(i)
    ENDDO

    M1234(0,RK4step) = m0_RK4 * (q(0) - g(0) - p0x * F0 &
           - r_Xtr0 * Ytmp(0)) * dt
    M1234(1:n,RK4step) = (q(1:n)-g(1:n)) * dt

  enddo !RK4step

    Y = Y + (V + (M1234(0:n,1)+ M1234(0:n,2)+ &
               M1234(0:n,3)) / 6.d0) * dt
    V = V + (M1234(0:n,1) + 2.d0 * (M1234(0:n,2)+ &
               M1234(0:n,3)) +  M1234(0:n,4)) / 6.d0

  !END MAIN RK4 ROUTINE
```

This concludes the discussion of the method of the time domain analysis of a 1D approximation of the cochlea model. This structure can address and analyze, and model several different cochleae. It needs a proper set of parameters, and can be used for different mammalian ears, insofar as data are available.

We have not (yet) gone into the fine structure of the CP, not because we would want to suggest that it does not have a function, but because we choose to limit the discretization grid to cover at least ten longitudinal cell "layers" in the Organ of Corti ($\approx 100\,\mu$m). Moreover, these layers are not as separate as some modelers do want us to believe, there are several interesting longitudinal couplings. At the moment we take a global view looking at combined effects, and more in particular: stay away from a model with too many free parameters.

Even with this simplification, several relevant cochlea model properties can be analyzed by specifying the $g(i)$ function. This is worked out in the following section.

Table 5.3 Minimum set of parameters required for a proper cochlea model analysis

Model specification		
Name	Symbol	Units
Frequency map		
Frequency at base	$f_{\mathrm{G}}(0)$	Hz (kHz)
	κ_{G}	1/m
Apex correction	A_{p}	Hz
Middle ear		
Stapes area	A_{st}	m^2
Eardrum area	A_{TM}	m^2
ME resonance frequency	$f_{\mathrm{c,me}}$	Hz
ME quality factor	$Q_{\mathrm{3dB,me}}$	-no-
ME transformer ratio	n_{t}	-no-
Damping coupler	d_{me}	$\mathrm{Ns/m}^5$
Mass coupler	m_{me}	$\mathrm{kg/m}^4$
Stiffness coupler	s_{me}	$\mathrm{kg/s}^2\mathrm{m}^4$

The index G of $f(0)$ and κ refers to the Greenwood-map, see Sect. 5.3.5.1

5.3.5 Selection of a Specific Cochlea Model or Structure

A set of specific parameters that covers all necessary constants should be available for any mammal for which computation are made. Table 5.3 gives a shortlist of a minimum set.

Given these parameters, we can address the next interesting point: the selection of the g-function, or of the CP damping and stiffness parameters along the cochlea. We use the simplification that the relevant mass per unit area is constant.

The versions with the active NL damping option are the ones that can generate spontaneous oscillations or otoacoustic emissions. All nonlinear versions will generate distortion products, among which are DPOAEs. The damping point in a standard passive system characterizes the point where energy is dissipated. Formally, changing the sign of the damping term to negative transforms a dissipater into an emitter. In other words, changing from passive to active involves changing the sign of the damping term for small amplitudes. At larger amplitudes damping has to become positive again, stabilizing the properties of the oscillator.
We will discuss several of the specification options following the column numbering in Table 5.4.

5.3.5.1 Linear Model (#0)

The first column gives the combination for linear damping and linear stiffness, although it also contains the option to introduce nonlinear stiffness. This linear case is "identical" to the linear model in Chap. 3, and in particular the analysis in Sects. 3.1–3.5 does apply.

Table 5.4 A selection of parameters for the g-function: the damping + stiffness terms in the transpartition branches

Model #	0	1	2	3	4	5	6
Damping:	L	vdP	NL1	NL2	NLG	LZ	NLZ
Passive	p		p	p	p		
Active		a	a	a		*	*
Stiffness:	L NL	L	L	L	NLG	LZ	NLZ

The "values" in the table indicate linear (L) or nonlinear (NL) elements, and active vs. passive is denoted with a or p. Column # 1 (vdP) refers to the classical van der Pol oscillator with parabolic damping term. NLG gives the option to select Furst and Goldstein's nonlinear model (Furst and Goldstein 1982), and LZ or NLZ links to a rather different version following Zweig's specifications for local feedback (Zweig 1991)

One issue that is treated similarly for all model versions is the frequency map. Presently fairly accurate data concerning the frequency map are available, both from observations of BM-responses, and from HRP-labeled auditory nerve data. Of course one has to be aware of the fact that if such data have been obtained from almost intact preparations, then the cochlea would have been operating nonlinearly, and the linear approximation should be regarded with proper caution!

One of the first generally accepted estimates of the map was the Greenwood map Greenwood (1961, 1990), which uses three parameters: an initial value of f_c at the base (in Greenwoods original paper at the apex), the exponential decay towards the apex, and the asymptotic apical term. In the formula the propositions imply:

$$\begin{aligned} f_c(x) &= 165.4\,(10^{0.06(35-x)} - k) \\ &= 20{,}822\,(10^{-0.06x} - k.10^{-2.1}) \\ \text{and for } k &= \;\; 0.8 \text{ or } 0.85\;: \\ f_c(34) &= 57.6 \text{ or } 49.3 \text{ Hz}, \\ f_c(35) &= 33.1 \text{ or } 24.8 \text{ Hz}. \end{aligned} \tag{5.69}$$

In the original equation, x was taken to increase from apex to base, just opposite from the notation that we are using. In the current formulation, the apical limit has become sensitive to the length of the cochlea in combination with the basal boundary condition. The constants are $f_c(0) = 165.4 \times 10^{2.1} = 20{,}822.6$, the exponential factor is $0.06\,\text{mm}^{-1}$. The apical limit parameter originally had the value $k = 1$. In the (1990) study, Greenwood suggest that this original value "may sometimes be better replaced by a number from about 0.8 to 0.9" to obtain a lower frequency limit at the apex. For the human cochlea, we will use his map, with $k = 0.85$ (see also Sect. 3.3).

Note that the value of $f_c(0)$ depends critically on the value 2.1, which is the product of 0.06 and 35. For instance, for a small variation in cochlear length, between 34.9 and 35.1 mm, the initial frequency changes from 20,537 to 21,112 Hz.

The frequency map seems to provide rather direct information about the CP-parameters. Local oscillators have local resonance frequencies, and it has been tempting to link the one directly to the other and derive CP parameters from insufficiently detailed information and intelligent interpretation. Figure 8.1 indicates that only for the "symmetric" case (middle column), the resonance frequency of a simple oscillator is independent of the damping. This is true for all band-pass filters with symmetric input–output characteristic, independent of the order of the system. However, as soon as the symmetry does not apply, the peak frequency starts to deviate from the resonance frequency, depending on the damping, and correlated to the phase behavior. This is indicated in the left-hand and right-hand columns in Fig. 8.1.

Summarizing this point: the map is very important for the estimation of the local mass and stiffness ratio, but for the full values independent additional information is required, either about the response phase properties or about one of the two parameters: mass or damping (see also Sect. 8.5.1).

5.3.5.2 van der Pol Model (#1)

Model #1 uses the classical quadratic (parabolic) damping term. Small amplitude negative damping specifies an oscillator that generates a limit cycle oscillation if undriven and uncoupled. The relative scaling of the active part is characterized by the parameter ε.

Some Relevant Properties of the Parabolic Damping Term

Two important properties of a parabolic damping profile, as in the van der Pol oscillator, deserve specific attention.

The first is of theoretical nature and leads straightforwardly to an exponential power-law output–input relation, with the power $1/3$. This follows from the consideration that near resonance, the high-amplitude behavior is limited by the damping term, which is of the family $d.v$. For large amplitude, and also for large velocity v, the damping coefficient d is proportional to v^2, and the input $s(t)$–output $v(t)$ relation approaches:

$$s(t) \propto v^3(t) \qquad \text{or} \qquad v(t) \propto \sqrt[3]{s(t)}, \tag{5.70}$$

(where the Zero-phase cubic root term is considered).

The second is practical: an odd-order nonlinearity can only generate odd-order harmonic and intermodulation distortion. This provides an accuracy check on any applied method. By contrast, an even-order nonlinearity can generate both even-order and odd-order products. This is the reason to limit the analysis to odd-order nonlinearity. It does not imply that we assume that the real system is completely odd order. Obviously, there are some even order distortion products (e.g., difference tones) although they tend to be much smaller than the odd-order components.

The (un)coupling refers to the interaction with neighboring elements within Corti's organ. Specifications as well as results of the Model #1 have been presented in Duifhuis et al. (1985). Their equation (1) equals[26] (5.30). For both $d(x)$ was specified as

$$d(x) = -d_1(x) + d_2(x)\, w_{\mathrm{t}}^2.$$

Note that the dimensions of $d_1(x)$ and $d_2(x)$ are quite different: the definition is such that the product $d_2/d_1.w_{\mathrm{t}}^3$ is dimensionless. The fact that the parabolic relation is in the velocity w_{t} rather than in deflection w can be considered as treating the system as a Rayleigh oscillator rather than a van der Pol oscillator (see Definitions 9.12 and 9.13). For a critical comparison with general van der Pol oscillators, as formulated in (5.31), it is useful to apply the following scaling factors: first, divide all terms by m:

$$\frac{\varphi}{m} = w_{tt} - \frac{d_1}{m} w_{\mathrm{t}} + \frac{d_2}{m} w_{\mathrm{t}}^3 + \frac{s}{m} w$$

and then scale the time by 1 /(angular resonance frequency), $\theta = t\,\sqrt{s/m}$, which brings us at[27]

$$\frac{\varphi}{s} = w_{\theta\theta} - \frac{d_1}{\sqrt{sm}} w_\theta + \frac{d_2}{m}\sqrt{\frac{s}{m}}\, w_\theta^3 + w,$$

which is of the form of the general Rayleigh equation (9.29). This can also be written as

$$f(\theta) = \ddot{w} - \varepsilon(1 - \alpha\dot{w}^2)\dot{w} + w \tag{5.71}$$

with

$$\varepsilon = \frac{d_1}{\sqrt{sm}} \quad \text{and} \quad \alpha = \frac{d_2\, s}{d_1\, m}. \tag{5.72}$$

Integration transforms the Rayleigh equation to a regular van der Pol equation ($\dot{w} = x$), and then scaling by $u = x\sqrt{3\alpha}$ can alter the appearance of the NL damping term[28]:

$$\ddot{x} + (3\alpha x^2 - \varepsilon)\dot{x} + x = \dot{f}, \tag{5.73}$$

$$\ddot{u} + (u^2 - \varepsilon)\dot{u} + u = \sqrt{3\alpha}\,\dot{f}. \tag{5.74}$$

---

[26]The only difference is the use of the symbol y instead of w for the transmembrane deflection.

[27]Note that the scaled time parameter θ is dimensionless.

[28]This operation and the connected discussion originates from Broer and Takens (2009).

The applied Broer–Takens transform moves the ε-parameter from without to within the bracket term, thereby modifying its role. The equation now has a Hopf-bifurcation at $\varepsilon = 0$. For negative ε all solutions converge to zero, for positive ε the resulting oscillator is an active van der Pol oscillator.

Initially[29] (Duifhuis et al. 1985) we started with a network model where all branches were active with[30] $\varepsilon = 0.05$. The reason to start analyzing the van der Pol oscillator has primarily been that there was not a direct biophysical natural source of spontaneous otoacoustic emissions. In general, a sensor must pick up a tiny amount of energy in order to detect or measure (estimate) some parameter accurately. Lossless detectors are unphysical. So at some point, a sensory system has to dissipate some data-related energy, and it is quite surprising that the auditory system appears to have a modification that also allows generation of energy. If this amount of generated energy is large, then the ear would really be transformed from a sound detector to a sound producing device. This line of thinking is consistent with the idea that low-level damping is reduced in order to enhance the sensitivity to weak sounds. Spontaneous emissions would arise at points where this reduction is too strong. Apparently, this mechanism requires intact OHCs, but for the time being, that is where our speculations pause.

The use of the constant ε parameter, independent of x, presented us with the surprising result that the undriven response of the cochlea showed chaotic behavior (op cit., Figs. 1–3). Each point tries to oscillate at its own resonance frequency, which implies that the resonances frequencies of adjacent points differ, so that they cannot synchronize, at least not continuously (Fig. 5.9). The consequence of the breaking of synchronization at irregular times is represented in the generated unpredictable response. If a stimulus is presented to this model cochlea, then the part(s) driven strongly enough will synchronize to the stimulus, and the chaotic behavior disappears (op cit., Figs. 5 and 6, and Fig. 5.10, this section).

The characteristics of the classical (parabolic) damping profile of the van der Pol oscillator are important enough to readdress them in detail, including the point of bifurcation, even though we have become aware of important limitations. For instance, in (5.71) the parabolic damping term leads to a power of $1/3$ in the output–input ratio. If this power extends from the threshold to the upper range of hearing (0–120 dB), then the internal range is nicely compressed to 40 dB, but the characteristics of the responses in the frequency domain show too much overlap, or decrease of selectivity. This result is apparent in the BM response profile as a function of level (see Fig. 5.10). Therefore, we decided to go on investigating modified van der Pol oscillators, with a general even-order nondecreasing damping term, i.e., with a function that grew more slowly than quadratic and might even saturate. Some examples are given in the next subsection (Fig. 5.11).

---

[29] Actually, this study was preceded by an analytical approach by van Netten and Duifhuis (1983) where the analysis focused on the properties of single driven van der Pol oscillators, and the application to CMs.

[30] Note that the definition of ε (5.72) is similar to the definition of δ or $1/Q$ for a linear second order system.

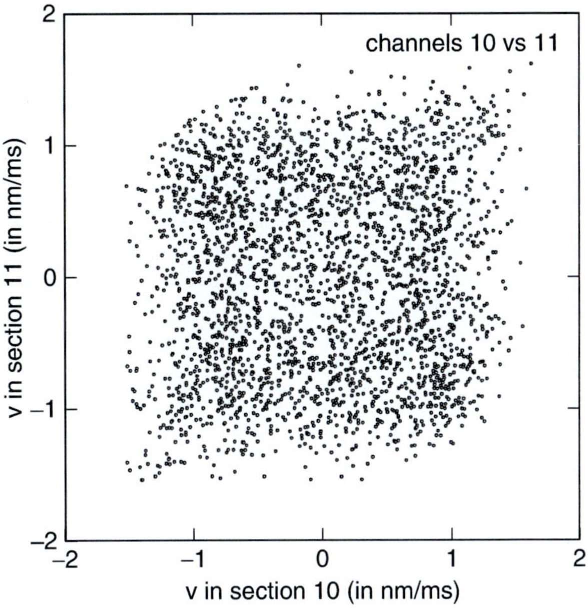

Fig. 5.9 As an example, the *dot display* showing the (lack of) correlation between the responses of two adjacent branches of the cochlea model is presented. Data points are collected over 100 ms at the sampling interval of 2.5 μs. Within this interval there are some rather weak correlation bands, but the overall impression is that of an evenly filled field. After Duifhuis et al. (1985, Fig. 3b)

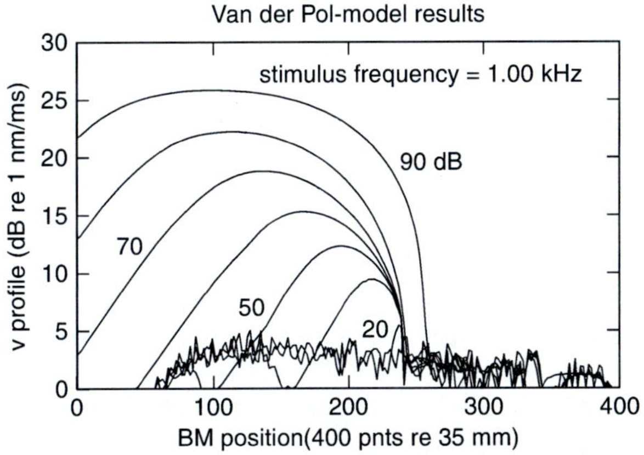

Fig. 5.10 BM excitation in a van der Pol cochlea for a 1 kHz tone as a function of level. At the apical end, the response remains chaotic for all levels, at the basal end the response synchronizes to the stimulus as soon as the stimulus is present, even at a level below the background chaotic level. The peak of the excitation profile moves basalward with level (indicated for some examples). Only at the highest level (90 dB) an extension in apical direction starts to develop

5.3.5.3 Modified Models (#2 and 3)

This concerns model versions that contain NL oscillator elements, such as defined and described in Sect. 9.2, and are able to generate spontaneous otoacoustic emissions if modeled as van der Pol oscillators. For that reason, we suggested to

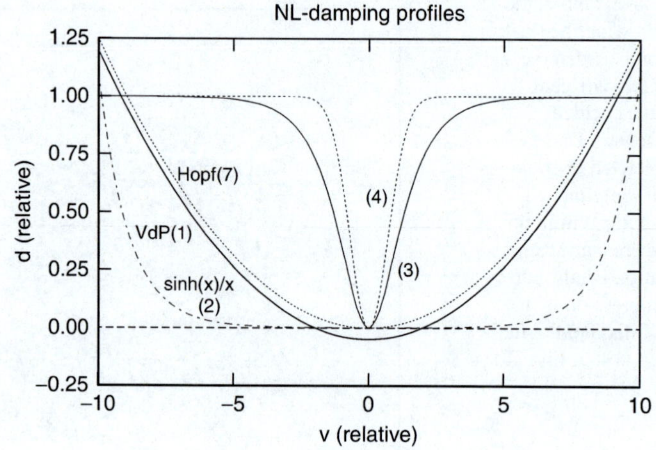

Fig. 5.11 Scaled damping profiles as a function of velocity. The standard VDP profile (1) and the $\sin h(x)/x$ profile (2) continue to increase with increasing velocity; the more recently introduced profiles (3,4) move from low values at low velocities to a saturation level, scaled to 1, at high velocities. The values around 0 may be several orders of magnitude smaller. The Hopf-profile (7) proposed by Duke and Jülicher can be considered as a version of the Van der Pol profile: scaled in such a way that its limit value for small deflections is exactly zero. For this limit, they introduced the term "critical oscillator". They also allow either velocity or deflection as the independent parameter

use a damping function that explicitly includes a dB-scale approximation over some amplitude range:

$$d(x,v) = d_0(x)\left(\chi \frac{\sinh(\alpha v)}{\alpha v} - \frac{\gamma}{\cosh \beta\, v}\right). \tag{5.75}$$

In this equation, the positive part provides the output–input behavior, and the negative part accounts the active properties. The parameter $d_0(x)$ scales the bracket part to order 1, and carries the dimension of damping; the bracket term is dimensionless (Fig. 5.12).

The hyperbolic sine function is constant for small amplitudes and increases exponentially at large amplitudes. Over that large amplitude range, the output–input ratio becomes logarithmic . The parameter α determines where $\alpha \times v$ is of order 1, which is where the damping function changes from approximately constant to exponentially increasing. This $\sin h(x)/x$-function approaches 1 if the argument (x) value goes to 0. The parameter χ was introduced to allow adjustment of this level. Generally, it is smaller than 1.

The additional term $-\gamma/\cosh \beta\, v$ accounts for active behavior as long as it is stronger than the $\sin h$-damping part. Similarly to the way in which α determined a characteristic width of the positive damping term, β defines the width of the active damping section, or the velocity range over which negative damping can occur. But this role of β is strongly affected by the value of γ.

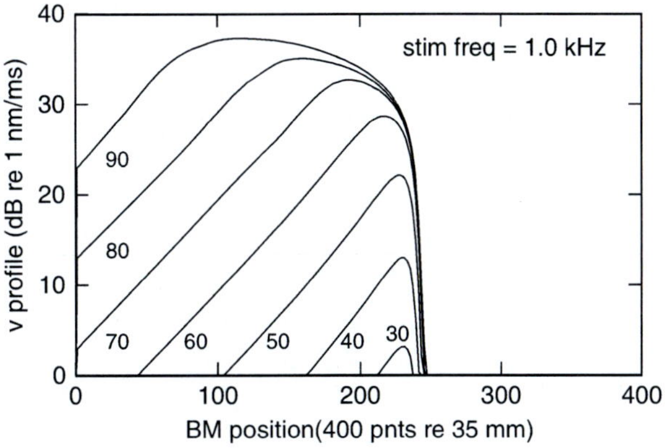

Fig. 5.12 BM excitation in a NL cochlea with $\sinh(\alpha v)/(\alpha v)$ damping (profile #2 in Fig. 5.11) for a 1 kHz tone as a function of level (30–90 dB; passive model version). The peak of the excitation profile moves basalward with level. The extend of the relatively linear low-level range depends on α, and can be greater than for the VDP model

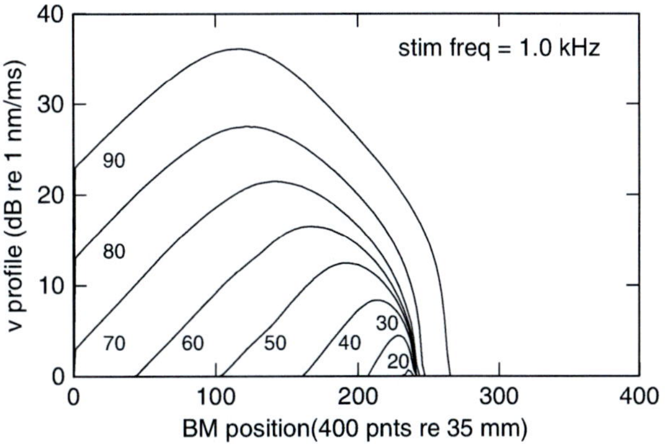

Fig. 5.13 BM excitation in a NL cochlea with damping profile #3 for a 1 kHz tone as a function of level (30–90 dB, passive model version). First, the peak of the excitation profile moves basalward with level. At the top level, this effect saturates. As in Fig. 5.12, the extend of the relatively linear low-level range depends on α, and can be greater than for the VDP model

A Hopf-bifurcation arises at $d(x, v) = 0$, and analyzing (5.71) it is clear that this implies that the bifurcation appears at $\gamma \to \chi$. As long as $\gamma \leq \chi$, spontaneous oscillations will not be generated. The γ-parameter, $\gamma(x)$, is a random vector (fixed for each ear) of values between a specified ranged.

The damping function specified above (5.75) does have limitations around velocity 0, but it does not approach a constant value at high SPLs. As indicated in Fig. 5.10, the continuing increase of damping with level does not match the real data. Instead, at high levels, the ear appears to linearize and approach a constant damping value. This brings us to the current trend that assumes two specific extreme values: one for low-level stimuli (low damping) and one for high-level stimuli (high damping), connected by a transition range (Figs. 5.13 and 5.14).

The profile that covers the range from low damping to high damping, usually over the mid-level range from about 30 to 60 dB, can be modeled in several ways. The derivative of the output–input relation must be nonnegative over the entire range, and low-level and high-level asymptotes approach 1.

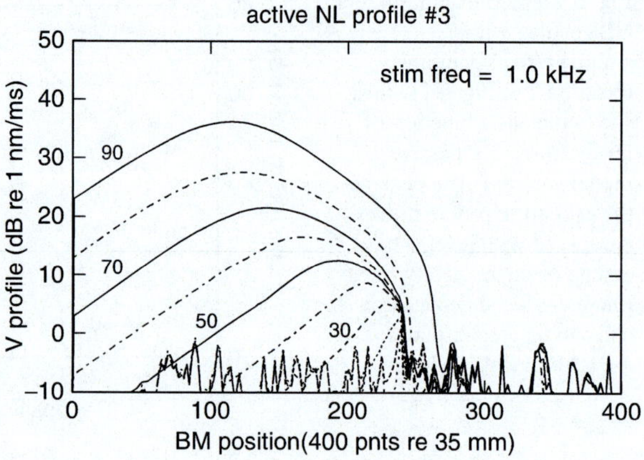

Fig. 5.14 As Fig. 5.13 but for the active model version. In this case there is again some low-level noise. In the basal end it disappears at high stimulus levels due to NL entrainment. At the lowest levels, the system acquired an enhanced sensitivity, and a nonlinear peak response (difference in vertical scale accounts for much of the "shape change")

5.3.5.4 Furst and Goldstein Model (#4)

The nonlinear model proposed by Furst and Goldstein (1982) was also very suited for time domain analysis. The g-function proposed by Furst and Goldstein (1982) contains a nonlinear damping profile, similar to the van der Pol nonlinear profile, but lacking its characteristic near-zero negativity. An additional nonlinear correction is applied to the stiffness term. Together, these provide

$$g_{FG}(x,t) = d(x,t)\, V(x,t)(1 + a\, V(x,t)^2) + \frac{Y(x,t)}{C(x,t)\,(1 + |b\, V(x,t)|^{\delta})}. \quad (5.76)$$

Values for the parameters a, b, and δ are presented by Furst and Golstein. The values for $d(x,t)$ and $C(x,t) = 1/s(x,t)$ are similar to the values proposed for the previous model versions. The boundary values at base and apex have to be updated to physically realistic values.

Furst and Lapid (1988) demonstrated that this model, with a proper middle-ear coupling, will generate DPOAEs, but neither SOAEs nor CEOAEs. This had to be expected because the bidirectional coupling is capable of reflecting the effects of frequency change in the nonlinear transforms, but the system lacks sources for generation of triggered responses. Moreover, it demonstrates an important difference between narrow-band noise behavior (Furst and Lapid) and active oscillator behavior.

5.3.5.5 Zweig's Linear and Nonlinear Models (#5 and 6)

Both the linear and nonlinear versions of the cochlea models proposed by Zweig (1991) have properties that can directly be implemented in a time domain model. At a CP-point, a feedback loop is specified. Feedback delay is tuning-frequency dependent, in other words, it is determined by the frequency map.

Zweig introduces the feedback in the local stiffness, such that the local CP-impedance (specific acoustic impedance) is:

$$Z_{sa,CP}(x,\omega) = \mathrm{i}\omega m + d(x) + \frac{s(x)}{\mathrm{i}\omega}\left(1 + \rho \mathrm{e}^{-\mathrm{i}\omega\tau(x)}\right). \tag{5.77}$$

Obviously, $\rho = 0$ is the standard case, and $\rho = 0.1416$ follows Zweig's adaptation. In a practical implementation and analysis of Zweig's proposal by van Hengel (1993), he found that he had to modify the proposed values at the boundaries at the base and at the apex in order to avoid (too) strong reflections from those points. Smooth transition from an active parameter set in the center to a passive set at the boundaries was sufficient (see his Fig. 1, op. cit). van Hengel adapted Zweig's model parameters at base and apex accordingly.

The time domain representation of the g-function part of (5.77), i.e., the damping-velocity and stiffness-deflection term is

$$g_{zw}(x,t) = d(x,t)\,V(x,t) + s(x)\,\left[Y(x,t) + \rho Y(x,t-\tau(x))\right]. \tag{5.78}$$

The nonlinear version of the model differs from the linear version (5.77) in that both damping and stiffness terms in the CP "impedance" will saturate at high levels.[31]

In collaboration with the ENT-group in Leiden[32] (Schneider et al. 1999 Schneider 2004; de Kleine 2001), and later with the Medical Physics group in Oldenburg (Mauermann et al. 1999a,b), Peter van Hengel analyzed model predictions for DPOAE group delays, in particular for the model version with Zweig's nonlinear damping (and stiffness) parameters. The exact formulas and parameter sets were adapted per occasion, where it should be noted that Schneider's data concerned measurements from guinea pigs, whereas Mauermann et al.s data where from humans.

The general nonlinear extension of (5.78) is

$$\begin{aligned} g_{zw,nl} = {} & V(x,t)\,d(x,t)\left[1 + \frac{\mu_1\beta|V(x,t)|}{1+\beta|V(x,t)|}\right] \\ & + s(x)\left[Y(x,t) + \left(\rho + \frac{\mu_2\beta|V(x,t)|}{1+\beta|V(x,t)|}\right)Y(x,t-\tau(x))\right] \end{aligned} \tag{5.79}$$

[31]The quotes denote that the *impedance* concept is borrowed from linear signal analysis, and formally only is approximately applicable in the nonlinear case.

[32]This collaboration was supported by NWO, the Netherlands Organization for Scientific Research, from 1996–2001.

with

$$\mu_1 = \mu_1(x, V(x,t)) \qquad \text{and} \qquad \mu_2 = \mu_2(x, V(x,t)). \tag{5.80}$$

The nonlinear extensions depending on μ_1 and μ_2 tend to vanish for large $|V|$. The criterion is set by the parameter β, a reciprocal velocity of the order of 0.01 ms/nm. Parameters μ_1 and μ_2 depend on the other partition parameters and are related to the formulation in Mauermann et al.'s equations (2a) and (2b).

A further discussion of model predictions and data is postponed to Chap. 6.

The order of presentation of the next subsection and sections is somewhat arbitrary, e.g., because of the link between cochlear amplifier behavior and nonlinear oscillator properties. Nevertheless, we first give a brief extension to 2- and 3-D approaches of analysis of the cochlea.

5.3.6 Two- and Three-Dimensional Approximations

The introduction into 2D and 3D model cochlea's has been given in Sect. 3.6. The first successful 2D model that uses boundary conditions all around the 2D structure is the model by Lesser and Berkley (1972). One important difference between the original 2D setups and our current cochlea model is the behavior at the stapes/oval window. In linear models, it has been customary to prescribe the pressure as determined directly by a driving force. Now we would also want to take the interaction with the entire system into account. Lesser and Berkley obtained a solution of the boundary value problem using a Fourier series analysis, in which they found a practical truncation limit after a certain number of terms.

The solution tends to approach the WKB(J) solution presented in Sect. 3.3.1. The same result had been obtained by Steele and Taber (1979), and had been confirmed by several others. Of these studies, several have addressed and estimated effects of boundary conditions, such as solid walls, scalae curvature, smoothness of scalae cross-section (irregular shapes in some bats), etc. However, very few have addressed the properties of the nonlinear cochlea. One of the first explicit numerical studies was the PhD-study by Diependaal (1988).

As several of the other 3D studies, he effectively reduces the 3D problem to a lower degree model using the integral equation for the behavior at the center of the BM. For the rectangular box model (Fig. 3.1), Diependaal derives the following set of equations:

$$\frac{16\rho}{\pi^2 b}\int_0^L \beta(\xi)\, G^{3\mathrm{D}}(x,\xi)\ddot{u}_1(\xi,t)\mathrm{d}\xi - m(x)\ddot{u}_1(x,t) = g(x,t) + \frac{8\rho A_s \ddot{u}_s(t)}{\pi b h}(L-x) \tag{5.81}$$

with

$$G^{3D}(x,\xi) = \sum_{n=0}^{\infty} \left\{ \frac{\cos(n\pi b_1(x)/b) + \cos(n\pi b_2(x)/b)}{b^2 - n^2\beta^2(x)} G_n(x, xi) \right\}, \quad (5.82)$$

$$g(x,t) = d(x,t)\dot{u}_1(x,t) + s(x,t)u_1(x,t), \quad (5.83)$$

where the g-function is similar to the one used before. The box-width b consists of three parts, the ridges $b_1(x)$ and $b_2(x)$, and the actual BM-width $\beta(x) = b - b_1(x) - b_2(x)$, and the function G_n follows from

$$G_n(x,\xi) = -\frac{b^4\left[\cos(n\pi b_1(\xi)/b) + \cos(n\pi b_2(\xi)/b)\right]}{\delta_n L[b^2 - n^2\beta^2(x)]} \times \sum_{k=0}^{\infty} \left\{ \frac{\coth(\gamma_{n,k}h)}{\gamma_{n,k}} \cos[(k+1/2)\pi x/L] \cos[(k+1/2)\pi\xi/L] \right\}, \quad (5.84)$$

with

$$\delta_n = \begin{cases} 2, & n = 0 \\ 1, & n \geq 1 \end{cases}$$

and

$$\gamma_{n,k}^2 = [(k+1/2)\pi/L]^2 + (n\pi/b)^2. \quad (5.85)$$

Nonlinear and/or active behavior is introduced in the damping and or stiffness term in the g-function (5.83), as in the 1D model.

A software version of Diependaal's time domain 3D model is available through the URL. A major difference with the 1D model is that the matrix equation that describes the longitudinal coupling between BM-sections (or CP-sections) is no longer three-diagonal, but a completely filled $N \times N$-array. This means that the solution time increases dramatically. The program uses matrix solution tools from standard libraries (IMSL). The ordinal program has been slightly modified and updated from F77 to f90, but except for some input/output properties this should not have changed the behavior.

Diependaal concludes

> We have calculated responses on a pure tone and on a tone burst in a nonlinear, active model, the modified van der Pol oscillators model (Duifhuis et al. 1985). In all these cases, the 2D response is close to the 3D solution; the 1D solution deviates considerably from the multidimensional solutions. However, from simulation studies in earlier, passive models we learn that for realistic values of the parameters the 2D model response may deviate significantly from the 3D response (Diependaal 1988). So the appropriate dimensionality of the cochlea depends on how much information one wants to obtain from the simulation

studies. If one is only interested in global behavior, a 1D approach is often satisfactory. For a more detailed evaluation of the cochlear model, a 2D, sometimes a 3D, approach is necessary.

The solution of the matrix equation also provides an error matrix, from which the relative accuracy in each cell can be estimated. The errors maximize near stapes and helicotrema. It is not yet clear to what extent this is due to the use of the at the time common but physically incorrect boundary conditions: stapes velocity follows the stimulus linearly and apical "termination" is a shortcut rather than minimal reflection.

Tentatively, we conclude that the most prominent differences in response profiles generated by 1D through 3D models is that the peak shape moves basalward and is broader.

5.4 Cochlear Amplifier

The concept of a "cochlear amplifier" (CA) process was introduced in the same year by Davis and by Neely (1983). Davis aimed to describe an active process that below about 60 dB might provide additional energy to a narrow segment of the basilar membrane, near the "apical foot" of the traveling wave envelope. This would be particularly effective in the tip of the tuning curve. This argument might seem similar to Gold's (1948) argument favoring active processes in the cochlea, but in fact Davis presented a significantly updated review with many new arguments.

Neely analyzed the power flow around the point of maximum excitation. To model his findings, he promoted a negative damping element in an additional local impedance term (see also Kim et al. 1980; Neely and Kim 1986), and argues that "this negative damping provides a convenient means of modeling active mechanical behavior [in a linear cochlea]". Although I would dispute the end of the sentence—which I placed in brackets—the major idea is not disputed. The discussions around 1980, such as at the 5th ISH (1980) where the option to use the van der Pol oscillator surfaced, appear to have provided a fruitful basis for these developments. The use of negative damping in one form or another has appeared in many studies since.

A recent follow-up discussion about the unequivocal experimental evidence of cochlear amplification, at the 10th International Workshop on the Mechanics of Hearing (Cooper and Kemp 2008), evaluated the progress of 25 years of further development. It has not yet brought a complete consensus, but that is one of the thrills that keeps scientific research exiting.

In the next subsection, references will be made to contributors to that discussion, as well as to some of the original remarks from Davis. The transcripts of the workshop are available in Cooper and Kemp (2008, pp. 467–476).

5.4.1 Definitions of CA

The above discussion about the existence of a cochlear amplifier started with a proper definition of the concept. At the occasion, it was proposed by Shera as:

> I define it here as *a process*, so in other words it's not a protein, it's not prestin, it's not the transduction channel, it's nothing like that. It is *a process* that provides cycle by cycle amplification of the vibrational response in the normal hearing organ. This vibrational response, in the mammalian pathway at least, might include, or definitely does include, but certainly is not limited to the vibrational response of the BM.

Shera continues his introduction emphasizing that in his opinion:

- Nonlinearity in itself is no evidence for power amplification.
- Physiological vulnerability (of the intact system) is definitely no evidence for power amplification.

Note that the definition does follow Davis' original definition rather closely.

The lively discussion that follows leaves a few open ends. Santos-Sacchi brings up Gold's original argument, based on viscous damping of BM-radial fibers, a point also addressed in footnote 10 in Chap. 3. Allen and Shera come back to Shera's definitions, Allen stating that not everybody is going to buy the power amplification definition.

Apparently, Allen's statement is based on controversies that played an important role during the development of the topic. In addition to the points mentioned above by Shera, it implies straightforward biophysical points of view, such as:

- A cochlear amplifier does not simply generate amplitude amplification (either deflection, velocity, or pressure), but the properly integrated (averaged) combination of, e.g., pressure and volume velocity.

This means that measurements of deflection (or velocity, or pressure) alone do not give sufficient information for an answer. Obviously, a proper combination provides the required power measure rather than an amplitude measure only.

Most discussants appear to agree on the point that the occurrence of distortion products in a nonlinear system are no proof of power amplification. This includes DPOAEs.

How about SOAEs? How natural are these, and are they a valid proof for power being emitted by the cochlea? The discussion does not appear to share the current conviction that in young normal hearing listeners there always is a "fingerprint" or specific individual SOAE pattern. With more sensitive measurement setups, in particular using specialized probe microphones and connectors, the percentage of normal hearing people showing emissions rapidly approaches 100.

If one considers the damping profiles that we have been testing in this chapter (Fig. 5.11), it may be clear that profiles 1 and 2 (VdP and $\sinh(x)/x$) are not very helpful because frankly, they focus on different properties. Profiles 3 and 4, however, are of the category that looks promising: there is an upper limit reached at high levels (e.g., 80 dB) and a lower horizontal asymptote, on average at a 30 to

50 dB lower damping value (0.001 is not distinguishable on the scale of Fig. 5.11). Spontaneous emissions are modeled by adding an amount of jitter on the low-level value. Whenever the net value of $d(0) < 0$, we are dealing with a general van der Pol oscillator, with its limit cycle oscillation behavior and with the entrainment property. The spontaneous emissions can disappear either if the dip-value in curves 3 and 4 retracts, and the damping reduction around $v = 0$ is less, or if the jitter disappears.

5.4.2 *Summarizing*

The discussion about a cochlear amplifier, or even *the* cochlear amplifier, is still continuing. Although some of the historically proposed arguments remain disputable, I believe that it may be useful to interpret the mechanism that reduces the high-level damping (normalized at 1 in Fig. 5.11) by a few of orders of magnitude as an amplification mechanism, even though the amplification is indirect: the process decreases the effective damping to a value near zero, aiming at enhancing the system's sensitivity. Strictly spoken, no net power has to be added to the signal, but the net dissipation at the sensor decreased.

Allowing this definition, the use of a NL cochlea model with a van der Pol type damping profile can be considered to belong to the category of models that include cochlear amplification.

References

ANSI S (2004) Acoustical Terminology. American National Standards Institute, Inc.

de Boer E (1980) Auditory physics. Physical principles in hearing theory. I. Physics Reports 62:87–174

Broer HW, Takens F (2009) Dynamical Systems and Chaos. Appl. Math. Series 172, Springer, New York

Butcher JC, Wanner G (1996) Runge-kutta methods: some historical notes. Appl Numer Math 22:113–151

Cai H, Manoussaki D, Chadwick R (2005) Effects of coiling in the micromechanics of the mammalian cochlea. J R Soc Interface 2:341–348

Cooper NP, Kemp DT (eds) (2008) The Biophysics of Hearing, Mechanics of Hearing, World Scientific, Singgapore

Dallos P (1973) The Auditory Periphery. Academic, New York

Davis H (1983) An active process in cochlear mechanics. Hear Res 9:79–90

Diependaal R (1988) Nonlinear and active cochlear models: Analysis and solution methods. PhD thesis, TU-Delft, Netherlands

Diependaal RJ, Duifhuis H, Hoogstraten HW, Viergever MA (1987) Numerical methods for solving one-dimensional cochlear models in the time domain. J Acoust Soc Am 82:1655–1666

Duifhuis H, Hoogstraten HW, van Netten SM, Diependaal RJ, Bialek W (1985) Modelling the cochlear partition with coupled Van der Pol oscillators. In: Allen JB, Hall JL, Hubbard AE, Neely ST, Tubis A (eds) Peripheral Auditory Mechanisms, Springer, New York, pp 290–297

Duke TAJ, Jülicher F (2008) Critical oscillators as active elements in hearing. In: Manley GA, Fay RR, Popper AN (eds) Active Processes and Otoacoustic Emissions, SHAR, vol 30, Springer, New York, chap 3, pp 63–93
Eguíliuz VM, Ospeck M, Choe Y, Hudspeth AJ, Magnasco MO (2000) Essential nonlinearities in hearing. Phys Rev L 84:5232–5235
Furst M, Goldstein JL (1982) A cochlear nonlinear transmission-line model compatible with combination tone psychophysics. J Acoust Soc Am 72:717–726
Furst M, Lapid M (1988) A cochlear model for acoustic emissions. J Acoust Soc Am 84:222–229
Gold T (1948) Hearing. II. The physical basis of the action of the cochlea. Proc Royal Soc London, Series B, Biol Sc 135(881):492–498
Greenwood DD (1961) Critical bandwidth and the frequency coordinates of the basilar membrane. J Acoust Soc Am 33:1344–1356
Greenwood DD (1990) A cochlear frequency-position function for several species—29 years later. J Acoust Soc Am 87:2592–2605
Guckenheimer J, Holmes P (1983) Nonlinear oscillations, dynamical systems and bifurcations of vector fields. Springer-Verlag, New York
Harris FJ (1978) On the use of windows for harmonic analysis with the discrete fourier transform. Proc IEEE 66:51–83
van Hengel PWJ (1993) Comment on: Shera & Zweig 'Periodicity in otoacoustic emissions'. In: Duifhuis H, Horst JW, van Dijk P, van Netten SM (eds) Biophysica of Hair Cell Sensory Systems, World Scientific, Singapore, p 62
Inselberg A (1978) Cochlear dynamics: The evolution of a mathematical model. SIAM Review 20:301–351
Kern A, Stoop R (2003) Essential role of couplings between hearing nonlinearities. Phys Rev Lett 91(12):128,101–1–128,101–4
Kim DO, Molnar CE, Matthews JW (1980) Cochlear mechanics: Nonlinear behavior in two-tone responses as reflected in cochlear-nerve-fiber responses and ear-canal sound pressure. J Acoust Soc Am 67:1701–1721
Kinsler LA, Frey AR, Coppens AB, Sanders JV (1982) Fundamentals of Acoustics, 3rd edn. Wiley, New York
de Kleine E (2001) Boundary conditions of otoacoustic emissions. PhD thesis, University of Groningen, Groningen, Netherlands
Kohllöffel LUE (1990) Cochlear mechanics: Coiling effects (I,II) and the absorption equation (III). Hear Res 49:19–28
Lamb H (1895) Hydrodynamics, 2nd edn. Cambridge University Press, Cambridge
Lesser MB, Berkley DA (1972) Fluid mechanics of the cochlea. Part 1. J Fluid Mech 51:497–512
Magnasco MO (2003) A wave traveling over a Hopf instability shapes the cochlear tuning curve. Phys Rev Lett 90(5):058,101–1–058,101–4
Mauermann M, Uppenkamp S, van Hengel PWJ, Kollmeier B (1999a) Evidence for the distortionproduc frequency place as a source of distortionproduct otoacoustic emission (dpoae) fine structure in humans. i. fine structure and higher-order dpoae as a function of the ratio f2/f1. J Acoust Soc Am 106:3473–3483
Mauermann M, Uppenkamp S, van Hengel PWJ, Kollmeier B (1999b) Evidence for the distortionproduc frequency place as a source of distortionproduct otoacoustic emission (dpoae) fine structure in humans. ii. fine structure for different shapes of cochlear hearing loss. J Acoust Soc Am 106:3484–3491
Middleton D (1960) An Introduction to Statistical Communication Theory. McGraw-Hill, New York
Neely ST (1983) The cochlear amplifier. In: de Boer E, Viergever MA (eds) Mechanics of Hearing, Nijhoff/Delft Univ. Press, Netherlands, pp 111–118
Neely ST, Kim DO (1986) A model for active elements in cochlear biomechanics. J Acoust Soc Am 79:1472–1480
van Netten SM, Duifhuis H (1983) Modelling an active, nonlinear cochlea. In: de Boer E, Viergever MA (eds) Mechanics of Hearing, Delft Univ. Press/Nijhoff, pp 143–151

O'Connor KN, Puria S (2008) Middle-ear circuit model parameters based on a population of human ears. J Acoust Soc Am 123:197–211
Olson ES (1998) Observing middle and inner ear mechanics with novel intracochlear pressure sensors. J Acoust Soc Am 103:3445–3463
Olson ES (1999) Direct measurement of intra-cochlear pressure waves. Nature 402:526–529
Olson ES (2001) Intracochlear pressure measurements related to cochlear tuning. J Acoust Soc Am 110:349–367
Olson ES (2004) Harmonic distortion in intracochlear pressure and its analysis to explore the cochlear amplifier. J Acoust Soc Am 115:1230–1241
Peterson LC, Bogert BP (1950) A dynamical theory of the cochlea. J Acoust Soc Am 22:369–381
Puria S, Allen JB (1991) A parametric study of cochlear imput impedance. J Acoust Soc Am 89:287–309
Puria S, Allen JB (1998) Measurements and model of the cat middle ers: Evidence for tympanic membrane acoustic delay. J Acoust Soc Am 104:3463–3481
Rayleigh JWS (1894) The Theory of Sound, vol I, 2nd edn. MACMILLAN, London, [Dover edition: 1945]
Rayleigh JWS (1896) The Theory of Sound, vol II, 2nd edn. MACMILLAN, London, [Dover edition: 1945]
Schneider S (2004) Amplitude and phase characteristics of distortion product otoacoustic emissions. PhD thesis, University of Leiden, Leiden, Netherlands
Schneider S, Prijs VF, Schoonhoven R (1999) Group delays of distortion product otoacoustic emissions in the guinea pig. J Acoust Soc Am 105:2722–2730
Steele CR, Taber LA (1979) Comparison of WKB and finite difference calculations for a two-dimensional cochlear model. J Acoust Soc Am 65:1001–1006
Steele CR, Zais J (1985) Effect of coiling in a cochlear model. J Acoust Soc Am 77:1849–1852
Tinevez JY, Jülicher F, Martin P (2007) Unifying the various incarnations of active hair-bundle motility by the vertebrate hair cell. Biophys J 93:4053–4067
Tinevez JY, Martin P, Jülicher F (2009) Active hair-bundle motility by the vertabrate hair cell. In: Cooper NP, Kemp DT (eds) The Biophysics of Hearing, Mechanics of Hearing 10, 2008, World Scientific, Singapore, pp 415–424
Viergever MA (1978) Basilar membrane motion in a spiral-shaped cochlea. J Acoust Soc Am 64:1048–1053
Zweig G (1991) Finding the impedance of the organ of Corti. J Acoust Soc Am 89:1229–1254
Zweig G, Lipes R, Pierce JR (1976) The cochlear compromise. J Acoust Soc Am 59:975–982
Zwislocki J (1965) Analysis of some auditory characteristics. In: Luce R, Bush R, Galanter E (eds) Handbook of Mathematical Psychology, Vol. III, Wiley, New York, pp 1–97
Zwislocki JJ (1974) Cochlear waves: interaction between theory and experiments. J Acoust Soc Am 55:578–583

Part III
Results and Open Issues

Part III starts with a presentation of results obtained with the time domain analysis, and with some suggestions for new experiments (Chap. 6). It focusses on three groups of nonlinear effects: level effects, combination tones, and delay effects.

The final Chap. (7) contains 3 sections:

1. Presents applications where the use of NL time domain analysis is potentially useful. This includes hearing aids and automatic speech recognition (ASR).
2. Discusses the link between Hopf-bifurcation and van der Pol-oscillator.
3. The final section discusses some open issues, and concludes the book.

Chapter 6
Results

Abstract In this chapter, we present example results from the NL cochlea, analyzed in the time domain. The examples cover three categories: level effects, combination tones, and onset delay effects. The first two topics were originally addressed in a project by Marc van den Raadt, the last two by Peter van Hengel.

6.1 Spontaneous Emissions: SOAE

The nonlinear model cochlea with the negative damping near zero velocity is one of the few, general models of spontaneous oto-acoustic emissions. The emissions are generated within the cochlea and leave the cochlea through the middle ear and outer ear. Classical cochlea models used to define a stapedial input. This does not allow emission feedback, neither spontaneous nor evoked. In addition, the specialized coupling through closed or open ear canal entrance is relevant, as is the matching impedance at the helicotrema.

By the end of the 1970s the Van der Pol oscillator was proposed as a candidate descriptor of SOAE. Although the Van der Pol oscillator was well known as a single oscillator, and had been described extensively in the applied mathematical literature for more than half a century, the analysis of a few coupled oscillators turned out to be a different matter. Two limit cycle oscillators of comparable strength and different frequencies appear to be able to generate (quasi) chaotic responses. This implies that some time after an onset time the response phases can no longer be estimated reliably on the basis of the initial conditions. A fundamental reason of this property is that any experimental measure has a limited accuracy. Different behavior is seen if one oscillator is much stronger than a second, coupled oscillator. In that case, the weaker one will be entrained by the stronger one. Usually there is a transition region, where the response switches from one oscillator to the other in a (quasi)

H. Duifhuis, *Cochlear Mechanics: Introduction to a Time Domain Analysis of the Nonlinear Cochlea*, DOI 10.1007/978-1-4419-6117-4_6,

random way. We observed and reported this behavior in Duifhuis, Hoogstraten, van Netten, Diependaal, and Bialek (1985, Sect.3) and van den Raadt and Duifhuis (2011, Fig. 1b).[1]

The analysis of many coupled nonlinear processes lacks sufficient analytical tools and has to be addressed numerically. Around 1980 the required computation time was extremely demanding; nowadays, the computation demand is getting closer to real time sound processing. The remaining factor for computation of the mathematical core for a streaming stimulus is slightly over one order of magnitude (Fortran 2003, Intel i7 processor).

The essential biophysical concept of SOAE is that acoustic power is emitted by an ear that receives no net acoustic input from the external environment. Internal contributions, sometimes summarized as body noise, are not completely absent. Some of these can be (partly) controlled by well-trained subjects (e.g., timing of breathing noise), whereas other internal processes, such as blood flow, remain virtually uncontrollable.

Our practical way to model individual emission profiles is to define subject related quasi-random damping profiles. In terms of Fig. 5.11 and (5.75), this applies to the amount of negativity near $v = 0$. In this equation, the limit cycle amplitude is controlled by $\gamma = \gamma(x)$. In other words, the practical model involves the definition of subject- dependent $\gamma(x)$ profiles.
Summarizing:

- Insofar as the cochlear amplifier only reduces the damping to a very low value, it will not produce a limit cycle or a spontaneous emission.
- A nonlinear system with a negative damping term near zero (van der Pol oscillator) generates a local emission, which propagates through the auditory system.
- Stability of a van der Pol oscillator is guaranteed by a nonlinear damping term, which reaches positive values at increasing velocities.[2]
- Negative damping in a linear system is unstable.
- Negative damping in a selection of coupled linear systems requires specifications that depend on stimulus parameters and does not provide a general solution. Moreover, the normal cochlea is nonlinear to begin with.
- The alternative hypothesis of coherent reflection requires at least a thermal noise input to produce a coherent reflection with spontaneous emission properties.

Our current hypothesis is that a specific oscillator of the van der Pol family is the best candidate to represent the generation of spontaneous emissions.

---

[1]A general definition of chaos is still lacking, but a practical focus appears to involve three points. Following Lynch (2010, Sect. 7.3.II) the responses are:

1. Long-term aperiodic (often: almost periodic)
2. Extremely sensitive to initial conditions, and/or
3. Often have a fractal structure

A similar definition was given by Wiggins (2003, Chap. 30), and a more theoretical one by Broer and Takens (2009, their definition 2.12). We checked points 1 and 2 in particular. Point 1, e.g., is shown in the phase plane plot (Fig. 1b) referred to above.

[2]More precisely, the van der Pol oscillator has a parabolic deflection profile; the Rayleigh-oscillator has a parabolic velocity profile. In both cases the NL-term is a damping term.

6.2 Level Effects

6.2.1 The Nonlinear Residual Response Method[3]

The stimulus paradigm for the nonlinear residual response method was developed by Brass and Kemp (1991) as a tool to measure ear canal responses to tonal stimuli. It is one of the techniques in the field of stimulus frequency otoacoustic emissions (SFOAE). The technique uses precise timing information which in practice allows measurement of relatively weak nonlinear distortion effects even at the input frequency. Stimulus constructs were specified such that linear processing produces a zero output, whereas nonlinear processing leaves a distinct residue.

The paradigm requires accurate summation of a continuous background tone (carrier) and a properly synchronized probe tone. This accurate summation was realized in four exactly equal time intervals, labeled A, B, C, and D, of duration $(N + 1/2)$ cycles of a carrier frequency $T_b = 1/f_b$. Hence, each interval contained exactly the same odd number of semi-cycles of the background tone, and the background waveforms appearing in successive intervals differed in sign (or a π phase shift). In other words, the polarity of the carrier flips at each section transition, and the phase rotates by π. The probe was presented only during the intervals C and D. The probe shapes for C and D were identical (Table 6.1). The background frequency f_b was subject to the constraints specified above, the probe frequency f_p and both levels L_b and L_p were free.

The residual was obtained by addition and subtraction of ear canal pressure measured during those four intervals as defined by the rule

$$\text{Residual} = \text{A} - \text{B} - \text{C} + \text{D}.$$

Hence, the duration of the residual is that of a single segment.

The residual of a linear system must be equal to zero. Experimentally, however, the responses in the ear canal do not cancel completely: due to nonlinearity in the system a small residual component remains (Brass and Kemp 1991).

Table 6.1 Temporal structure of the ABCD stimulus (time proceeds downwards)

A	+Background	↓
B	−Background	
C	+Probe +background	
D	+Probe −background	t

Duration of the four blocks is identical: $T_{\text{block}} = (N + 1/2)/f_b$

[3]Part of the contents of this section has been presented at the 6th International Symposium on the Mechanics of Hearing (Duifhuis and van den Raadt 1997). Our analysis was initially prepared by M.P.M.G. van den Raadt, and is updated with additional results by the author.

6.2.1.1 Specification of the Analysis

For optimal comparison between model results and data, we aimed to stay close to the stimulus parameters used by Brass and Kemp. Therefore, duration of a single section was kept at approx. 40 ms. Obviously this limits the maximum spectral resolution to (at best) 25 Hz. Originally, probe tones were shaped with cosine-shaped onsets and offsets of 1 ms. Brass and Kemp claim that with this duration of the window no significant high frequency components or other transients are detectable in the residual. Their choice was motivated by the necessity to collect an optimum set of data over a limited period.

In our recent analysis, the following arguments favored a few adjustments.

1. Low-frequency parts of transient responses extend over considerably longer intervals than 40 ms.
2. Accurate computation does not require averaging.
3. Computation time is not an issue.

Therefore, section duration was an adjustable parameter, usually kept at 40 ms, but extended for low-frequency studies. We set the probe onset and offset to 4 ms, and placed the probe not in Sects. C and D (Brass and Kemp) but in B and C. This changes the sign of the residue, but also prevents cut-offs from increment responses over another section duration.[4]

The simulations presented have been obtained using a nonlinear or linear 1-D time domain cochlea model that has been presented in Chap. 5. The advantage of this model is that nonlinear (as well as linear-) parameters can be manipulated, in stiffness as well as in damping, that interaction between elements is taken into account, and that time effects (onset and offset transients) are computed properly. The classical Runge–Kutta method was chosen as time integration method with a time step of 2.5 μs.

6.2.1.2 Relevant Cochlea Properties and Model Parameters

The analysis was performed for a 1-D model with $N = 400$ BM points. Several nonlinear damping and stiffness parameters listed in Table 5.4 were tested, both from the passive and active groups.

It should be obvious that the coupling properties at the ear canal/middle ear side and at the helicotrema play a significant part, both in ear and in model. A simplified middle ear and canal coupler are assumed to cover the basic physical properties.

The apical coupling is also more relevant than generally acknowledged. It determines how fast the cochlea adapts to transients. Of course all transients

[4]The accuracy of the model computation was sufficient to obtain reliable results within a single run (see Sect. 6.2.2.1). Onset effects in a run were minimized by addition of two introduction sections, the response to which was discarded from further analysis.

contain low-frequency components, and the amount affecting the cochlear response depends strongly on the middle ear transmission properties. The cochlear reaction to these components will be optimized, i.e., show minimum reflection, by finding the optimum termination properties at the helicotrema. In other words, (in a linear approximation)[5] an optimum impedance match to the local characteristic impedance—if the gradient is sufficiently small—would prevent reflections.

6.2.2 *Results*

6.2.2.1 The Residual of a Linear Cochlea Model

Several properties were studied, both for equal parameters for probe and background ($L_b = L_p$, $f_b = f_p$), and for different values for probe and background. In the linear case with $N = 400$ there is a clear low-frequency low-level component in the residue (at approx. 100 Hz.). This was unexpected because theoretically, the residual response should approach zero for the linear model. The observed low-frequency component remains linearly related to stimulus level. This was checked by measuring the residue in absence of the probe. Figure 6.1 shows an example at $L_b = 80$ dB. An essential difference exists between our computation, based on a single set of intervals A through D, and the experimental data, which involve continuously repeated stimuli. In the first case, the transients are clipped at the end of interval D; in the experiments, the transient responses are continuously taken into account.

6.2.2.2 Nonlinear Damping

We present the results for two possible damping nonlinearities, viz. the power law nonlinearity and the exponential nonlinearity. These two have been used frequently in recent cochlear modeling. The comparison with other nonlinear models is left to the reader.[6]

Examples of results in Figs. 6.2 and 6.3 show that the residual has three main characteristics: an initial transient, a relatively stationary part, and a final transient which continues until (or even beyond) the end of the frame. Furthermore, at the start of the time section a much smaller transient can be present. In the presented results this is clear in Fig. 6.3, but definitely not in Fig. 6.2. The two large transients in the residue, especially for model #3, were not expected. They can vary considerably with (relative) level, (relative) frequency, and nonlinear characteristics. Apparently, this residue displays a strong contribution from the high-frequency

[5]The brackets reflect the limitation of linear analysis to nonlinear systems.

[6]The required tools are available through the URL.

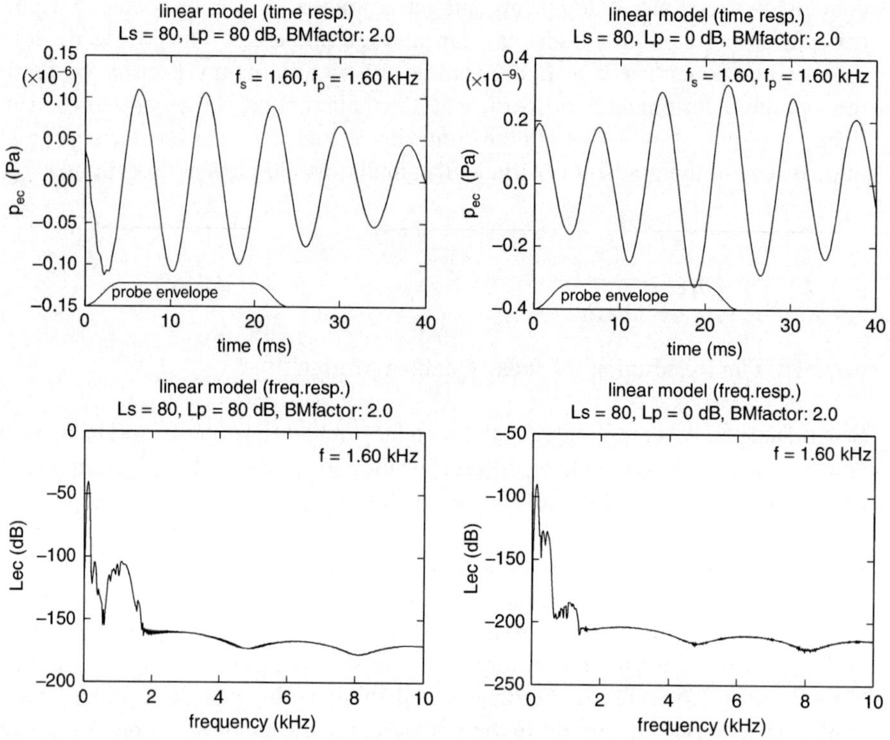

Fig. 6.1 The residue of the linear model shows a low frequency response at about 125 Hz. The *top row* gives two time responses (*left*: $L_b = L_p = 80$ dB, *right*: $L_b = 80$, $L_p = 0$ dB), with the probe envelope indicated at the *bottom* of the figures. The *bottom row* gives the related amplitude spectra. The response to the background only is 50 dB lower than the response to equal background and probe (*left column*). Nevertheless, the responses remains at low levels, but differ from zero. Stimuli and response parameters are defined for a plugged ear canal, near the plug

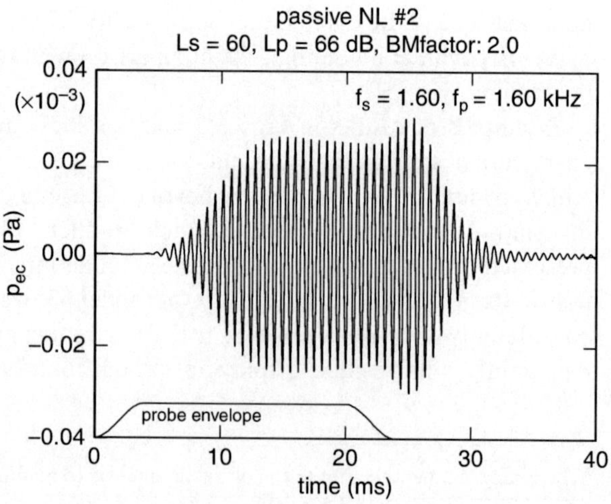

Fig. 6.2 The residual of nonlinear model #2 shows the three main characteristics for the example: $L_b = 60$, $L_p = 66$, probe envelope indicated at the *bottom* of the figure. Both onset and offset responses extend over a longer intervals then the stimulus onsets and offsets. The plateau value is not completely constant for this stimulus duration

Fig. 6.3 The residual of nonlinear model #3. Other parameters as in Fig. 6.2. The transient responses are much stronger than for model #2

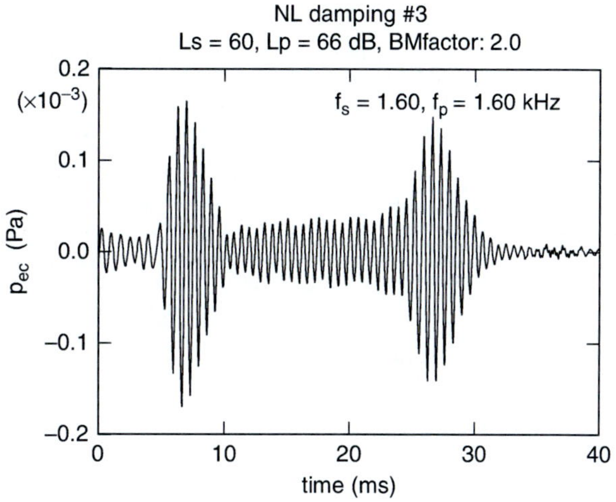

part of the cochlea. That is the part that responds fastest, and because of the spectral sensitivity, provides strong transient responses. The cochlear response near resonance grows more gradually, but in that range the phase gradient is larger and vectorial summation can then lead to a weaker result. It is also the place where phase behavior plays an important role. In the simple, straightforward model cochlea two parameters are available to control the phase behavior, e.g., the gradients of $Q(x)$ and of $s(x)$ over the response range x.[7]

The results here are at a "typical" level, but it is noted that the profile can depend rather strongly on level. This does particularly apply to the transient part of the responses. The plateau response may help to distinguish between different models.

The BM-factor that scales m, d, and s, without changing the frequency map, and the d-term by itself, changing the quality factor and phase. One might expect the Q-factor to present the following effects: A higher Q implies a greater frequency selectivity, and, therefore, a slower but stronger response ($\Delta f \times \Delta t$ remains constant). However, it also implies a faster phase transition at resonance, which will reduce the net residual response. This may lead to opposite effects, a point that deserves further exploration. For the results presented in this section, the value 2 has been used.

Comparison of Figs. 6.2 and 6.3 might also lead to hypotheses about the response delay. Many experimenters only consider the group delay, a property that we will address in Sect. 6.4. But two other delays are relevant in signal processing (Papoulis 1962) and also in auditory processing (e.g., Ruggero 1980), namely the phase delay, and the signal-front delay. The definitions are presented in Sect. 8.2.1.The signal-front delay may be the least well known, and its direct measure does come with the experimental difficulty: how to measure the real front

[7]Obviously, any alternative independent set of two from $m(x)$, $d(x)$, $s(x)$, $\omega(x)$ and $Q(x)$ can be used.

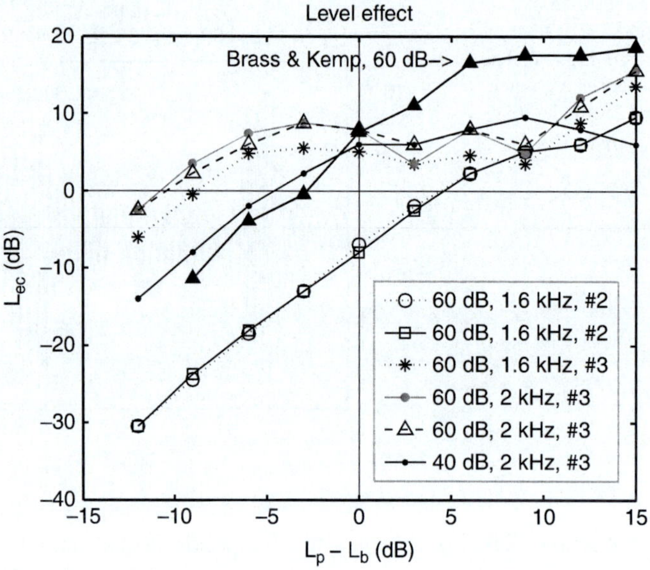

Fig. 6.4 Effect of the level on ear canal residue. Input variable: $L_p - L_b$; other parameters: frequency and model type. Levels are based on the peak-to-peak amplitude values. Note that the entries for 60 dB, 1.6 kHz and #2 and for 60 dB, 2 kHz and #3 are repetitions with slightly different initial conditions. The filled triangles are 60 dB points read from Brass and Kemp's Fig. 3a

delay, can you be absolutely sure that the onset is not hidden in the noise? For a linear system, this delay can also be measured in the frequency domain. It is determined by the asymptotic behavior ($\omega \rightarrow \infty$) of the phase vs. (angular) frequency ratio (8.40), stressing that the high-frequency part determines the apparent onset delay.

Brass and Kemp showed level effects (their Fig. 3(a)) of the residual phase and residual level at background levels from 40–60 dB. The results are presented against the probe level—background level (actually the latter is denoted as "stimulus level"). We present only some level results for two different model types, both passive, viz., the expanding NL model #2, and the saturating NL model #3. In Fig. 6.4 we present results for models #2 and #3, and the 60 dB data from Brass and Kemp. The Brass and Kemp's data differ from both models. The difference with #2 is most obvious: the saturation at higher $L_p - L_b$ is insufficient, and the response around the 40 dB background dropped out of the figure, implying that the nonlinear effect decreased considerably (actually, the upper limit of this set is just above the −40 dB base line at $L_p - L_b = 15$ dB). The results for model #3 do show saturation, but start at a lower $L_p - L_b$-value, viz. at −5 dB instead of +5 dB. The order of magnitude of the residual effect does fit.

6.2.2.3 Nonlinear Stiffness

Because the question about the location of the source of auditory nonlinearity was not completely solved, we also investigated the impact of a nonlinear stiffness, both separately and in combination with a nonlinear damping function. We choose functions similar to those proposed by Furst and Goldstein. At low levels the discontinuity in the derivative at $y = 0$ [y = partition displacement] can lead to instability in the response. Stiffness nonlinearity does not seem to provide transients above the average continuous response level.

An alternative nonlinear stiffness characteristic, quite different from the one mentioned above is the term associated with the Duffing equation. It gives an odd nonlinear stiffness element, instead of the symmetrical, even element mentioned above. Tentative results indicate that it may produce larger onset "delays".

Along this line we started to test the effect of an asymmetric nonlinear stiffness term as assumed to be relevant for hair bundle motion. The primary effect appears to be characterized by overall stiffness reduction, which leads to a basalward shift of the characteristic frequency.

6.2.3 Discussion

We started the analysis of SFOAEs. The technique, introduced by Kemp and Souter (1988) and expanded by Brass and Kemp, was simulated in a transmission-line cochlea model. Brass and Kemp's method opens the possibility to do temporal measurements at the stimulus frequency.

A nonlinear model cochlea generates a reproducible residual result. A linear cochlea model displays a low-frequency oscillatory response, because the apical part of the cochlea, often beyond the region of resonance, is relatively slow, and it is stimulated by transients in the stimulus. Part of this effect is exaggerated by the use of a too simple middle ear; the appropriate match will reduce high-frequency as well as low frequency transients.

With a straightforward specification of the acoustic coupler, it was possible to predict the levels of the residue. Also, global frequency effects and level effects follow the experimental data. In this macroscopic model, there is—so far—no apparent need for a second filter.

For the nonlinearities tested, we have not yet been able to mimic every detail of the experimental data. Open questions are source of strong initial and final transients and for the "delay" of the residual. As mentioned above, both middle ear properties and cochlear parameters involving the phase characteristics directly affect this result. At least one of those has to be modified to explain the data. The "delay" caused by the traveling wave is in fact the gradual onset profile of the coupled cochlea elements, and thus formally there is no delay. The time necessary for the residue to reach 50% of the stationary response can be used for a comparison, physically it is determined by the onset properties at the sites where the nonlinearity is prominently present. Likewise, travel time through the fluid back to the ear canal is negligible. Relevant envelope effects are determined by the same onset properties, which incorporates the coupling with all elements in the system.

6.3 Time Effects

In Sect. 6.2 the attention was focused on Level Effects, but the applied technique for measuring those effects also displayed some clear "time effects". These were mentioned briefly in Sect. 6.2.1.1 and are rather obvious in the onset and offset phenomena displayed in Figs. 6.2 and 6.3. Although the delays were noted above, they were not studied systematically. This point will be taken up here.

Most experimental studies that followed have concentrated on distortion product otoacoustic emissions (DPOAEs), assuming that the measurement would be simpler, and more reliable because the DPOAE(s) can be spectrally separated from the input stimulus components, a technique that seems less demanding than the ABCD technique.

In this section we want to stress the power of a straightforward time domain analysis method. The primary-tone phase variation (PTPV) technique introduced by Whitehead et al. (1996) will be used to demonstrate this point. We support their view that this technique provides a valuable tool for reliable measurement and interpretation of DPOAE delay.

6.3.1 DPOAE Delays Measured with Phase Rotation

We make the following assumptions for the cochlear nonlinearity:

- The net damping term is of the type: odd-order (following Sect. 5.4).
- Potential stiffness NL has not more than a weak effect on the overall behavior of a section.

The real cochlea also generates even-order distortion products, in particular the difference tone for a two-tone stimulus, but these are almost always significantly weaker than the odd order components. It is very convenient to neglect the even components for the moment, because it helps to validate the procedure. An odd order nonlinearity does only generate odd harmonics and odd intermodulation products. For instance, if the damping function would be modeled as

$$d = d_0(1 + \alpha v + \beta v^2) \tag{6.1}$$

then the β-term generates the family $m\,f_1 \pm n\,f_2$ where $m + n =$ odd and f_1 and f_2 are the frequencies of the stimulus components. Similarly, the α-term generates this family with $m+n$ = even[8] containing the difference tone $f_2 - f_1$. Since α and β are independent parameters, so is the behavior for even and odd-order nonlinearities. It will be useful to discern these two different families (the odd-order DPs and the even-order DPs). The β-term generates an odd NL family, the α-term generates the even family of intermodulation products.

In Sect. 9.1 the Bennett function is introduced which provides a proper estimate of amplitude and phase of the distortion product at the generation site. This is presented in (9.13). The basic result is that for combination tones of the family $m f_1 \pm n f_2$, $m > n$ the dependence of the DP-phase on the primary phases[9] is $\phi_{DP} = m\phi_1 \pm n\phi_2$, except for a possible additional factor π, or sign change. Details are discussed in Sect. 9.1.3.

The nonlinear relation described above is the velocity response to a local pressure source. In several studies, this nonlinearity is approximated by a cubic *conductance* term as the first odd order term after the linear one in the commonly employed Taylor expansion. This remains unfortunate, and can lead to errors, in particular at the high-level end, where the normal nonlinear behavior is compressive (saturating). Compressive and expensive nonlinearities do predict opposite phases for $2f_1 \pm f_2$. The behavior for other values of m and n is given in Fig. 9.3 for values of m and n from 0–9.

On the other hand, a cubic *damping* asymptotes (at high levels) to the saturating power 1/3, and is a member of the νth-law nonlinearities, which provides a more efficient fit to compressive nonlinearities.

The ear canal measure of the returned "by-effect" of the DP-generation, or the net DPOAE, is the summed result of all cochlear places involved in generation and processing of the component. The academic discussion about the location of this point has not yet reached a really scientific level: discussion has been limited to the question whether the primary source is at the f_1-place, f_2-place, or the $2f_1 - f_2$-place (e.g., Withnell et al. 2003). In fact all these approximations are oversimplifications of the more likely full story stated above: *the summed result of all cochlear places involved.*

With respect to intermodulation product generation site, the nonlinearity is most effective in the area where the partition response to both stimulus tones is approximately equal.[10] A very important difference between a local effect and the summed result seen in the ear canal is that a summation does not necessarily enhance the response, it can also reduce the summed response, because the phases of the generated DPs rotate rather quickly over the places of generation. If we consider the phenomenon that the phase rotation ($\partial\phi/\partial\omega$) is maximum near the point of resonance (see Fig. 8.1), and additionally that the maximum response is reached *before* (basalward of) that point of resonance, it follows that the maximum

[8]Obviously, if $m+n$ = odd, then $m-n$ = odd as well, and similarly: $m+n$ = even $\Leftrightarrow$ $m-n$ = even. See also Sect. 4.1.

[9]Each primary component $i, i \in (1, 2)$, is defined as $y_i(t) = \cos(\omega_i t + \phi_i)$.

[10]Nonlinearity in the damping term implies that the relevant variable is the partition velocity.

interaction[11] occurs before the f_1-place and near the f_2-place.[12] In this area all intermodulation products and overtones are generated, at strengths that decay with increasing distance from the primary frequencies. Although in principle all distortion products might propagate in both cochlear "directions," local "impedances" are such that only the low-frequency intermodulation products propagate efficiently to their own "places," whereas the high-frequency components remain weaker – and often much weaker– than the primaries, and are only traceable with sensitive frequency analysis. At high levels only, the high frequency distortion components may reach a perceptible level. Note also that the local source effects can be modified significantly by phase effects when these lead to interference and reduce the net contribution, which is the vector sum of all local cochlear contributions. High-frequency-side distortion products match neither apical nor basal impedance, but the basal mismatch will be smaller so that the basal component will be stronger.

The phase dependence of the response components, primaries as well as DPs, was used by Whitehead et al. (1996) to design a method to measure the time profile of the ear canal response precisely. The phases of the components are systematically changed in steps $\Delta\phi_1$ and $\Delta\phi_2$ between tests in such a way that the phase of the generated $2f_1 - f_2$ is constant in all cases, whereas the primaries could be modified. The steps in the phases changes of the primaries $\Delta\phi_1$ and $\Delta\phi_1$ were selected to null both the f_1 and the f_2 primary sums. These steps are efficiently noted in the complex plane, and the complex representation of the sine wave. The step counter is denoted by k, and N steps complete one or more times 2π.

$$s_{10}(t) = |s_1|\,\mathrm{e}^{\mathrm{i}(\omega_1 t - \phi_1)} \quad \text{and} \quad s_{20}(t) = |s_2|\,\mathrm{e}^{\mathrm{i}(\omega_2 t - \phi_2)}, \tag{6.2}$$

$$s_{1k}(t) = |s_1|\,\mathrm{e}^{\mathrm{i}(\omega_1 t - \phi_1 + k\Delta\phi_1)} \quad \text{and} \quad s_{2k}(t) = |s_2|\,\mathrm{e}^{\mathrm{i}(\omega_2 t - \phi_2 + k\Delta\phi_2)}, \tag{6.3}$$

$$s_{1k}(t) = s_{10}(t)\,\mathrm{e}^{-\mathrm{i}k\Delta\phi_1} \quad \text{and} \quad s_{2k}(t) = s_{20}(t)\,\mathrm{e}^{-\mathrm{i}k\Delta\phi_2}. \tag{6.4}$$

The steps in the primaries now have to fulfill the conditions

$$\sum_{k=0}^{N-1} \mathrm{e}^{-\mathrm{i}\,2\,\Delta\phi_1} = 0 \qquad \text{and} \qquad \sum_{k=0}^{N-1} \mathrm{e}^{-\mathrm{i}\,\Delta\phi_2} = 0 \tag{6.5}$$

in such a way that the phase for the selected DP, here $2f_1 - f_2$, does not rotate. The phase of the fundamental of this DP is,

$$\phi_{2f_1 - f_2} = \pm(2\phi_1 - \phi_2) \tag{6.6}$$

[11]Maximum nonlinear interaction occurs if both inputs to the nonlinearity have a similar strength; otherwise, the nonlinear response is dominated by the strongest components, and the response to the weak component follows the behavior presented in Table 4.1: linear for a weak A_2 and quadratic for a weak A_1.

[12]Because of the asymmetric excitation patterns the area of optimum interaction will be closer to the f_2-place than to the f_1-place.

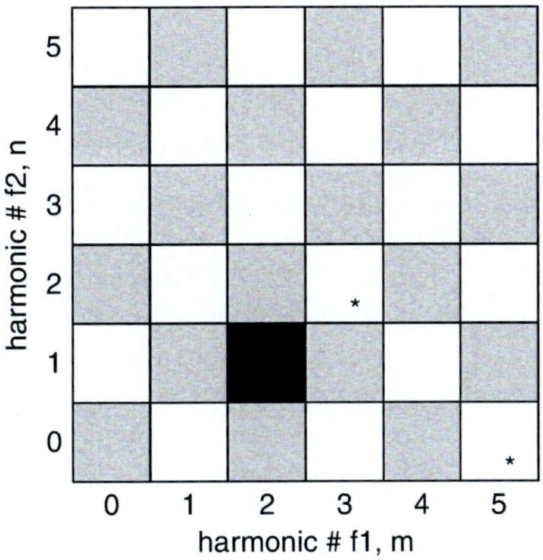

Fig. 6.5 Summation results of checked intermodulation products for N = 12. Initial phase conditions for f_1 and f_2 are 0. The figure presents global results in a tabular format with mf_1 along the horizontal axis and nf_2 along the vertical axis. Gray scale code used: (0) *white*: net intermodulation (harmonic) adds to zero, except for the marked blocks (*), where an $mf_1 + nf_2$ component can be present, depending on the initial phases; (1) *light gray*: irrelevant even intermodulation combinations (do not appear in an odd NL system); (2) *black*: maximum response to intermodulation product

and the variation by rotation step k:

$$k\Delta\phi_{2f_1-f_2} = \pm(2k\Delta\phi_1 - k\Delta\phi_2) = 0 \tag{6.7}$$

must be zero for all k, or

$$2\Delta\phi_1 = \Delta\phi_2. \tag{6.8}$$

More generally, the ratio of the phase steps has to match the ration m/n, where the counters define the intermodulation products. This condition can easily be matched. For example, for $2f_1 - f_2$ the combination of 12 steps ($N = 12$) of $\pi/6$ for f_1 coupled to 12 steps of $\pi/3$ for f_2 is appropriate to also rotate the phases of the primaries such that these cancel. The verification that other relevant DPs are also suppressed remains more elaborate, but is also rather straightforward (see Fig. 6.5). The proof is left to the interested student.

6.3.2 DPOAE Analysis Details

In our numerical analysis the α-term from (6.1), which represents even-order DPs, is omitted for two reasons:

1. The odd-order distortion products are the most prominent set.

2. The omission provides an accuracy test of the analysis because an odd order nonlinearity should produce no even-order distortion products. [Note that the odd-order nonlinearity arises from an even damping term: the constant + the quadratic elements in the bracket term in (6.1).]

Although it is clear that amplitudes and phases of the DPOAEs carry relevant information, it is not immediately obvious how that is done, and what experimental results can effectively be compared with the computed model results at this stage. For instance, the psychophysical cancellation technique is not modeled at this point, and, therefore, level data are not directly compatible.[13] A similar problem applies to AN data. At the stage of the auditory nerve more information processing is involved than the CMs, processing that involves additional nonlinear behavior and adaptation.

The most direct comparisons not suffering from these additional effects are:

- Direct measurements of DPOAEs, both in human subjects and in animals.
- Direct measurements on the BM or cochlear partition, which are only available from animal studies.

At relatively low primary levels (<40 dB SPL), the $2f_1 - f_2$ DPOAE is 20–40 dB below the primary level for the frequency ratio with maximum DP, which occurs at about 1.15 for man. Wilson (1980) reports a reduction of approximately 55 dB at the ratio 1.3 (for man).

Basilar membrane velocity studies (Nuttall et al. 1990; Robles et al. 1990; Ruggero et al. 1992) match these numbers rather well. It is noted, however, that the studies mentioned tend to be limited to the basal part of the cochlea.

An other observation is that the strength of the DP drops markedly with small increases of the frequency ratio f_2/f_1.

In general, some phase uncertainty is introduced by the hardware. For low-pass filters required before sampling (Nyquist criterion) the effect tends to be negligible below 8 kHz. Use of bandpass filters leads to stronger effects. Also the use of FFT-analysis usually involves phase shifts related to window shape, position, and signal duration.

The method treated in this section avoids such phase uncertainties. It was introduced by Whitehead et al. (1996).

The method is applicable for the reliable assessment of DPOAE travel times, DPOAE levels, and DPOAE phases. The travel time assessment requires accurate measurement of onset delay. This is by no means a trivial procedure because the auditory system is locally band limited and frequency selective. In term of CMs: the cochlea couples many band-limited sections, which follow the cochlear frequency map and frequency selectivity properties. Hence, the impulse response in such sections will build up in time with velocities related to the local resonance frequencies. The overall impulse response seen reflected at the ear canal (CEOAE) will contain contributions over the full frequency range, including very fast high-frequency parts, and slow low-frequency ones.

[13] The cancellation stimulus consists of three components, where a third component is adjusted in frequency, amplitude, and phase, to minimize the $2f_1 - f_2$ DP. But this component plays a role in a nonlinear process and will affect the nonlinear interactions.

The PTPV method was introduced to operate on narrow band signals. For such signals, it enables separation of spectral components in the response by stimulus component preselection. The method has virtually no effect on the stimulus envelope.[14] Even so, the exact distinction between travel delay and rise time (for a linear system, the convolution of the impulse response with the stimulus) remains hard to establish experimentally because of measurement noise (or computation noise). Choosing a practical threshold level, e.g., at a fixed percentage of the peak response value and above the noise, provides a number, and such a number can be used in a verifiable comparison with reasonable model predictions. Here we use primarily the *50% from the peak height* onset envelope criterion. In exceptional cases the peak and the 90% match may fall after the displayed 40 ms window.

On similar grounds, Whitehead et al. used a different optimum criterion, based on correlation between the expected waveform and the measured result. For the long term response, it is usually straightforward to determine the amplitude and phase response. After matching this part to a continuous signal of the same amplitude and phase, the correlation between response and match is computed, and then the response delay was determined based on three different criteria: fit by eye, fit at a correlation criterion, and fit at the normalized RMS (envelope) response. The correlation criterion was set at 75%, the envelope (amplitude) criterion at 63%. In addition, the stimulus onset was shaped to suppress splatter, but whereas we used a sinusoidal shape with a 5 ms half-height delay, Whitehead et al. used 0.7 ms. This causes more splatter, and a difference in expected delay of about 4 ms. This point will be addressed in the discussion.

Different cochlea models, and different stimulus levels, produce different effects on the DPOAE onset shape, and thus on the definition and assessment of delay. Some of these are explored next.

Ear canal responses have been computed for several frequency combinations, e.g., for $f_1 = 1$ and $f_2 = 1.2$ kHz. Figures 6.6 and 6.7 show results for 30 dB (SPL) and 80 dB (SPL) stimuli, respectively. The upper part of the figures show a single trace of the total response for ϕ_1 and ϕ_2 equal to zero, the lower trace shows the PTPV response optimized for $2f_1 - f_2$.

The sinusoidal stimulus onset with overall rise time of 10 ms is clearly present in the top panel. Analyzing the $2f_1 - f_2$ response (lower panel) it is clear that stimulus onset shape and response onset shapes are not identical. At best they are quite similar in shape, but onset delay of the response may vary from somewhat shorter to significantly longer than the original 10 ms. In other cases, however, the shape may be much more "deformed" and envelope variations in the response continue over more then 100 ms. Similar differences in experimental results have been reported by Whitehead et al. They investigated human and rabbit responses and found significantly shorter delays for rabbits than for humans. In this section we

[14]The critical reader might want to know how well the envelope of the two-tone stimulus represents the power flow into the cochlea. The waveform shape depends on the primary phases, and the relative phases change over the conditions. Here we neglect any possible small effect due to these variations.

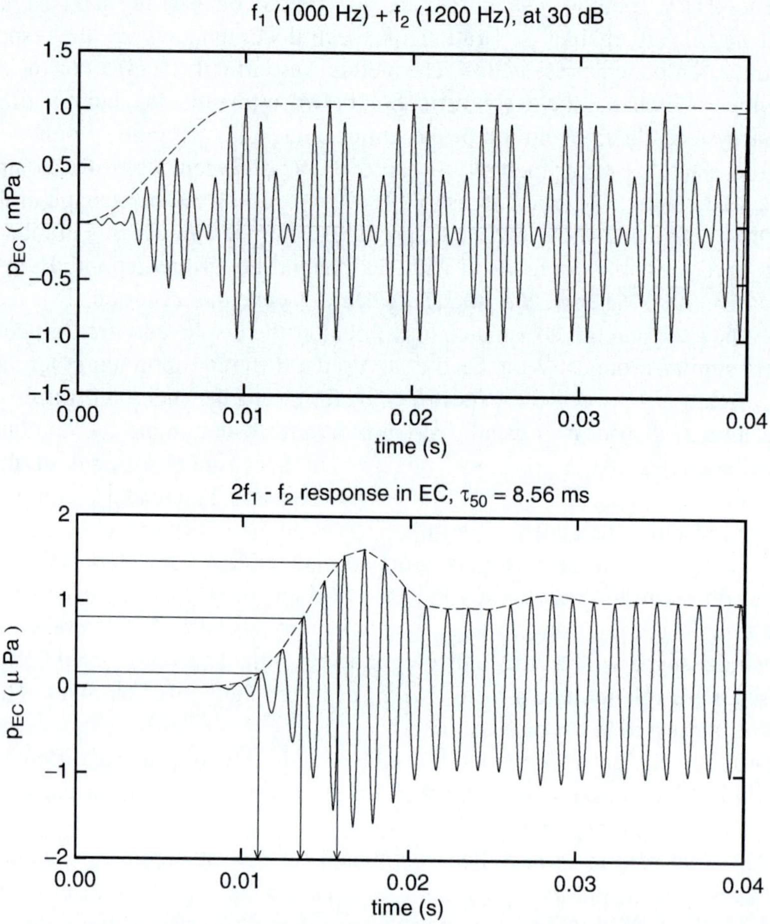

Fig. 6.6 Stimulus $f_1 + f_2$ and PTPV response for model #4 with linear stiffness. Stimulus parameters: $f_1 = 1$ kHz, $f_2 = 1.2$ kHz, $L_1 = L_2 = 30$ dB. Time plots over the first 40 ms. The response onset is delayed by approx. 8.5 ms at 50% of the response peak amplitude. Note the overshoot in the response onset and a smaller peak around 28 ms. Whereas the envelope of the stimulus is smooth (half a circle of a sinusoid), the envelope of the response consists of simple straight lines. The 50% point of the stimulus slope is by definition at 5 ms; the reading of the indicated 50% point from the response gives the delay + 5 ms. As before, stimulus and response parameters refer to a plugged ear canal point just after the plug

focus on human parameter model data, in Sect. 6.4 guinea pig data are addressed, which also turn out to have somewhat shorter delays.

We will first consider examples where stimulus and response onset shape are rather similar.[15] We allow, however, for a difference in onset slope. This

[15]Note that we distinguish here stimulus and response. In fact the stimulus is derived from the recording of a single track at stimulus onset. Hence, it also will contain any reflections, but these

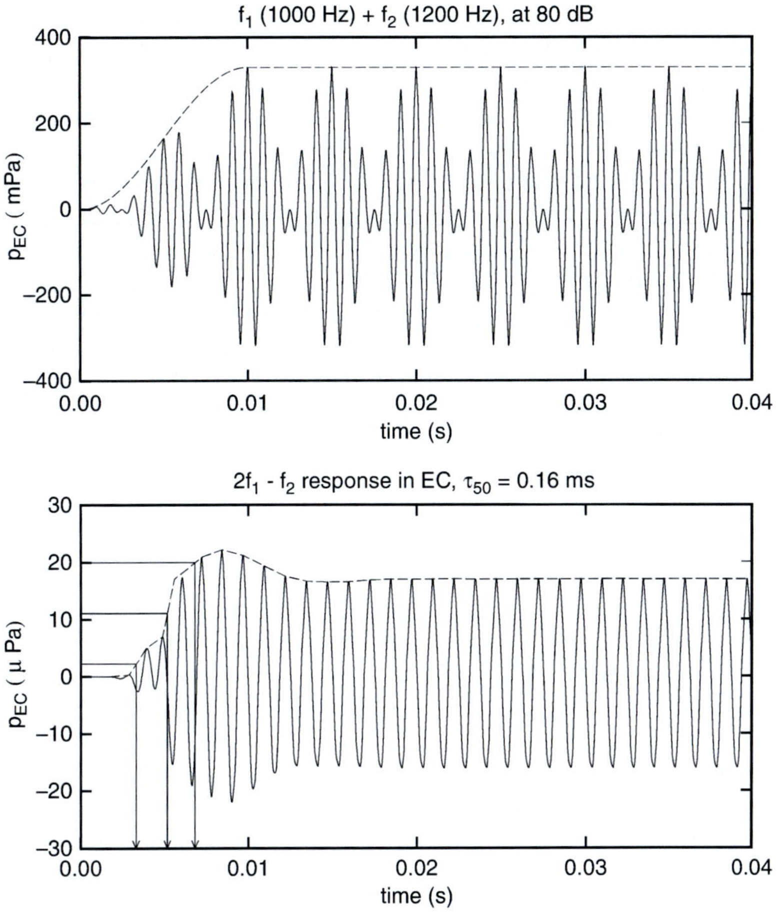

Fig. 6.7 As Fig. 6.6 but at stimulus level of 80 dB. Note that the delay is much shorter. Moreover, the onset shape is no longer smoothly sinuoidal

occurrence appeared to be dependent on stimulus level and on model version/type. For the time being, we focus on the delay at half height, and leave the additional points, low (10%), and high (90%) levels, as additional information for the reader. It is clear that the information is a quantitative measure of the shape change.

The first step is to map the response onset envelope (shape) on the stimulus onset, or vice versa, and make the comparison (optimum time shift). At 30 dB (SPL) the match occurs at a delay of approx. 8.6 ms. Note that at this low stimulus level, the DP response is very weak: more than 100 dB below the primary level. At the higher

are supposed to be at a low level. In general they are not visible in the time record, but they might shows some low-frequency low-level peaks in the spectrum. The PTPV response of the DP, on the other hand, only contains components that come back from the cochlea.

level (80 dB SPL), the DP level is close to 35 dB (SPL), or 45 dB below the primary level. But more relevant for this section is the dramatic decrease in delay. Because of the change in response rise time the definition of delay is problematic, as indicated above. The value of approximately 0.3 ms (or even shorter) follows from a threshold taken at the top.

The reason to display at least 40 ms of the responses is the occasional irregular behavior between 25 and 40 ms after stimulus onset. The 30 dB response shows two somewhat unexpected peaks, the first at ≈17 and the second at ≈28 ms after onset. Additional results for several other conditions are presented in the appendix to this chapter, in Chap. 10, together with the response spectra. These spectra are computed over an interval of either 40 or 50 ms, starting 10 ms after stimulus onset, and after down-sampling the original signals by a factor of 4. Given a starting sampling frequency of 400 kHz, this left us with 100 kHz, or a $dt = 0.01$ ms. The total time interval of 5,000 points has a duration $T = 40$ or 50 ms, so that the frequency resolution is $df = 1/T = 25$ or 20 Hz. The signal edges of the samples were suppressed using a Kaiser–Bessel window ($\beta = 3.5$). Computing the FFT in double precision provided a maximum accuracy of 180 dB or more below the highest primary peak level. Since the levels of the distortion products are lower than the primary levels, the noise floor was higher for the DPOAE estimates, but usually this did not affect the results in any significant way.[16]

Figure 6.8 shows the relation between DPOAE delay and primary levels over the range from 30 to 80 dB (SPL), or more, for several models. A selection was made that limits the presentation to the 2 + 2.4 kHz cases. The differences between model (4) versions d and $d + s$ remained rather constant from 0 to 50 dB, but disappeared at higher levels. The differences between (3) d and $d + s$ are hardly significant. At high levels, there are clear differences in behavior of models with and without nonlinear stiffness. Nonlinear stiffness produces significantly more distortion and irregular behavior than linear stiffness. The reason that nonlinear stiffness primarily affects the high-frequency part of the related spectra is that the properties of the basal part of the cochlea (basalward with respect to the resonance point) are stiffness dependent, and cover the high-frequency mapped part of the cochlea.

In addition, there is a change in the phase of the DPOAE component with level, as shown in Fig. 6.10. At levels below 60 dB (SPL), the phase is relatively constant, whereas above 60 dB (SPL) there can be a marked increase with level.

6.3.3 Discussion

DPOAEs have occupied many auditory researchers over the last three decades, and one of the major topics has been the DP delay. Most studies focused on the

[16] The computation noise is best observed and checked in a linear model version, when distortion products should be absent.

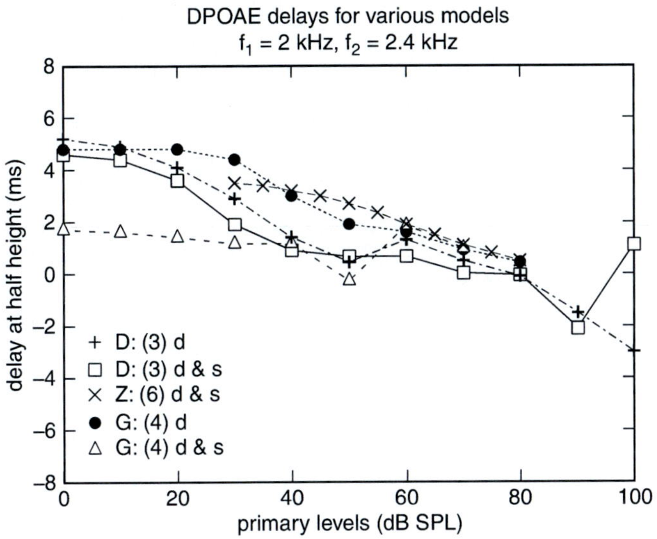

Fig. 6.8 Delays at half-height vs. level for several models. The numbered labels refer to Table 5.4 and the presence of the letters *d* and/or *s* indicate that in all cases nonlinear damping functions were used, and two cases also include nonlinear stiffness. At high levels the inclusion of stiffness nonlinearity can dramatically affect the behavior. At the low levels (10–20 dB) there is a significant difference between the $G(4)$ $d\&s$ version whereas from about 30–60 dB three models score 2 ms lower than the remaining 2; from 60–80 dB the differences are not significant, and above 80 dB the computations are less accurate

group delay, based on measurement of $\partial\phi/\partial\omega$ (e.g., Moulin and Kemp 1996; Schneider et al. 2000). However, the biophysical interpretation of the data remains problematic. The major problem with this method is that the common interpretations rely on system conditions (linearity) which appear not to be fulfilled. We come back to this approach in Sect. 6.4.

Therefore, a direct method such as the PTPV method of the response delay as proposed by Whitehead et al. (1996) is to be preferred if one is interested in the actual response delay, and a proper model analysis helps the interpretation of the underlying mechanisms.

Measurement of amplitude and phase of a distortion product can be estimated efficiently (Whitehead et al. 1996) by eye by mapping the DPOAE signal to a comparison tone as well as by analyzing the PTPV result.

The DPOAE is not generated at one or two points, but it is generated over a cochlea range, i.e., at many points. At these points, levels and phases depend on sites of generation and, also, because of the nonlinearity, on overall level. Obviously, intermodulation can be generated only at sites where both primary components are:

1. Present
2. Subject to the local nonlinearity

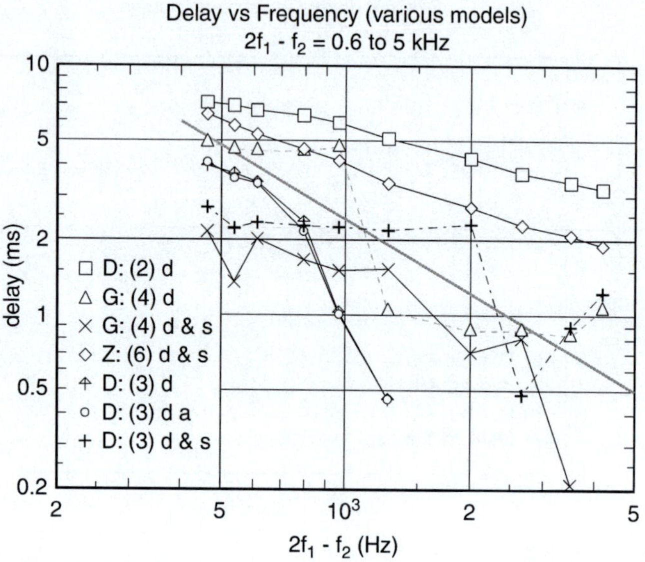

Fig. 6.9 Delays at half-height vs. $2f_1 - f_2$-frequencies for several models. The numbered labels refer to Table 5.4 and the presence of the letters *d* and/or *s* indicate that in all cases nonlinear damping functions were used, and two cases also include nonlinear stiffness. Average level: 45 to 50 dB. Obviously, some models produce more variation than others. Almost all models show a decrease of delay with increasing frequency. The model (3) with additional label *a* represents results for an active version of the model. The results do show that the difference between active and passive damping is marginal at this level. A few additional points for this model attained negative delays, which are excluded for the logarithmic plot. The *heavier gray line* represents average group delay date measured in guinea pig (Schneider et al. 1999). One should keep in mind that the front delay or even its equivalent at 50% is different from the linear group delay measure

This is in the range where nonlinear damping is relevant. The extent of the range increases with level, primarily so in the basal direction. This increase is the primary reason for the decrease of DP delay with level. The precise interpretation is complicated by the fact that the integrated summation of many contributions with different phases must deal with phase interference. The net result in the ear canal is the strong dependence of phase on level (Fig. 6.10). This type of phase change does not occur at a single point nonlinearity for which the phase can be determined precisely (Duifhuis 1989). Therefore, the basis for the observed phase change is that it is derived from an integrated contribution of many adjacent points.

At a single point, the generation of two-tone intermodulation distortion products results from cross terms of the primaries (see Sect. 9.1). They arise as product terms that can be decomposed into the intermodulation products, or sum terms. This leads to the result that $2f_1 - f_2$ and $2f_1 + f_2$ are generated at the same strength, which does not apply to $2f_1 - f_2$ and $2f_2 - f_1$. The components $2f_2 - f_1$ and $2f_2 + f_1$ form a different pair, with a different amplitude factor. At the site of generation, only low-frequency side components (relative to the local tuning frequency) feed

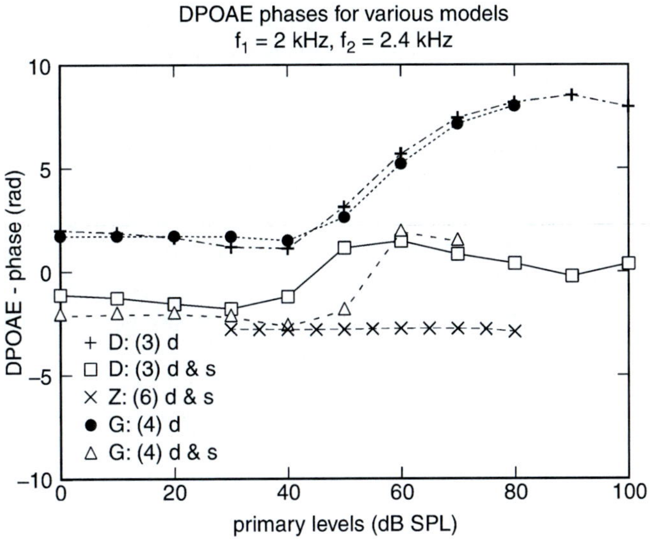

Fig. 6.10 Phases of $2f_1 - f_2$ for several models with respect to the primary phases as a function of primary levels. Model references as before. At low levels (<40-dB) the phases are independent of level, and we recognize two populations that differ by approx. π. At higher levels the Zweig model (6) remains constant. All other models show some changes in the mid-level range, tending again to constant phases at levels above 60 dB. The two NL-d only models [D:(3)d and G:(4)d] predict a phase increase over the full range (except 100 dB). Note that a reliable determination of the increase may require a finer sampling (smaller level steps), in order to exclude 2π-ambiguities

back effectively into the system. The high-frequency side components meet with an input impedance mismatch (strictly speaking, the term impedance applies to linear systems only). The mismatch increases with frequency distance. This partly explains why the $2f_1 - f_2$ component is much stronger than the other components, except at very high levels where the impedance matched range of the $2f_2 - f_1$ component begins to overlap the expanding excitation site. The other factor follows directly from the generation formulas (9.14)–(9.17).

Low-frequency DPs propagate to their corresponding tuning sites. Since these sites also have nonlinear properties they also generate distortion products. At low levels this concerns primarily generation of overtones, thereby slightly modifying the input levels. That effect is of the order of the effect generated by a single tone stimulus, and its contribution to the ear canal pressure is relatively weak (about 40 dB down), as can be concluded from the data from Brass and Kemp (1991) in an experiment that has some similarity with PTPV: a linear addition and subtraction of stimuli at different phases (Sect. 6.2.1). At high levels the generation of second order intermodulation products (e.g., between f_1 and $2f_1 - f_2$) will complicate the situation.

The delay values presented in the model results, partly summarized in Figs. 6.8 and 6.9 agree fairly well with the results from Whitehead et al. This may be most obvious when comparing their Fig. 6 to Fig. 6.9. Note that their data (single subject) are presented at linear scales, whereas we use log scales. The difference in onset shape may have introduced some additional differences: splatter generates more high frequency during the onset and thereby a shorter decay. But here it seems to be within the uncertainty range. Only the "extremely" low delay at low frequencies for model (4), nonlinear stiffness case, appears to fall beyond the acceptable range.

Another point that we have not discussed in detail, but which is relevant in the comparison with the results from Whitehead et al., is their observation of local minima in the DPOAE level vs. frequency plots in Fig. 9 for the human subject, and in Fig. 11 for rabbits. These plots show similar local minima of -5 to -10 dB. Those dips are probably related to similar dips in the excitation profile that we have reported for single spontaneous acoustic emissions (van Hengel et al. 1996, Fig. 6). The latter study explains that the minima arise from the "impedance" mismatch at the cochlea–middle ear transition. This causes a reflection of the outgoing wave, which leads to an interference pattern. The reason that such interference is not observed for external sources, or in general, waves traveling in apical direction, is that most of these waves are completely dissipated before they reach a source of significant reflection. Those are only expected at the helicotrema, where optimum termination for the power flow cannot be matched over the complete frequency range (Duifhuis 2000).[17] Very low frequencies (below $\approx$100 Hz) will generate additional apical reflections. In both cases, viz., the input side (stapes, oval window + the parallel RW) and the apical end (helicotrema) even a small "impedance" mismatch causes measurable reflections. Optimization of the matches minimizes these reflections, but optimization is limited to a limited part of the frequency range, and for nonlinear systems: also for a limited part of the level range.

6.3.3.1 In Conclusion

Time domain methods of analysis are superior in accuracy both in experiments (Brass and Kemp 1991; Whitehead et al. 1996), and in modeling (e.g., Duifhuis and van den Raadt 1997; van Hengel and Duifhuis 2000) because the impact on the signals is minimum. The results appear to be sufficiently accurate to claim that the main effects of DPOAE generation are covered. This means, among other things, that a global definition of cochlear partition parameters appears to be sufficient. The differences between rather different models remain small, and do no require an analysis on a smaller scale, which would imply detailed involvement of the CP fine structure.

The disadvantage of the PTPV measurement mentioned by Whitehead et al., viz.,

> that it requires the subject to be very still and quiet during measurements because complete removal of the primary tones depends critically on cancellation, which is sensitive to very small variations of signal phase and amplitude

[17] Details were presented at the 3rd European Bioph. Congress, Munich 2009.

is an important experimental problem. But it plays no role in modeling. There, one just has to be certain that the accuracy is sufficient. One practical example: the algorithm that steps in time can drift away if the next value is computed by addition of a (fixed: floating point with bounded accuracy) increment. This is remedied by using the integer step numbers and computing the resulting time as the product of the (accurate) number of steps and the (constant) step size.

Spectral analysis requires averaging over a time window. Bandpass filtering techniques also affect the waveform; at best, it is only smoothed marginally through convolution with the filters impulse response. The time analysis approach involves repetition in time and averaging, operations that can be performed very precisely. These operations do not have to affect details of the waveform, and therefore are suited to investigate the time course of nonlinear properties of the cochlea.

6.4 DPOAE Group Delay Measurements

Our involvement in group delay measurements resulted from a cooperation with the group at the University of Leiden around 1990, when Sandra Schneider and Peter van Hengel cooperated on the project. Experimental studies took place in Leiden, model studies in Groningen. Biophysically, the guinea pig model is very similar to the human model, except for the model parameters. We decided to set up the model in such a way that easy transformation from one to the other set is possible by changing just one single parameter. As yet, parameter sets for human ears and for guinea pig have been implemented, other extensions are still open. Details are presented on the program codes on the URL, including these parameter sets.

As explained in Sect. 8.2.1, the group delay concept is defined in linear signal analysis, and its applicability to nonlinear systems is unclear. This question is hardly discussed seriously, because the linear approach appears to match the data reasonably well. To obtain the group delay for the DPOAEs one has to measure phases over some frequency range. Most commonly, one of the primary frequencies is fixed, and the other is varied (swept) over a certain range, including the range where the DP amplitude is maximum. At that point, the stability of the phase vs. DP-frequency slope is checked, and measured, and that slope is $-\tau_{gr}$. Interestingly, the f_1-sweeps and the f_2-sweeps tend to give different results.

In this case, the model analysis procedure was very similar to the experimental one. Computations of amplitudes and phases had to be done for every sweep point and then the optimum slope for the set was determined (Fig. 6.11).

Note that the group delay measured for the guinea pig on average is shorter than the predicted (model) values for human, except for the Zweig model. Note that the slopes also differ; in particular, the slope of the guinea pig data is shallower than that of Zweig's models and of a linear constant Q cochlea model. Model results for the variable Q case follow the data slope more closely, but the difference in size is still a factor of 2.

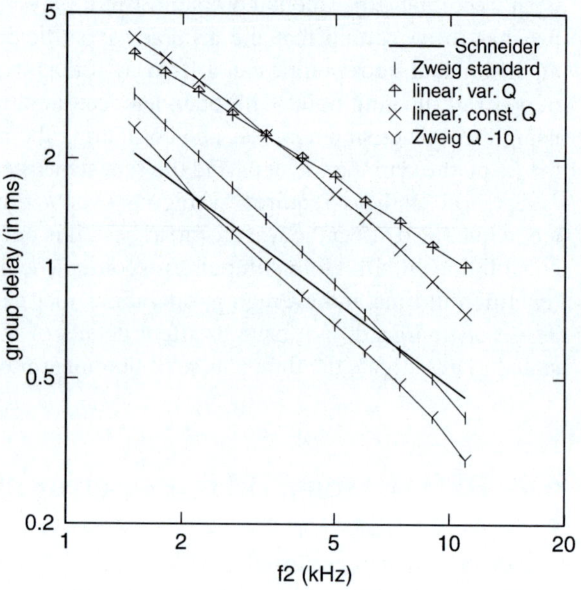

Fig. 6.11 Group delay measurements of $2f_1 - f_2$ for Zweig models in linear and nonlinear versions. The *heavier line* represents average guinea pig data collected by Sandra Schneider (same average line as in Fig. 6.9 measured using f_1-sweeps). The versions with constant-Q show a steeper slope than the models with systematically changing Q (higher at the fast hf-end, lower at the slow lf-end). The tested variable Q models shows a gradient that is more in line with the data, but the difference of approx. a factor 2 over the entire line

6.5 Summary

6.1 *Spontaneous Emissions:*
We hypothesize that an oscillator from the van der Pol family is still the best candidate to represent the generation of spontaneous emissions.

6.2 *Level Effects:*
Although some major effects can be reproduced by the proposed nonlinear model cochlea, not every detail of the experimental data is covered. This may be due, at least in part, to the use of a very simple middle ear model.

6.3 *Time Effects:*
Time effect are most accurately measured in the time domain, experimentally as well as in models. These methods typically do not require averaging over a time window. For auditory data processing, the front delay is probably more important than the group delay.

6.4 *Group Delay Measurements:*
Although the group delay is widely used, its possible limitations in a nonlinear environment deserve further analysis. See also the previous point.

The discussion of applications and perspective is continued in Chap. 7.

References

Brass D, Kemp DT (1991) Time-domain observation of otoacoustic emissions during constant tone stimulation. J Acoust Soc Am 90:2415–2427

Broer HW, Takens F (2009) Dynamical Systems and Chaos. Appl. Math. Series 172, Springer, New York

Duifhuis H (1989) Power-law nonlinearities: A review of some less familiar properties. In: Wilson JP, Kemp DT (eds) Cochlear Mechanisms: Structure, Function and Models, Plenum, New York, pp 395–403

Duifhuis H (2000) Theoretical biophysics of the cochlea. Europian Biophys J 29:246, (abstract)

Duifhuis H, van den Raadt MPMG (1997) Usefullness of the nonlinear residual response method. In: Lewis ER, Long GR, Lyon RF, Narins PM, Steele CR, Hecht-Poinar E (eds) Proceedings of the International Symposium on Diversity in Auditory Mechanics, World Scientific, Singapore, pp 226–232

Duifhuis H, Hoogstraten HW, van Netten SM, Diependaal RJ, Bialek W (1985) Modelling the cochlear partition with coupled Van der Pol oscillators. In: Allen JB, Hall JL, Hubbard AE, Neely ST, Tubis A (eds) Peripheral Auditory Mechanisms, Springer, New York, pp 290–297

van Hengel PWJ, Duifhuis H (2000) The generation of distorion products in a nonlinear transmission line model of the cochlea. In: Wada H, Takasaka T, Ikeda K, Ohyama K, Koike T (eds) Recent Developments in Auditory Mechanics, World Scientific, Singapore, pp 409–415

van Hengel PWJ, Duifhuis H, van den Raadt MPMG (1996) Spatial periodicity in the cochlea: The result of interaction of spontaneous emissions? J Acoust Soc Am 99:3566–3571

Kemp DT, Souter M (1988) A new rapid component in the cochlear response to brief electrical efferent stimulation: Cm and otoacoustic observations. Hear Res 34:49–62

Lynch S (2010) Dynamical systems: With applications using Maple^{TM}, 2nd edn. Birkhäuser (Springer), Boston

Moulin A, Kemp DT (1996) Multicomponent acoustic distortion product otoacoutic emission phase in humans. II. Implications for distorion product otoacoustic emssion generation. J Acoust Soc Am 100:1640–1662

Nuttall AL, Dolan DF, Avinash G (1990) Measurements of basilar membrane tuning and distortion with laser doppler velocimetry. In: Dallos P, Geisler CD, Matthews JW, Ruggero MA, Steele CR (eds) The Mechanics and Biophysics of Hearing, Springer, Berlin, pp 288–295

Papoulis A (1962) The Fourier Integral and its Applications. McGraw-Hill Book Company, New York

van den Raadt MPMG, Duifhuis H (2011) A generalized van der P0l-oscillator cochlea model. In: Dallos P, Geisler CD, Matthews JW, Ruggero MA, Steele CR (eds) The Mechnics and Biophysics of Hearing, Springer, Berlin, pp 227–234

Robles L, Ruggero MA, Rich NC (1990) Two-tone distortion products in the basilar membrane of the chinchilla cochlea. In: Dallos P, Geisler CD, Matthews JW, Ruggero MA, Steele CR (eds) The Mechanics and Biophysics of Hearing, Springer, Berlin, pp 304–311, (disc: 312–313)

Ruggero MA (1980) Systematic errors in indirect estimates of basilar membrane travel times. J Acoust Soc Am 67(2):707–710

Ruggero MA, Robles L, Rich NC, Recio A, Brown AM, Evans EF (1992) Basilar membrane responses to two-tone and broadband stimuli [and discussion]. Phil Trans R Soc Lond B 336(1278):307–315

Schneider S, Prijs VF, Schoonhoven R (1999) Group delays of distortion product otoacoustic emissions in the guinea pig. J Acoust Soc Am 105:2722–2730

Schneider S, Prijs VF, Schoonhoven R, van Hengel PWJ (2000) F1- versus f2-sweep group delays of distortion product otoacoustic emissions in the guinea pig; experimental results and theoretical predictions. In: Wada H, Takasaka T, Ikeda K, Ohyama K, Koike T (eds) Recent Development in Auditory Mechanics, World Scientific, Singapore, pp 360–366

Whitehead ML, Stagner BB, Martin GK, Lonsbury-Martin BL (1996) Visualization of the onset of distortion-product otoacoustic emissions, and measurement of their latency. J Acoust Soc Am 100:1663–1679

Wiggins S (2003) Unknown Title, 2nd edn. No. 2 in Texts in Applied Mathematics, Springer, New York

Wilson JP (1980) The combination tone, $2f_1 - f_2$, in psychophysics and ear-canal recordings. In: van den Brink G, Bilsen FA (eds) Psychophysical, Physiological, and Behavioural Studies in Hearing, Delft University Press, Delft, pp 43–50(2)

Withnell RN, Shafer LA, Talmadge CL (2003) Generation of DPOAEs in the guinea pig. Hear Res 178:106–117

Chapter 7
Applications and Perspective

Abstract In this chapter we present examples of areas where an efficient NL time domain cochlea model is potentially useful. This includes the front-end application for auditory brain processing, ASR models, and hearing aids. Next we address the relation between Hopf bifurcation models and van der Pol models for CMs. We conclude with our general conclusions.

7.1 Modeling Level Effects

The study of perceptual level effects is a relatively simple and straightforward example of potential application of a better understanding of the operation of the auditory front end.

7.1.1 Pure-Tone Masking Level Effects

The classical nonlinear pure-tone masking level effect, published by Wegel and Lane (1924), still provides an important set of data that cannot be accounted for by linear models.

Potentially, there is more than one possible source of nonlinearity. First, the CP response to a tone is nonlinear, as has been shown in BM responses and detailed OC responses. This has been a main issue of the present book. Next, the response in the nervous system deals with nonlinear coding of signal strength. Finally, it is not unequivocally clear how the processing across all channels may affect the overall behavior.

H. Duifhuis, *Cochlear Mechanics: Introduction to a Time Domain Analysis of the Nonlinear Cochlea*, DOI 10.1007/978-1-4419-6117-4_7,

The first full quantitative analysis of this problem originated in the CBG[1] at the MIT during the 1960s. Elements appear in several theses, but the first publication in a journal was by Siebert (1965) and a more complete story was presented in the conference paper Siebert (1968). He starts with a linear filterbank model of CMs, that does not deviate significantly from the later gamma-tone filterbank models. It is followed by separate transduction channels that model the transformation from BM response to AN response. That step includes a compressive nonlinearity, neural adaptation effects, and it uses a Poisson process as a model for spike generation. In a later study, Siebert also discussed the time information that is available in the precise synchronization of the neural spikes to the stimulating waveform, at least for sufficiently low frequencies. In this study (Siebert 1970), he concludes that there is enough information in the rate coding, so it is unlikely that the additional time coding, which may appear to be more precise, is also used.

Treating the nerve responses in the different channels as stochastically independent with respect to the involved neural noise, Siebert estimated the psychophysical thresholds of frequency discrimination and amplitude discrimination, assuming that processing beyond the auditory nerve is optimum. If that is the case, then some interesting theorems from system analysis apply (e.g., the Cramér–Rao inequality).

During the last two decades the technique was picked up again, updating some of the original assumptions and providing a biophysical alternative for the gamma-tone filterbank that had become popular among experimental psychologists.

As noted above, the front end of Siebert's model did not deviate significantly from the gamma-tone filterbank. Siebert starts in the frequency domain with filter shapes that match the tip of the neural tuning curves, whereas the gamma-tone filters start in the time domain format with impulse responses. We note that an approximate relation between those two forms had been analyzed and presented already by Goldstein et al. (1971).

A revival of Siebert's approach was promoted by Heinz, Colburn, and Carney in two detailed new papers on the limits of auditory performance (Heinz et al. 2001a,b).

We advocated the use of the time domain method, with proper nonlinear properties, as the front end, thereby basically substituting the linear filterbank with a nonlinear cochlear model. A good start would be the analysis of (global) level effects response for pure tones, and combine these results to estimate pure tone masking data. This might be a successful approach to obtain convincing estimates for Wegel and Lane's data. It is clear that their results from almost a century ago always have been confirmed: asymmetry, different behavior above *vs* below masker frequency, but also that so far no nonlinear mechanism has achieved the status of "explanatory proof without any doubt."

The CMs study presented in this book does not provide an analytical formulation, or model, of the stimulation of the IHC's, the sensory cells driving the auditory nerve fibers. It fact it does not even address details of the 3-D coupling in detail.

[1]The Communications Biophysics Group at the Electrical Engineering Dept., headed by W. M. Siebert.

The brief sidetrack in that direction addresses the canal fluid motion; the 3-D fine structure in the OC remains unaddressed, primarily because the number of relevant free parameters still exceeds the available reproducible data. This point will be mentioned in more detail in the final summary.

Detailed incorporation of all properties of the CMs, including 3-D local mechanical coupling interactions and mechanisms like cellular adaptation, can give a better understanding of these steps. Some of these I have employed before treating the nonlinear CM behavior into account, and that did not help to understand nonlinear masking aspects (Duifhuis 1973, App. A).

7.1.2 Conclusion

The interested student should be able to address this topic in more detail and work it out to an interesting model of level representation and amplitude discrimination.

7.2 Applications

Auditory processing techniques are used in numerous applications in information technology and in biomedical engineering. The first category contains technical speech processing as used in the early days of telephone communication and in ASR, the second involves developments for auditory diagnosis and supporting devices as hearing aids and auditory implants.

Some of these techniques go back to the time that van der Pol started in electronics, the spectral analysis revived at the second half of the previous century, with a revival of the Fourier analysis, and the even earlier FFT algorithm (often cited to Cooley and Tukey 1965), but going back even farther than Fourier, viz. to Gauss around 1805 (referenced by Heideman et al. 1984).

Real-time analysis (i.e., only with a continuous time window delay) became available in hardware and software at the seventies, as long as the number of bits representing the amplitude remained limited, and the bandwidth (time sampling) remained between a few hundred and a few thousand Hertz (sampling frequency about 8 kHz, or slightly higher).

For some practical applications, (quasi) real-time streaming is essential—as in hearing aids; for others, focusing on more detailed analysis might be more appropriate and the real-time condition might be dropped. For a long time, the time windows used for FFT analysis of audio signals, in particular of speech, were between 20 and 40 ms, providing spectral resolutions from 50 to 25 Hz (slightly broader due to proper windowing). This delay is just audible: it is equivalent to the delay of sound in air over a distance of the order of 10 m, a distance over which many normal hearing observers have no problems combining audio- and visual speaker information.

The disadvantage of using amplitude spectra only, which is common practice, is that part of the sensitive timing information is averaged away over the FFT window. Its envelope smoothens the underlying fine structure.

Moreover, the basic theory of signal analysis is a theory of linear signal analysis. Its (common) properties rely on the system to be linear.

Obviously, a nonlinear system does not meet this criterion, so that we have to be suspicious about any result of linear analysis. Our choice was to go to back to a sufficiently accurate time analysis and check its accuracy and reliability, and using additional spectral analysis only in case of reasonably stationary signal–response situations.

So far applications have found their way in the following projects:

- van Hengel (1996), "Emission from Cochlear modeling." Besides the construction of the 1-D cochlea, it addresses spatial periodicity, 2nd filter, and phase behavior in the cochlea. Later on, phase data are compared to experimental DPOAE group delay data from Sandra Schneider. Finally Van Hengel addresses some issues of the numerical 3-D approach. As an example, the cupula motion in lateral line is modeled successfully. Suggestions were made for additional steps toward a 3-D mammalian cochlea analysis.
- Andringa (2002), "Continuity Preserving Signal Processing." This study focuses on cochlear processing as a front end for speech processing by studying the cochleogram. In this case the cochleogram is not a filterbank approximation but the output of a realistic cochlea model.
- The time domain model has been used by S Verhulst at the Technical University of Denmark for her model studies (Verhulst 2010). This study focuses on CEOAEs responses.
- The time domain model is being used by B Epp (Univ. Oldenburg) in a more general approach of auditory processing, with specific interest in nonlinear cochlea dynamics, e.g., in CTs and related OAEs (Epp 2011; Epp et al. 2010).

For details, in particular of the last two studies, the interested reader is referred to the presentation by the investigators. Some investigators used the software in a UNIX/Linux environment, others in Windows, for which we have developed a specific GUI. [The graphics structure for the UNIX/Linux environment differs substantially form the Windows version.]

7.3 The Hopf Bifurcation

In Chap. 5 we introduced a brief reference to the Hopf bifurcation that might be considered an alternative to the van der Pol oscillator model of spontaneous otoacoustic emissions.[2] References were given to studies by Magnasco (2003) and

[2]Note that the Hopf bifurcation alternative did surface about 20 years after the first proposals of the van der Pol oscillator.

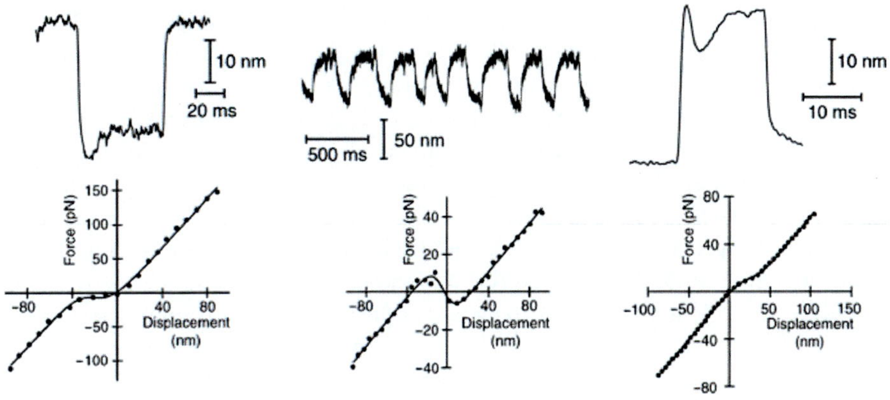

Fig. 7.1 As an example, results from experiments Tinevez et al. (2007, 2009) are presented. These concern measurement of hair bundle deflection in response to forces applied with a flexible glass fiber. The *left column* shows a transition in force response data for a small negative displacement; the *middle column* shows spontaneous oscillations at ~12 Hz, and the *right-hand column* has a positive transition. The *middle column* depicts the negative stiffness around zero, where the force–displacement curve has a negative slope. Reprinted from Biophysical Journal, Vol. 93, J-Y Tinevez, P Martin and F Jülicher, Unifying the Various Incarnations of Active Hair-Bundle Motility by the Vertebrate Hair Cell, 4053–4067, Copyright (2007), with permission from Elsevier

Kern and Stoop (2003). These represented an example of studies with roots in Hudspeth laboratory,[3] and later expansions in several European laboratories, in particular in France, Germany, and the United Kingdom. In addition a Swiss branch developed somewhat independently from the others. Some of this work will be discussed in the following section. Claims and results will be related to our studies. The interested student should find the open ends, and may find them intriguing enough to try to find the proper answer(s).

Technically a Hopf bifurcation is a feature of a much wider class of nonlinear functions, but the cochlear models that have been implemented and called Hopf bifurcation models are the same as the van der Pol oscillator models, as will be shown in this section (see Theorem 7.1 on page 183).

The basic idea starts almost simultaneously with the discovery of nonlinear behavior in hair bundle stiffness behavior, which has been confirmed for both mammalian and nonmammalian vertebrates. For "large" driving forces the responding deflection behaves linearly, indicating a constant stiffness, except that the asymptotes do not go through zero. Instead, around zero there is a nonlinear transition from one asymptote to the other that can be accompanied by a low or even negative stiffness (e.g., Hudspeth (Fig. 5, 2008), Tinevez et al. (Figs. 2 and 3, 2009), examples of which are shown in Fig. 7.1).

---

[3]Howard Hughes Medical Institute and Laboratory of Neuroscience, The Rockefeller University, New York.

Originally the phenomenon was introduced as a possible alternative for OHC body motility (Martin and Hudspeth 1999), a feature that is available in mammalian cochleae but lacking in nonmammals. Since nonmammals do show spontaneous emissions, another source was supposed to be available. Martin and Hudspeth "report physiological evidence that amplification can stem from active movement in the hair bundle" formally leaving room for other options. In practice, alternative options were not taken seriously.

The first proposals with the Hopf bifurcation were presented a little later (Eguíliuz et al. 2000; Martin et al. 2001) by the same group. Eguíliuz, Ospeck, Choe, Hudspeth, and Magnasco return to the use of the term *essential nonlinearity* that had been introduced about 30 years before, to describe the CT data: at low levels, CTs appeared to decrease linearly rather than disappearing completely.[4] They provide the basis for subsequent work, the details of which will be given in Sect. 7.3.1. Somewhat later, Magnasco (2003) presented the Hopf-bifurcation as a mathematical tool in a biophysical context aiming at the shape of cochlear tuning curves. Although some of his specifications remain uncertain he does pay attention to the energy flow and dissipation. Mathematically, he starts from the results of Eguíliuz et al., referring to a (local) response R which occurs as the output of an input F. This will be discussed in more detail in Sect. 7.3.1. The two papers are still cautious about the biophysical interpretation of the required source of nonlinearity, but the option that nonlinear hair bundle stiffness fulfills this role is presented in some detail.

After $\approx$2003, the developments converged toward nonlinear stiffness in line with the hair bundle stiffness results.

A very important point in the arguments used by all investigators in this group is the assumption that energy loss within the cochlea is basically accounted for by viscous loss. This loss would occur either only in the cochlear fluid and its boundary layers, or possibly also in additional viscous elements in cochlear partition. This appears to be one of the implicit points going back to the 1948 study by Gold.

The explicit point that is promoted in many of the papers dealing with the topic concerns the occurrence of active behavior. This reference relies on their interpretation of Gold's (1948) paper, probably without carefully analyzing his arguments (e.g., Martin and Hudspeth 1999; Eguíliuz et al. 2000; Martin et al. 2001; Martin and Hudspeth 2001; Magnasco 2003; Duke and Jülicher 2003; Kern and Stoop 2003; Chan and Hudspeth 2005; Martignoli et al. 2007; Kern et al. 2008; Hudspeth 2008; Hudspeth et al. 2010; Maoiléidigh and Jülicher 2010, and many others).

Basically, Gold excluded the possibility of sharp cochlear tuning on the basis of viscous damping of the BM radial fibers. The role of viscosity was just one of the three major points that prevented his success during the fifties. Obviously Gold was a very talented scientist, who had positioned himself in the auditory science field in a very short time, but his ideas about the cochlea where to a large extent based on incorrect starting points:

[4]Unfortunately, Eguíliuz et al. assumed a cubic distortion generator for the $2f_1 - f_2$ component.

1. Gold concludes on the basis of psychophysical data that the frequency selectivity of a point in the cochlea should be more than an order of magnitude greater than expected to be achievable. This assumption contains two underlying points: (a) he overestimated the psychophysical frequency selectivity probably by an order of magnitude, (b) he underestimated the possibility of cochlear frequency resolution because of data so far were based on dead cochleae. One might argue that the last point is exactly what his statement wanted to promote, but I am not so sure. Let us consider the next point:
2. Gold effectively assumes the cochlea to be modeled properly by the Helmholtz-type model. This implies parallel 2nd order filters, which are much too simple to model the cochlea. This was about 25 years after Wegel and Lane (1924) had published a transmission line model with fluid coupling, and many auditory scientists had switched or began to switch from the Helmholtz type to the newer ideas, which do produce a type of tuning not considered by Gold.
3. Finally, the limitation of the Helmholtz model at the time was assumed to be due to viscous drag of radial BMy fibers in viscous cochlear fluid. The current conviction is that viscous losses at that point are negligible, one reason being that the membrane fibers do not vibrate independently in the cochlear fluid. Many current studies neglect the fluid viscosity completely, possibly except at some specific boundaries.

The argument misses one very important point, elementary in physics, biophysics, and engineering.

The cochlea is a fundamental element in a biological measuring device that is able to do much more than auditory signal detection. Any measuring device needs to consume some input energy.[5] The optimum sensory device minimizes the loss of input power. Just like the visual system is able to detect optical photons, the auditory system can detect stimuli at the Brownian noise background level (de Vries 1948, 1952). This implies that a sensitive ear will be optimized to pick up the acoustic energy, thereby involving the hair cell, but at the same time minimizing viscous losses.

Scientific history may give the answer to the question if the share that Gold received in the recognition for the experimental discovery of SOAE and other OAEs by Kemp and co-workers is justified. In this context, the recent chapter from Kemp (2008, Chap. 1) is illuminating.

7.3.1 *Details of the Hopf Bifurcation*

Eguíliuz et al. (2000) present the following generic equation to describe a useful Hopf bifurcation:

$$\dot{z} = (\mu + \mathrm{i}\omega_0)z - |z|^2 z, \tag{7.1}$$

[5] An equivalent argument is that the quantities involved are bounded by the uncertainty relation, rewritten in the proper dimensions.

where $z = z(t)$ is a complex variable in time, and ω_0 the natural frequency of oscillation.[6,7] As noted below, we must assume that the parameters are dimensionless. A related mechanical equation would be either

$$d\dot{w} + s_0(1 - \gamma_1 w^2)w = 0, \quad w \in \Re \tag{7.2}$$

with d and s_0 representing damping and stiffness per unit area and w the across partition deflection (neglecting the inertia term!), or

$$\dot{v} + d_0(1 - \gamma_2 v^2)v = 0, \quad v \in \Re \tag{7.3}$$

with d_0 is the damping per unit area, v the across partition deflection, and where the stiffness term is neglected.[8] Obviously (7.2) places the nonlinearity in the stiffness, whereas (7.3) implies a nonlinear damping term. The occurrence of the term $\mathrm{i}\omega_0 z$ in the right hand side of (7.1) suggest a close connection to linear systems analysis. In nonlinear analysis the use is only by approximation, and can be avoided.

Let us consider the solutions described by the authors. Equation (7.1) has characteristic solutions for the different parameter settings. First, the exact solution $z = 0$ is stable because it implies that $\dot{z} = 0$. For small deviations from zero, stability of the solution depends on the sign of μ. This follows from the consideration that for small values of z, with $|z|^3 \ll 1$, the cubic term can be neglected, yielding the solution: $z(t) = z_0 \mathrm{e}^{\mu t + \mathrm{i}\omega_0 t}$. This solution expands for positive μ, whereas it is stable for a negative value (Eguíliuz et al. 2000).

As is a common approach in the analysis of (non)linear dynamical systems, one tries to find periodic solutions by subjecting the system to an oscillatory force. A trial solution of the form $z(t) = R\mathrm{e}^{\mathrm{i}(\omega t + \phi)}$ gives

$$\dot{z} = \mathrm{i}\omega \mathrm{e}^{\mathrm{i}(\omega t + \phi)} = \mathrm{i}\omega z \tag{7.4}$$

$$= (\mu + \mathrm{i}\omega_0)z - R^2 z + F\,\mathrm{e}^{\mathrm{i}\omega t} \tag{7.5}$$

where $R = |z|$ (as already implied in the trial solution) and where a stimulus term $F\,\mathrm{e}^{\mathrm{i}\omega t}$ has been added. After inserting the trial solution $z(t)$, multiplication by $\mathrm{e}^{-\mathrm{i}(\omega t + \phi)}$, and some reshuffling this yields:

$$-F\mathrm{e}^{-\mathrm{i}\phi} = \mu R - R^3 + \mathrm{i}(\omega_0 - \omega)R.$$

---

[6]Note that the dimensions of these parameters are puzzling. Obviously, the dimension of $|z|^2$ must equal that of T^{-1}, and the same applies to the dimension of μ. This leaves $\mathrm{T}^{-1/2}$ as dimension for z. This is a dimension that can not be associated with a biophysical quantity, in particular not to either damping $[\mathrm{M\,T}^{-1}]$ or stiffness $[\mathrm{M\,T}^{-2}]$. In other words, the formula can only be meaningful if it represents properly scaled dimensionless quantities.

[7]A slightly different formula was presented later by Martignoli et al. (2007) in an application study. Basically it is limited to the same insufficient initial and boundary conditions.

[8]The simplifications that discard either inertia or stiffness imply that neither (7.2) nor (7.3) is a good cochlea model.

The quadratic modules of both sides give their equation (1):

$$F^2 = R^6 - 2\mu R^4 + [\mu^2 + (\omega - \omega_0)^2]R^2, \tag{7.6}$$

which at the bifurcation point ($\mu = 0$) simplifies to

$$F^2 = R^6 + (\omega - \omega_0)^2 R^2. \tag{7.7}$$

Eguíliuz et al. (2000) argue that near resonance the response to stimulus relation is close to a cubic root, whereas at sufficient distance form resonance the linear relation will become dominant (their Fig. 2).

These results were placed in more specific contexts of biophysical properties of the ear in later studies. Magnasco (2003) discussed the relevance for cochlear tuning curves, and Duke and Jülicher (2003) aimed to derive a proper wave equation for the cochlea. A substantial part of their analysis is carried out in the frequency domain, which, unfortunately, may not be considered as valid for nonlinear systems analysis (see our example in Sect. 5.1.3.2). Magnasco's cochlea model appears to be of the Helmholtz type, where the effect of fluid coupling between adjacent points in the cochlea is lacking. In this sense, his analysis also is of the gamma-tone filterbank type (see Sect. 3.4).

Further points of agreement and disagreement will be summarized after the discussion of the direct link between Van der Pol oscillators and Hopf bifurcations.

7.3.2 Relation Between Hopf Bifurcation and Van Der Pol Oscillator

Let us start with the one end of the discussion, presenting the conclusion as a theorem:

Theorem 7.1. *There is no formal distinction between a properly defined Hopf bifurcation cochlea model and a general Van der Pol oscillator.*

Proof. The proof of this theorem was presented by Broer and Takens (2009) in their Sect. *1.3.1 A Hopf bifurcation in the Van der Pol equation.*[9] We follow their arguments. Just like the mathematicians quoted in Sect. 7.3.1, they use a somewhat loose formulation of the differential equations, scaling from physical to dimensionless quantities with relevant values around 1. The physical Van der Pol/Rayleigh equation, already partly modified by a scaled time differentiation

[9] See also: Bronshtein et al. (2003), Handbook of Mathematics, p. 831.

(5.71), where the link with the original parameters is added in (5.72), is reduced to the dimensionless form (5.73) or (5.74). They consider the special case with $\varepsilon = 1$:

$$\ddot{x} = -x - \dot{x}(x^2 - 1). \tag{7.8}$$

The equation is recognized to describe a damped harmonic oscillator, where the damping coefficient is represented by the term $(x^2 - 1)$. The damping term is negative for $|x| < 1$, increasing the energy, whereas for $|x| > 1$ one encounters the common positive damping. Broer and Takens define the energy of a solution $x(t)$ by

$$E(t) = \frac{1}{2}\left(x(t)^2 + \dot{x}(t)^2\right).$$

If the damping term is zero, then the energy term describes the conserved sum of potential and kinetic energy. For the van der Pol oscillator, a similar conservation is present, which follows from considering the derivative of $E(t)$. After differentiation and some reshuffling, one obtains

$$\dot{E} = -(\dot{x})^2(x^2 - 1).$$

If the damping term, $d(x)$ is zero, one would just be dealing with a harmonic oscillator. But this damping term $d(x) = x^2 - 1$ is (a) nonlinear and (b) characterizes a limit cycle amplitude and period.[10]

For the comparison with the Hopf bifurcation, the damping term is modified to $(x^2 - \mu)$, modifying the differential equation to[11]

$$\ddot{x} = -x - \dot{x}(x^2 - \mu). \tag{7.9}$$

Now the value of μ determines where the damping is negative. For $\mu > 0$, this is the interval given by $|x| < \sqrt{\mu}$.

The behavior of the solutions is:

- For $\mu \leq 0$ (7.9) has no periodic solutions; all solutions converge to zero.
- For $\mu > 0$ all solutions of (7.9) -except for the zero solution- converge to a fixed point solution, the amplitude of which depends on μ.

The transition of the behavior at $\mu = 0$ marks the bifurcation point. □

The bifurcation point occurs if the shape of the damping profile is allowed be shifted along the damping axis (vertical axis in Fig. 5.11). It differs from the original ε-parameter description, which quantifies a relative shift of the profile only.

---

[10] An estimate of the period is given in (9.33); for the case under consideration, $\varepsilon = 1$, the angular resonance frequency is still close to 1, and the maximum amplitude close to 2 (see Fig. 5.5).

[11] See Definition 9.16 and (9.31). In contrast to (7.8), (7.9) can be transformed to the general format with the ε-parameter. Actually, $\varepsilon \equiv \mu$, but the x-parameter will be scaled by a factor $\sqrt{\mu}$.

The extension proposed by Broer and Takens is appropriate for CMs because it allows careful consideration of the distribution of the bifurcation parameter over the cochlea.

In terms of (applied) mathematics it is straightforward to rewrite a 2nd order mechanical PDE into coupled first order equations, which are even more similar to the equations proposed as the Hopf bifurcation equation. However, it is noted that the background (bio-)physics is quite different. This is made clear in the following theorem:

Theorem 7.2. *The 2nd-order ODE's describing the uncoupled elements in a simple Van der Pol equation are—around the point of resonance—equivalent to the simple Hopf-bifurcation equation.*

Proof. The proof is most simply given for the Rayleigh equation, which is directly related to the van der Pol equation (see Definitions 9.12 and 9.13).
Let the simple version of the Rayleigh equation be given in by

$$\ddot{x} - \varepsilon\left(1 - \frac{1}{3}(\dot{x})^2\right)\dot{x} + x = s(t).$$

This notation follows the Definition 9.12 but adds the driving component $s(t)$.

The general interpretation is that the $\dot{x}$-parameter in this equation represents the velocity, whereas in the regular van der Pol equation the nonlinear damping term contains a deflection or displacement parameter. The above equation can be decomposed into the following coupled 1st-order equations[12]:

$$\dot{x} = -y, \tag{7.10}$$

$$\dot{y} = \varepsilon\left(y - \frac{1}{3}y^3\right) + x - s(t). \tag{7.11}$$

Application of the Broer–Takens transform: $u = \sqrt{\varepsilon}\,x$ and $v = \sqrt{\varepsilon}\,y$ yields a regular driven van der Pol–Rayleigh set:

$$\dot{u} = -v, \tag{7.12}$$

$$\dot{v} = \varepsilon v - \frac{1}{3}v^3 + u - s_1(t). \tag{7.13}$$

We compare this result to the hypothesized form by Eguíliuz et al. and find

$$\dot{z} = \mu z - |z|^2 z + \mathrm{i}\omega_0 z + F\mathrm{e}^{\mathrm{i}\omega t} \tag{7.14}$$

and recognize practically complete correspondence between the terms. A minor difference arises at the nonlinear term, but that difference is virtual because the van der Pol–Rayleigh set contains real elements only, and the above bifurcation equation

[12]Addition of this second dimension makes it possible to represent oscillatory solutions as 2-D rotations.

has been defined for a complex z; the additional factor of $\frac{1}{3}$ is a simple scaling factor. A more substantial difference appears within the next right-hand term, where v is compared to $\mathrm{i}\omega_0 z$. If the analysis was linear, then

$$u = -\int v\mathrm{d}t = -\frac{v}{\mathrm{i}\omega} = \frac{\mathrm{i}v}{\omega} \tag{7.15}$$

and we see an essential difference between a factor of $1/\omega$ in the one case, versus ω_0 in the other case. The minor part of the difference is the constant appearing in the numerator,[13] the more serious difference is the occurrence of the ω-parameter in the other denominator only. In other words, around

$$\omega \approx \omega_0 \approx 1$$

the similarity is complete; for deviating values the systems will behave differently.

□

7.3.2.1 Comparison of Activation Patterns

A direct comparison of the activation patterns generated by cochlea models following the van der Pol model and the Hopf bifurcation models is discussed briefly. It has been presented to the 11th MoH symposium in 2011. It indicates some practical differences, and also leads to a number of practical questions.

In terms of the set of differential equations, the differences between the Hopf bifurcation and the van der Pol model focus on the damping term.[14] For the van der Pol model, the damping at 0-velocity is negative—causing a limit cycle response if undriven. In the Hopf bifurcation, the damping is reduced to zero. In both simple[15] shapes, damping increases quadratically. In the van der Pol model version in Sect. 5.3.5.2, we actually used the Rayleigh version

$$d(x,t) = -d_1(x) + d_2(x)\, w_t^2 = d_1(x)\left(a\, v(x,t)^2 - 1\right). \tag{7.16}$$

A fundamental difference between the oscillators and the specified Hopf bifurcation is that all oscillators have negative damping at zero velocity. This makes the (close to) zero solution unstable. That is why the oscillators find an equilibrium at the limit cycle, quite far from zero, and at a point where the dependence of d on v is

[13] It is noted that Martignoli et al. (2007) modified or scaled (7.14) by putting ω_0 before all terms on the right hand side. This reduces the difference.

[14] The model version of the Hopf-bifurcation model by Duke and Jülicher (2003) was modified to obtain reasonable middle ear coupling and proper apical termination.

[15] The more general damping form of the van der Pol-model or Hopf-bifurcation does not have to follow the original parabola. However, it remains an even order function with a minimum at zero.

considerable. For a zero or positive location of the bifurcation point, the above effect does not play a role. Around $v = 0$ a linear solution of an undamped oscillator arises, which is determined by the initial conditions. For larger values of v, the damping nonlinearity follows the same profile as for the Rayleigh oscillator.

For the undriven van der Pol- and Rayleigh model, the limit-cycle amplitude of the response for the small ε-case is independent of the actual average damping as well as of the resonance frequency. It is completely determined by the bracket term in (7.16): $\sqrt{a}v(x,t) = 2$. This means that the factor a must be considered the scaling factor that determines the limit cycle amplitude value. For the Rayleigh equation, its dimension has to match the dimension of $v(x,t)$: the product $\sqrt{a}v(x,t)$ is dimensionless.[16]

Although there are good arguments to let the a-parameter depend on x, we have used constant values so far. These are meant to approximate the mid-frequency range (1–2 kHz) values reasonably. But choice of profiles can significantly affect the longitudinal interactions, and thereby modify the activation patterns. Note that the choice of a constant factor for the velocity implies a frequency dependent factor for the deflection and vice-versa. More specifically, if we want to use similar factors for the common van der Pol version with the quadratic displacement term and for the Rayleigh format with the velocity term we should introduce a scaling difference of the order $\omega_0^2(x) = (2\pi f_0(x))^2$. This follows from the linear approximation of the time derivative by $i\omega$. This scaling factor does depend on the frequency-place map, which demonstrates the problem. It it starts with a constant van der Pol-a-factor, then the Rayleigh-a-factor will depend on x, and vice-versa.

The $d_1(x)$ profile was usually optimized to obtain equal limit cycle velocity responses. Remember that its value is related to ε (5.72a): $d_1 = \varepsilon\sqrt{sm}$.

7.3.3 Summary: Questions and Conclusions

1. Mathematically, there is overwhelming similarity between the Hopf bifurcation model and the van der Pol model of cochlear nonlinearity. The bifurcation parameter μ simply equals the van der Pol ε-parameter. This similarity involves scaled representations of the underlying (bio-)physical system. The involved issues are listed separately. The relation between the two has been presented in several text books (e.g., Guckenheimer and Holmes 1983; Broer and Takens 2009). This early awareness makes it puzzling why the development of the recent bifurcation models made no connections to the Van der Pol applications.
2. In physics as well as in applied mathematics, the van der Pol oscillator is most often interpreted as a 2nd order mechanical system—even though its origin lies in electronics. Nonlinearity appears in the damping factor, where a parabolic factor multiplies the standard damping term. The parabolic factor $(1 - x^2)$ can

[16]For the van der Pol-equation the proper the link is to $w(x,t)$ rather than to $v(x,t)$.

either involve the displacement directly, in which case one speaks of a van der Pol-oscillator), or x represents the velocity, in which case one refers to the Rayleigh equation. In both cases, the nonlinear term is the damping term.

3. The nonlinearity leads to a compression of the output (r) dynamics compared to the input (s) range: for the standard traditional shapes with the parabolic factor the output-to-input relation converges to

$$r = s^{1/3}.$$

This is in line with early predictions of the nonlinear relation required to model CT amplitudes properly (e.g., Smoorenburg 1972a,b).

4. The Hopf bifurcation equation first proposed by Eguíliuz et al. but taken over by many others is usually interpreted as representing a 1st order system, usually one that involves a linear damping and a nonlinear stiffness only (7.2). This equation lacks one of the essential mass elements in the system. It would also imply that a "dead" cochlea cannot have a frequency map, which contradicts von Békésy's classical observations.
5. In connection with the previous point, the proposed bifurcation model poses the cochlear damping completely in viscous loss. None of it is employed for information processing. This is (bio-)physically incompatible with stable information processing systems.
6. Both the single (i.e., uncoupled) van der Pol oscillators and the proposed Hopf bifurcations miss the difference between the classical Helmholtz model (parallel 2nd order filters) and modern cochlea models, by ignoring the fluid inertia in the scalae. The essential role of this fluid coupling was introduced by and Wegel and Lane and rapidly accepted in subsequent models (except by Gold) who treated the coupling as an essential part of CMs (see Chaps. 1 and 3). The present study follows that line. As mentioned in Sect. 3.4, for a linear cochlea only, there always exists a reversible transform from the transmission line model to the quasi-independent filterbank model, but the linear transmission line can not be mapped to a bank of parallel 2nd order filters.

   For the nonlinear transmission line, a reversible filterbank equivalent simply does not exist. It misses the possibility to properly transmit internally generated distortion products and other byproducts, amongst which onset/offset phenomena.
7. Any mathematical analysis can benefit from proper scaling to dimensionless quantities. The interpretation of the results of the behavior of the mathematical bare bones back to the underlying system needs proper definition and book-keeping of all these steps. One of the strong tools to be used in applied physics is the check of agreement of dimensions in all equations. This has not been done consistently with the published data.
8. There is little doubt that around zero deflection of the hair bundle its stiffness shows a remarkable behavior, which is best understood as a nonlinear term in the mechanical stiffness of the hair cell. However, in terms of energy balance the stiffness plays only a marginal role. Increase of sensitivity of the cochlea

can be obtained by an effective reduction of energy loss, a loss caused by damping. Local energy supply, able to account for SOAEs, requires a local negative damping—equivalent to locally active behavior.

9. It may well be possible that the cochlea deals with both a stiffness and a damping nonlinearity. This was first suggested by Kim et al. (1973) in a model study; they mentioned that the effect of the Duffing-equation part would be to increase frequency selectivity at higher levels, and do not report further exploration. Furst and Goldstein (1982), Furst et al. (1988) and Furst and Lapid (1988) have explored the effect of nonlinear stiffness in more detail, and obtain proper results if the stiffness nonlinearity does not appear to dominate the behavior.

   Usually, we have treated the damping term as the prominent one, primarily because it helped us to check accuracy of the simulation. Strong stiffness nonlinearity appears to lead to much more high-frequency distortion than has been observed experimentally.
10. In Zweig's model, where the distinction between stiffness term and damping term is less clear, the two become inseparable. This is also true in our implementation of his models, both in the linear and nonlinear version. The most puzzling result is the stability of the DPOAE phase as a function of level (Fig. 6.10); other characteristics so far did not seem to generate different behavior.

7.4 General Summary and Discussion

We conclude by discussing the major objectives and achievements of this book, basically following the order of presentation. The structure is summarized in the following list that also highlights some intriguing open questions.

7.4.1 Time Domain Analysis

The major goal is to show that the analysis of the healthy, nonlinear cochlea requires an approach that differs essentially from linear systems analysis. We assume that essential nonlinearity of the healthy cochlea is a given, well established fact over the last decades. Proper tools to analyze such systems are not well known, primarily because they do not appear in undergraduate signal analysis textbooks. However, there exists also a wealth of advanced nonlinear studies, in fields such as Dynamical Systems and Chaos theory, as well as in certain specializations of fluid mechanics. One of the strongest approaches of these problems, after discovery that full solutions do not exist, is the application of sufficiently accurate numerical analysis in the time domain. This way essential properties such as causality and power balance are easily kept under control. The results presented, e.g., in Chap. 10, show that the proposed technique works properly.

The model properly produces the family of OAEs, which most alternative cochlea models are unable to do at this moment.

7.4.2 Cochlea Parameters and Experimental Data

Obviously, the cochlea itself is a coiled 3-D fluid mechanics structure, and the OC (CP) also is a rather complicated anatomical structure. It is more elaborate than described in common 3-D models that just assume slices orthogonal to the length direction, whereas in fact the structure contains several different longitudinal connections such as the pillar cells stair case, Hensen's cell's, OHC top–bottom connection, and angles in the TeM radial fiber structure. In terms of a proper mechanical description, this leaves us for the time being with too many free parameters. This becomes apparent in a comparison of a number of nonlinear cochlea models (e.g., in Sect. 6.3.1 and Chap. 10), where significant differences might have been expected, but only marginal differences occur. More precise data are required along many lines of the CMs, and those data that are available should be used carefully within the experimental context for which they have been designed and optimized. This must reduce the number of free parameters and leave us with a solvable problem.

One point that needs direct improvement is the middle ear coupling, from ear canal to oval window. As indicated in Sect. 5.3.2, we have been using a crude approximation, which allows us to get reasonable quantitative results within a small frequency range. For a better representation of the high-frequency behavior, and across frequency interactions (time effects) a more comprehensive coupling is necessary. An initial impetus was presented in Sect. 5.3.2.2 as a promising expansion, but the implementation still needs to be done.

7.4.3 Comparison Across Mammals

The analysis in the text may appear to focus on the human cochlea, but much of the analysis is directly applicable to other mammals. For several mammals, the required parameters from outer ear up to the cochlea are equally well known as for humans. Presently our numerical analysis also contains a set of parameters for the guinea pig. It is expandable with data from other mammals, at least insofar as they have similar specifications. For instance, many bat families have a differently specialized cochlear structure, where part of the cochlea appears to be optimized for echo location. The outer and middle ears are specifically shaped to optimize hearing over the relevant input range, i.e., limited by boundary conditions of shape and size.

The comparison across mammals is very useful to underpin the general validity of assumptions that have been made on the basis of a very limited number of species.

7.4.4 Nonlinear Damping and/or Nonlinear Stiffness?

Since the analysis by Kim et al. (1973), and several related studies, no serious doubts about the primary source of cochlear nonlinearity appears to have arisen. They suggested a predominant damping nonlinearity. The question was addressed again by Furst and Goldstein (1982). A relatively weak stiffness nonlinearity term tends to have a positive effect on the cochlear response; a strong stiffness nonlinearity generates significantly more high-frequency distortion than has been observed so far (cf. Chap. 4). Stiffness nonlinearity can induce oscillation (Duffing oscillator); however, stiffness is not suited for power dissipation, nor for power generation.

7.4.5 Active Emission Sources or Filtered Noise?

1. DPOAEs do *not* require active cochlear sources. They are also generated by completely passive compressive nonlinear systems.
2. SFOAEs do *not* require active cochlear sources. Same argument as for DPOAEs.
3. CEOAEs do *not* require active cochlear sources. The energy splatter over the broad spectral range implies that coupling mismatches are bound to occur at certain frequencies. These mismatches are sources for reflection.
4. The cochlear amplifier does not require active sources (but does not exclude it either). It requires a nonlinear transition from high-level damping to low-level damping, the latter being significantly less than the former. The reduction is correlated to an intact cochlea, in particular to intact OHCs.
5. SOAEs *do* require active cochlear sources. The alternative hypotheses of passive filtering of thermal noise or of correlated reflections may not have been ruled out completely, but have been excluded largely. The validity of the choice between filtered noise and oscillator, proposed by Bialek and Wit (1984), strongly depends on stability and length of the analyzed time intervals. But although the first analyses may have used marginally sufficient data, the conclusion was confirmed in later, more extensive controls.

One of the colleagues who kept us awake on the active–passive issue is JB Allen, who over and over posed the question about the proof for active behavior in the cochlea. The discussion reports from the Mechanics of Hearing workshops are illuminating, without giving all the answers. These reports illustrate that several other colleagues are actively involved.[17]

[17]Presently the most recent MoH workshop was the 11th, which was held Summer 2011 in Williamstown, Massachusetts.

7.5 The End

Back to the beginning of this chapter:
The answer to the still open question of the proper explanation of the masking asymmetry which was reported by Wegel and Lane in 1924, should be found within the next decade, or at least before 2024! We assume that time domain analysis of the nonlinear cochlea will play a role.

References

Andringa TJ (2002) Continuity preserving signal processing. PhD thesis, University of Groningen

Bialek W, Wit HP (1984) Quantum limits to oscillator stability: theory and experiments on acoustic emissions from the human ear. Phys Letters 104A:173–178

Broer HW, Takens F (2009) Dynamical Systems and Chaos. Appl. Math. Series 172, Springer, New York

Bronshtein IN, Semendyayev KA, Musiol G, Muehlig H (eds) (2003) Handbook of Mathematics. Springer, Berlin

Chan DK, Hudspeth AJ (2005) Mechanical responses of the organ of Corti to acoustic and electrical stimulation in vitro. Byophys J 89:4382–4395

Cooley JW, Tukey JW (1965) An algorithm for the machine computation of comples Fourier series. Math Comp 19:297–301

Duifhuis H (1973) Consequences of peripheral frequency selectivity for nonsimultaneous masking. J Acoust Soc Am 54:1471–1488

Duke T, Jülicher F (2003) Active traveling wave in the cochlea. Phys Rev Let 90(15):158,101–1–158,101–4

Eguíliuz VM, Ospeck M, Choe Y, Hudspeth AJ, Magnasco MO (2000) Essential nonlinearities in hearing. Phys Rev L 84:5232–5235

Epp B (2011) Processing strategies of the auditory system for improving the detection of masked signals. PhD thesis, Univeristy of Oldenburg (Carl von Ossietzky Universität), Oldenburg, Germany, superviser: Prof. Dr. Jesko L. Verhey

Epp B, Verhey JL, Mauermann M (2010) Modeling cochlear dynamics: Interrelation between cochlea mechanics and psychoacoustics. J Acoust Soc Am 128(4):1870–1883

Furst M, Goldstein JL (1982) A cochlear nonlinear transmission-line model compatible with combination tone psychophysics. J Acoust Soc Am 72:717–726

Furst M, Lapid M (1988) A cochlear model for acoustic emissions. J Acoust Soc Am 84:222–229

Furst M, Rabinowitz WM, Zurek PM (1988) Ear canal acoustic distortion at $2f_1 - f_2$ from human ears: Relation to other emissions and perceived combination tones. J Acoust Soc Am 84:215–221

Gold T (1948) Hearing. II. The physical basis of the action of the cochlea. Proc Royal Soc London, Series B, Biol Sc 135(881):492–498

Goldstein JL, Baer T, Kiang NYS (1971) A theoretical treatment of latency, group delay and tuning. Characteristics for auditory nerve responses to clicks and tones. In: Sachs MB (ed) The Physiology of the Auditory System, National Educational Consultants, Baltimore, pp 133–141

Guckenheimer J, Holmes P (1983) Nonlinear oscillations, dynamical systems and bifurcations of vector fields. Springer-Verlag, New York

Heideman MT, Johnson DH, Burrus CS (1984) Gauss and the history of the Fast Fourier Transform. IEEE ASSP Mag pp 14–21

Heinz MG, Colburn HS, Carney LH (2001a) Evaluating auditory performance limits: I. One-parameter discrimination using a computational model for the auditory nerve. Neural Computation 13:2273–2316

Heinz MG, Colburn HS, Carney LH (2001b) Evaluating auditory performance limits: II. One-parameter discrimination with random-level variation. Neural Computation 13:2317–2339

van Hengel PWJ (1996) Emissions from cochlear modelling. PhD thesis, University of Groningen

Hudspeth AJ (2008) Making an effort to listen: mechanical amplification in the ear. Neuron 59:530–545, review

Hudspeth AJ, Jülicher F, Martin P (2010) A critique of the critical cochlea: Hopf—a bifurcation—is better than none. J Neurophysiol 104:1219–1229, review

Kemp DT (2008) Otoacoustic emissions: Concepts and origins. In: Manley GA, Fay RR, Popper AN (eds) Active Processes and Otoacoustic Emissions, SHAR, vol 30, Springer, New York, chap 1, pp 1–38

Kern A, Stoop R (2003) Essential role of couplings between hearing nonlinearities. Phys Rev Lett 91(12):128,101–1–128,101–4

Kern A, Heid C, Steeb WH, Stoop N, Stoop R (2008) Biophysical parameter modification could overcome essential hearing gaps. PloS Comput Biol 4(4):e1000,161–1–4

Kim DO, Molnar CE, Pfeiffer RR (1973) A system of nonlinear differential equations modeling basilar membrane motion. J Acoust Soc Am 54:1517–1529

Magnasco MO (2003) A wave traveling over a Hopf instability shapes the cochlear tuning curve. Phys Rev Lett 90(5):058,101–1–058,101–4

Maoiléidigh DÓ, Jülicher F (2010) The interplay between active hair bundle motility and electric motility in the cochlea. J Acoust Soc Am 128(3):1175–1190

Martignoli S, van der Vijver JJ, Kern A, Uwate Y, Stoop R (2007) Analogue electronic cochlea with mammalian hearing characteristics. Appl Phys Lett 91(6):064,108–1–3

Martin P, Hudspeth AJ (1999) Active hair-bundle movements can amplify a hair cell's response to oscillatory mechanical stimuli. P N A S 96:14,306–14,311

Martin P, Hudspeth AJ (2001) Compressive nonlinearity in the hair bundle's active response to mechanical stimulation. P N A S 98:14,386–14,391

Martin P, Hudspeth AJ, Jülicher F (2001) Comparison of a hair bundle's spontaneous oscillations with a response to mechanical stimulation reveals the underlying active process. P N A S 98:14,380–14,385

Siebert WM (1965) Some implications of the stochastic behavior of primary auditory neurons. Kybernetik 2:206–215

Siebert WM (1968) Stimulus transformation in the peripheral auditory system. In: Kolers PA, Eden M (eds) Recognizing Patterns, MIT–Press, Cambridge MA, chap 4, pp 206–215

Siebert WM (1970) Frequency discrimination in the auditory system: place or periodicity mechanism? Proc IEEE 58:723–730

Smoorenburg GF (1972a) Audibility region of combination tones. J Acoust Soc Am 52:603–614

Smoorenburg GF (1972b) Combination tones and their origin. J Acoust Soc Am 52:615–632

Tinevez JY, Jülicher F, Martin P (2007) Unifying the various incarnations of active hair-bundle motility by the vertebrate hair cell. Biophys J 93:4053–4067

Tinevez JY, Martin P, Jülicher F (2009) Active hair-bundle motility by the vertabrate hair cell. In: Cooper NP, Kemp DT (eds) The Biophysics of Hearing, Mechanics of Hearing 10, 2008, World Scientific, Singapore, pp 415–424

Verhulst S (2010) Characterizing and modeling dynamic processes in the cochlea using otoacoustic emissions. PhD thesis, Technical University of Denmark

de Vries HL (1948) Minimum perceptible energy and Brownian motion in sensory processes. Nature 161:63

de Vries HL (1952) Brownian movement and the transmission of energy in the cochlea. J Acoust Soc Am 24:527–533

Wegel RL, Lane CE (1924) The auditory masking of one pure tone by another and its probable relation to the dynamics of the inner ear. Phys Rev 23:266–285

Part IV
Basic Linear and Nonlinear Tools and a PTPV Response Collection

Part IV consist of three additional, supporting Chapters:

Chapter 8 describes Basic Linear Tools, from a number of subdisciplines that are useful in linear CMs.

Chapter 9 describes some of the tools available for the analysis of nonlinear systems. It consists of two main parts.

- The first presents information about Bennett's series expansions for compressive nonlinearities, where Taylor's expansion is very inefficient.
- The second connects to the field of applied nonlinear dynamical systems and chaos.

Chapter 10 contains a collection of PTPV responses, computed for different parameters of the Groningen NL time domain model. It is an addition to Sect. 6.3.

Chapter 8
Basic Linear Tools

Abstract This chapter presents some elementary tools which are used to analyze physical properties of (some of the constituents of) the cochlea.

Acoustic signals depend strongly on the properties of the sound carrying medium. Vacuum, or an incompressible nonabsorbing wall, blocks propagation. Up to the middle ear the medium is air, with propagation properties that depend on temperature and humidity. The cochlea (and the connected vestibular system) are fluid filled with perilymph or endolymph, the acoustic properties of which approach those of water. In water, we can distinguish bulk waves and surface waves (in which the compressibility of the medium is essential), from fluid displacement waves (which depend on the mobility of the fluid but do not require compressibility). In the latter case, the mobility is primarily constrained by properties at the boundaries and fluid inertia.

The displacement waves will turn out to dominate acoustic fluid motion in the cochlea. This can be deduced from the comparison of properties of sound waves in water and the dimensions of the cochlear ducts. It turns out that the velocity of sound in the fluid is so high, ≈ 1.5 km/s—and consequently its wavelength is so large with respect to the cochlear dimensions—that for most mammals the cochlear fluid can be considered as incompressible. This implies that the velocity of sound within the cochlear fluid is practically infinite. For frequencies above about 10 kHz, this approximation needs to be refined. Then propagation begins to become noticeable. For larger cochleas than the human case, the frequency boundary will be shifted downward, for tinier cochleas it will be higher. Note that some bat species have a high-frequency limit of their echo location frequencies range that extends beyond 100 kHz, and in such cases the phase effects due to compressibility have to be taken into account.

H. Duifhuis, *Cochlear Mechanics: Introduction to a Time Domain Analysis of the Nonlinear Cochlea*, DOI 10.1007/978-1-4419-6117-4_8,

8.1 Fluid Mechanics

First, it should be noted that fluid mechanics treats a fluid (liquid or gas) as a continuous medium. This means that even the smallest volumes or points in a fluid(theoretically infinitely small) contain many molecules. This means that points in the fluid only can be very small compared to the sizes of the object under consideration, as long as they are be well above the molecular sizes.

The basic formulas for the motion of fluid particles are derived from three conservation laws: conservation of mass, conservation of mechanical momentum, and conservation of energy. The last case, obviously, must include all additional internal sources as well as all losses: primarily viscous and other damping losses and sensory power dissipation.

Conservation of fluid mass in a volume V_0 is expressed in an equation that equalizes the net influx of fluid to the associated increase of mass m (=density × volume), or $m + \Delta m = (\rho + \Delta\rho)V_0$:

$$\frac{\partial}{\partial t}\int \rho\,\mathrm{d}V = -\oint \rho\mathbf{v}\cdot\mathrm{d}A. \tag{8.1}$$

The left-hand integral, which represents the total mass, is a 3-D volume integral; the right-hand shows the related 2-D surface integral, and represents the net influx. Using a Green's formula to convert the surface integral to a volume integral leads (after some rewriting) to the equivalent differential equation:

$$\frac{\partial\rho}{\partial t} + \nabla\cdot(\rho\mathbf{v}) = 0, \tag{8.2a}$$

$$\text{or:}\quad \frac{\partial\rho}{\partial t} + \rho\nabla\cdot\mathbf{v} + \mathbf{v}\cdot\nabla\rho = 0. \tag{8.2b}$$

Note that the second term in (8.2a) is nonlinear, because both density and velocity are acoustic parameters. For many practical cases, the acoustic variations in pressure are small compared to the atmospheric pressure. Therefore, the nonlinear term is usually dropped, so that a linearized version of the equation can be applied. Writing the total density ρ as $\rho_{\text{tot}} = \rho_0 + \rho_a = \rho_0(1+\xi)$ where $\rho_a/\rho_0 = \xi \ll 1$, we obtain $\partial\rho/\partial t = \rho_0\partial\xi/\partial t$ and $\nabla\cdot(\rho\mathbf{v}) \approx \rho_0\nabla\cdot\mathbf{v}$ and (8.2) becomes, after division by ρ_0:

$$\frac{\partial\xi}{\partial t} + \nabla\cdot\mathbf{v} = 0. \tag{8.3}$$

For an incompressible fluid, the density is constant and the mass conservation condition simplifies to

$$\nabla\cdot\mathbf{v} = 0, \tag{8.4}$$

the form of which has been used most frequently in cochlear modeling.

8.1.1 Conservation of Momentum

Next we turn to the conservation of momentum, or the fluid equivalent of Newton's law: force = mass × acceleration. The fluid version, called Euler's equation, and presented by him in 1755, is of the form:

$$\frac{\partial \mathbf{v}}{\partial t} + (\mathbf{v} \cdot \nabla)\mathbf{v} = -\frac{\nabla p}{\rho}. \tag{8.5}$$

In this equation, $\mathbf{v}$ represents (as before) the particle velocity (at position x, y, z, and time t), p is the pressure at the same point, and ρ is the fluid density at that point. The difference with Newton's law occurs primarily in the second left-hand term, which represents internal motion within an accelerating mass.

If the fluid is in a gravitational field, then the gravitational acceleration term must be added:

$$\frac{\partial \mathbf{v}}{\partial t} + (\mathbf{v} \cdot \nabla)\mathbf{v} = -\frac{\nabla p}{\rho} + \mathbf{g}.$$

The internal friction in the fluid is characterized by the fluid viscosity. If we assume a constant dynamic viscosity η,[1] then adding the viscous loss terms leads to the well known Navier-Stokes equation:

$$\frac{\partial \mathbf{v}}{\partial t} + (\mathbf{v} \cdot \nabla)\mathbf{v} = -\frac{\nabla p}{\rho} + \frac{\eta}{\rho}\Delta\mathbf{v} + \frac{1}{\rho}\left(\zeta + \frac{1}{3}\eta\right)\nabla(\nabla \cdot \mathbf{v}) + \mathbf{g}. \tag{8.6}$$

The term ζ represents the so-called second or bulk viscosity. This equation becomes simpler if the compressibility of the fluid can be neglected (see 8.4), in which case $\nabla \cdot \mathbf{v} \approx 0$ so that the second viscous term can be dropped. Even with this simplification the equation remains quite complex, in particular because the quadratic term in $\mathbf{v}$ makes it nonlinear. Fortunately, for the normal sound parameters in air, the general consideration $(\mathbf{v} \cdot \nabla)\mathbf{v} \ll \partial\mathbf{v}/\partial t$ applies, and the second left-hand term, representing convective acceleration, can be dropped (cf. Sects. 3.6 and 5.3.6).

8.1.2 Conservation of Energy

The last conservation law regards conservation of energy, or balance of power.[2] Besides the mechanical potential and kinetic energy terms it covers the contributions of all additional local energy sinks and sources, except possible chemical terms

---

[1] In general the viscosity depends on temperature. Hence, a constant viscosity implies the assumption that temperature gradient effects are neglected.

[2] Obviously there is a fundamental difference between power P, expressed in W, and energy E, expressed in Ws or J. Balance of power ($\Sigma P = 0$) implies conservation of energy ($\Sigma E =$ constant).

which will be omitted from the present study. We will still be dealing with six different energy terms. These apply to a specific volume and its surface area. The analysis involves volume and surface integrals, or it is expressed as the analysis of variations of small parts, now expressed in differential equations. We follow the second approach and analyze energy conservation for a unit of mass. We follow the convention that thermodynamical quantities considered per unit of mass are written in lower case symbols and also called specific quantities, whereas the related regular quantities appear in standard upper case notation. However, the difference between regular volume and specific volume will be indicated by V and $1/\rho$, respectively.

The combination of the following terms determines if energy conservation holds. If the summed variation remains in balance, then conservation applies, else it is broken. The terms are:

1. Changes in kinetic energy and internal energy; a term relevant only for nonstationary movement, because changes are zero for stationary movement.
2. Net flux of kinetic and internal energy in the unit of mass.
3. Work done per unit of time by forces on the mass (e.g., gravity).
4. Work done per unit of time by surface forces (involving surface tension).
5. Net (external) heat flow into the unit of mass.
6. Effect of an internal heat source within a unit of mass.

We briefly consider some specific examples.

8.1.2.1 The Adiabatic Case

For the adiabatic case the sum of terms 1, 2, and 3 must be conserved. In this case, it includes 4 insofar as no irreversible losses are concerned. It assumes independent conservation in 5 and 6. The energy referred to in the first term is

$$\frac{1}{2}\rho v^2 + \rho u$$

containing a kinetic energy term and an internal energy term. The change in kinetic energy is

$$\frac{\partial}{\partial t}\left(\frac{1}{2}\rho v^2\right) = \frac{1}{2}v^2\frac{\partial \rho}{\partial t} + \rho \mathbf{v}\cdot\frac{\partial \mathbf{v}}{\partial t} \tag{8.7}$$

and the change in internal energy follows from one of the thermodynamic relations (in this case, the important "first law"):

$$\mathrm{d}u = T\mathrm{d}s - p\,\mathrm{d}\frac{1}{\rho} = T\mathrm{d}s + \frac{p}{\rho^2}\mathrm{d}\rho. \tag{8.8}$$

In a unit volume, the change in density in (8.7) will have thermodynamical consequences, in this case, characterized by the enthalpy equation

$$\mathrm{d}h = T\mathrm{d}s + (1/\rho)\mathrm{d}p. \tag{8.9}$$

In both s is the specific entropy. With some additional steps which involves fairly straightforward equations of continuity and motion, the kinetic energy change can be written as

$$\frac{\partial}{\partial t}\left(\frac{1}{2}\rho v^2\right) = -\frac{1}{2}v^2\nabla\cdot(\rho\mathbf{v}) - \rho\mathbf{v}\cdot\nabla\left(\frac{1}{2}v^2 + h\right) + \rho T\mathbf{v}\cdot\nabla s, \tag{8.10}$$

and the internal energy term becomes

$$\frac{\partial(\rho u)}{\partial t} = -h\nabla\cdot(\rho\mathbf{v}) - \rho T\mathbf{v}\cdot\nabla s. \tag{8.11}$$

Combining the two gives

$$\frac{\partial}{\partial t}\left(\frac{1}{2}\rho v^2 + \rho u\right) = -\left(\frac{1}{2}v^2 + h\right)\nabla\cdot(\rho\mathbf{v}) - \rho\mathbf{v}\cdot\nabla\left(\frac{1}{2}v^2 + h\right)$$

or

$$\frac{\partial}{\partial t}\left(\frac{1}{2}\rho v^2 + \rho u\right) = -\nabla\cdot\left[\rho\mathbf{v}\left(\frac{1}{2}v^2 + h\right)\right]. \tag{8.12}$$

(see Landau and Lifshitz (2004, Chap. I)).

8.1.2.2 The General Case

In the general case all terms are relevant, and all may interact. Important additions are terms 5 and 6 from the list, which deal with viscosity and with heat flux and/or sources.

In a fluid, viscosity is a major cause of internal dissipation. External dissipation can be due to loss in a moving structure or to a heat flux. Such effects lead to a change in *entropy* S, and are expressed, e.g., in the first law of thermodynamics, which states that the change in internal energy or *enthalpy* U is equal to the sum of the work done on a system and the heat supplied to the system. The formula describing the system with volume V is:

$$\mathrm{d}U = T\mathrm{d}S - p\mathrm{d}V, \tag{8.13}$$

where $T\mathrm{d}S$ represents the supplied heat and $p\mathrm{d}V$ the work done. This relation is frequently presented in so-called specific values, as shown in the above section in (8.8).

Dropping all terms that are already covered by mass and momentum conservation, one is left with the conservation of the heat function or enthalpy terms

$$H = U + pV \tag{8.14}$$

or

$$dH = TdS + Vdp, \tag{8.15}$$

which appeared above in specific units (8.9).

Rearranging and multiplying by ρ give per unit of mass, $\rho T ds = \rho dh - dp$ and taking the variation in time, one can obtain[3]

$$\rho T \frac{ds}{dt} = -\nabla \cdot \mathbf{q} + \Phi + (\sigma' \cdot \nabla) \cdot \mathbf{v}, \tag{8.16}$$

which states that the entropy at a "point" can be changed by a heat flux ($\nabla \cdot \mathbf{q}$), by a heat source (Φ), and by viscous effects in the last term.[4] The last term reflects the viscous effects through the stress tensor σ' which contains the dynamic viscosity and the relevant internal fluid velocities.

On a millisecond scale, the heat exchange between nearby points in air is rather slow. This implies that for the mid- and high-frequency ranges one may assume that sound waves are adiabatic. Only for very low frequencies can significant equalization take place. Thus, the general sound pressure wave is associated with an adiabatic temperature wave. We will show later that this temperature wave is rather small (compared to the environmental temperature of $\approx$300 K). In practice, it is so small that highly specialized detectors are required for its measurement.

When we apply fluid mechanics to the cochlea we will make plausible that additional practical considerations lead to simplifications of the basic formulas.

Proper combination of the continuity equations leads to the wave equation. This point is addressed in Sect. 8.2 after a first description of the applicable boundary conditions: the mechanical properties of the cochlea.

For a thorough discussion of the fundamentals, we refer to textbooks on fluid mechanics or fundamental acoustics, such as Morse and Feshbach (1953), Landau and Lifshitz (2004), Kinsler et al. (1982), and Beranek (1986), or one of the real classical studies as Lamb's "Hydrodynamics" (1895) and Rayleigh's "Theory of Sound" (1894, 1896).

8.2 Wave Equation, Traveling, and Standing Wave Solutions

Application of the equations for conservation of mass and momentum leads to the acoustic wave equations. We consider both waves in air and waves in a water-like medium. The first is necessary to understand common external sound and the properties of outer ear and middle ear, the last is relevant for the analysis of the

[3]Alternative formulae have been proposed, and differences may be very subtle, if they have been worked out at all.

[4]The vector $\mathbf{q}$ denotes the heat flux density due to thermal conduction. It is related to temperature variations across the fluid, and can be expressed as $\mathbf{q} = -\lambda \nabla T$, where λ in this context denotes the thermal conductivity (e.g. Landau and Lifshitz 2004, Chap. V).

Table 8.1 Acoustic properties of some liquids and gases. The top three rows (or less) give the common descriptions for each column. It is followed by the SI units for the data, and the last row in the header lists the standard symbols. Values are given in the body of the table. Humidity effects are computed after Davis (1992)

	Temp.	Density	Bulk modulus	Ratio of specific heats	Speed	Specific acoustic imped.	Coefficient of shear viscosity
	°C	kg/m^3	Pa		m/s	Pa s/m	Pa s
Liquids	T	ρ_0	B	γ	c	$Z_{sa} = \rho_0 c$	η
			$\times 10^9$			$\times 10^6$	
Water	20	998	2.18	1.004	1,481	1.48	0.001
Seawater	13	1,026	2.28	1.010	1,500	1.54	0.001
Ethanol	20	790			1,150	0.91	0.0012
Glycerin	20	1,260			1,980	1.11	1.2
Gases			Relative humidity %				
Air	0	1.293	0	1.402	331.3	428	0.000017
Air	10	1.247	0	1.402	337.3	420	0.000017
Air	20	1.204	0	1.402	343.3	413	0.000018
Air	20	1,199	50	1.402	344.0	412	
Air	20	1,194	100	1.402	344.6	411	
Air	30	1.165	0	1.402	349.1	406	0.000018
Oxygen	0	1.430		1.40	317.2	453	0.000020
Hydrogen	0	0.090		1.41	1,269.5	114	0.000009

cochlea. It is useful and necessary to separate the two media because they allow very different simplification steps in the analysis. Obvious differences are specific mass, or density, and compressibility, whereas air is highly compressible, water is almost incompressible. Some medium parameters are listed in Table 8.1.

Oscillatory motions in a compressible fluid are called sound waves as long as the amplitudes remain small enough. An additional constraint regards the oscillation frequencies: sound should be within the audio frequency range, or between 20 Hz and 20 kHz. Ultrasound applies to frequencies above that range, and infra-sound to lower frequencies.

Since the oscillation amplitudes are small, the velocities also remain small. If that is the case, then the quadratic term in Euler's equation (8.5) can be neglected. To quantify this statement, consider the sound pressure of a loud sound, for example, a sinusoidal signal, or pure tone, at 1 kHz presented at 100 dB sound pressure level (SPL). This is 100 dB above the standard reference SPL value $p_0 = 20\,\mu\text{Pa}$ which means that the effective amplitude value[5] at the 100 dB level corresponds to $p = 10^5 \times 20.10^{-6} = 2\,\text{Pa}$. The atmospheric pressure of 1 atmosphere (historical

[5]For the effective amplitude we take the r.m.s.-value of the oscillation. For a sinusoid this is $\sqrt{2}$ below the peak value of the oscillation.

units) corresponds to 101.3250×10^3 Pa, so that the pressure amplitude of the 100 dB tone is still $\approx$50,000 times smaller than the atmospheric pressure, and also much smaller than the commonly reported atmospheric pressure variations.

The velocity amplitude follows from the relation

$$Z_{\rm sa} = \frac{p}{v} \qquad \text{Pa s/m}. \tag{8.17}$$

as will be shown below. Values of $Z_{\rm sa}$ are listed in Table 8.1. Application of these values gives $v_{\rm air} = 2/340 \approx 5.9 \times 10^{-3}$ m/s, and $v_{\rm water} = 2/1.5 \times 10^6 \approx 1.3 \times 10^{-6}$ m/s at the 100 dB SPL. These results also imply that density variations due to the acoustic oscillations are small.

It is useful to reconsider pressure and density, and separate the average values from the acoustic variations. Let the total pressure, which is the pressure used so far, be written as $p_{\rm tot}$, and the average atmospheric pressure (averaged over a couple of minutes) be $p_{\rm atm}$. Then the acoustic pressure $p_{\rm a}$ is defined as the difference between total and atmospheric pressure:

$$p_{\rm a} = p_{\rm tot} - p_{\rm atm}. \tag{8.18}$$

In a similar way, we can separate an acoustic density fluctuation from the total density by subtracting the average value ρ_0:

$$\rho_{\rm a} = \rho_{\rm tot} - \rho_0 \tag{8.19}$$

Note that variations in $p_{\rm tot}$ and $\rho_{\rm tot}$ are identical to the variations in the acoustic values $p_{\rm a}$ and $\rho_{\rm a}$. Insofar as the effects of the averages are prominent, the total values can be replaced by the constant averages. This modifies Euler's equation (8.5) to

$$\frac{\partial \mathbf{v}}{\partial t} + \frac{\nabla p_{\rm a}}{\rho_0} = 0 \tag{8.20}$$

and the mass conservation equation (8.2) to

$$\frac{\partial \rho_{\rm a}}{\partial t} + \rho_0 \nabla \cdot \mathbf{v} = 0. \tag{8.21}$$

Equations (8.21) and (8.20) contain the unknown functions $\mathbf{v}$, $p_{\rm a}$, and $\rho_{\rm a}$. By combining the two equations, we can eliminate one, and moreover the variations in pressure and density are closely linked, especially for small variations. For small adiabatic fluid motion, we have

$$p_{\rm tot} \times V^\gamma \propto p_{\rm tot} \times \left(\frac{1}{\rho_{\rm tot}}\right)^\gamma = \text{constant}. \tag{8.22}$$

Taking the derivative, we obtain:

$$\partial p_{\rm tot} \left(\frac{1}{\rho_{\rm tot}}\right)^\gamma + p_{\rm tot}\, \partial \left(\frac{1}{\rho_{\rm tot}}\right)^\gamma = 0 \tag{8.23}$$

or

$$\partial p_{\text{tot}} \left(\frac{1}{\rho_{\text{tot}}}\right)^{\gamma} - p_{\text{tot}}\, \gamma\, \partial \rho_{\text{tot}} \left(\frac{1}{\rho_{\text{tot}}}\right)^{\gamma+1} = 0. \tag{8.24}$$

Applying:

$$\partial p_{\text{tot}} = \partial p_{\text{a}}, \quad \partial \rho_{\text{tot}} = \partial \rho_{\text{a}}, \quad p_{\text{tot}} \approx p_{\text{atm}} \quad \text{and} \quad \rho_{\text{tot}} \approx \rho_0 \tag{8.25}$$

yields, after division by the common term $(1/\rho_{\text{tot}})^{\gamma}$, the small amplitude approximation:

$$\partial p_{\text{a}} = \partial \rho_{\text{a}} \left(\frac{\gamma\, p_{\text{atm}}}{\rho_0}\right). \tag{8.26}$$

This additional constraint allows us to combine (8.20) and (8.21) to one equation. Taking the time derivative of (8.21) and the spatial gradient of (8.20), we arrive at:

$$\begin{aligned} \frac{\partial^2 \rho_{\text{a}}}{\partial t^2} + \rho_0 \frac{\partial \nabla \cdot \mathbf{v}}{\partial t} &= \frac{\rho_0}{\gamma\, p_{\text{atm}}} \frac{\partial^2 p_{\text{a}}}{\partial t^2} + \rho_0\, \nabla \cdot \frac{\partial \mathbf{v}}{\partial t} \\ &= \frac{\rho_0}{\gamma\, p_{\text{atm}}} \frac{\partial^2 p_{\text{a}}}{\partial t^2} - \nabla \cdot \nabla\, p_{\text{a}} = 0 \end{aligned} \tag{8.27}$$

or

$$\frac{\partial^2 p_{\text{a}}}{\partial t^2} - c^2 \Delta p_{\text{a}} = 0 \qquad \text{where} \qquad c^2 = \frac{\gamma\, p_{\text{atm}}}{\rho_0}. \tag{8.28}$$

This is the acoustic wave equation, which is applicable under the conditions discussed, such as small amplitude, and adiabatic waves.

Basically, the same equation can be derived for all related primary acoustic parameters: particle amplitude and velocity, and temperature. This is clearly true for the scalar quantities p_{a} and T_{a}, but for the vectors the analysis is a bit more complicated. That is a good reason to introduce an additional scalar parameter, the potential field $\boldsymbol{\phi}_{\text{a}}(x, y, z, t)$, the gradient of which will be equal to the particle velocity. The general wave equation is then

$$\frac{\partial^2 \boldsymbol{\phi}_{\text{a}}}{\partial t^2} - c^2 \Delta \boldsymbol{\phi}_{\text{a}} = 0. \tag{8.29}$$

The solutions of the wave equation are acoustic wave functions of the form:

$$\phi_{\text{a}}(\mathbf{x}, t) = f_1(\mathbf{x} - \mathbf{c}\, t) + f_2(\mathbf{x} + \mathbf{c}\, t). \tag{8.30}$$

The two components, f_1 and f_2, are practically independent as long as there are no precise boundary conditions and/or initial conditions: the only relation is that f_1 and f_2 travel in opposite directions. A sound wave requires a source, and the boundary conditions play an essential role in shaping the details of the wave. The constant c^2

in (8.29) reflects the velocity of sound, i.e., the speed with which the wave travels through the medium. Note that, although the speed of sound is a scalar quantity, it occurs as a vector in (8.30), thereby indicating the propagation direction. The magnitude depends on the specific heats of the medium $\gamma = c_p/c_v$, the density ρ_0 kg/m^3 of the medium and the atmospheric pressure p_{atm} Pa. Note also that the suggestion that might arise from the formula, viz. that c is proportional to the square root of atmospheric pressure, is false. This is so because ρ_0 also depends on p_{atm}. The term p_{atm}/ρ_0 equals rT, where r is the gas constant for air (= 287.068 J/kg/K) so that the dependency is completely characterized by the dependency on T. Hence, for the adiabatic condition a net dependency on p does remain, whereas for the isothermal condition the relation on p disappears completely.

From this point on, we will drop the use of the index "a" which so far specified the acoustic signals: because we will only deal with acoustic signals anyway. For one dimensional plane waves, a vectorial analysis is superfluous. Hence, we limit the analysis to behavior along that dimension. Choosing the x-direction as the direction of propagation of the acoustic wave, we obtain the following simplifications for the wave equation and its solution (particle velocity and amplitude are described completely by the scalar value of the x-components):

$$\frac{\partial^2\phi(x)}{\partial t^2} - c^2\Delta\phi(x) = 0. \tag{8.31}$$

The solutions of the wave equation are acoustic wave functions of the form:

$$\phi(x,t) = f_1(x - c\,t) + f_2(x + c\,t), \tag{8.32}$$

where f_1 and f_2 again are arbitrary functions in x and t, and where c is the magnitude of the speed of sound and now a scalar. Since audio signals by definition have a frequency spectrum that must be within the 20 Hz–20 kHz range, they can be considered as composed of elementary spectral components. Therefore, it is also useful to consider the solution for a single sinusoidal component of angular frequency ω. Such a component is also called a monochromatic wave, and can be written as:

$$\phi_1(x,t) = \Re\{\phi_1(0)\,\mathrm{e}^{-\mathrm{i}[\omega(t-x/c)+\varphi_1]}\} \tag{8.33a}$$

$$\phi_2(x,t) = \Re\{\phi_2(0)\,\mathrm{e}^{-\mathrm{i}[\omega(t+x/c)+\varphi_2]}\}. \tag{8.33b}$$

As before, ϕ_1 and ϕ_2 go in opposite directions. The complex notion of the oscillation will be used because it is parsimonious. The physical relevant part of the solution is always the real part, as indicated explicitly here (at times that specification may be dropped). The constant phase angles φ_1 and φ_2 are as yet arbitrary constants, depending on initial and boundary conditions. Let us check that ϕ_1 is a solution of (8.31). We obtain (for the complex version):

$$\frac{\partial^2 \phi_1(x,t)}{\partial t^2} = (-\mathrm{i}\omega)^2\, \phi_1(0)\, \mathrm{e}^{-\mathrm{i}[\omega(t-x/c)+\varphi_1]} = -\omega^2 \phi_1(0)\, \mathrm{e}^{-\mathrm{i}[\omega(t-x/c)+\varphi_1]},$$

$$\frac{\partial^2 \phi_1(x,t)}{\partial x^2} = \left(\frac{-\mathrm{i}\omega}{c}\right)^2 \phi_1(0)\, \mathrm{e}^{-\mathrm{i}[\omega(t-x/c)+\varphi_1]} = -\left(\frac{\omega}{c}\right)^2 \phi_1(0)\, \mathrm{e}^{-\mathrm{i}[\omega(t-x/c)+\varphi_1]}$$

and this meets the condition of (8.31):

$$\frac{\partial^2 \phi_1(x,t)}{\partial t^2} - c^2\, \frac{\partial^2 \phi_1(x,t)}{\partial x^2} = 0$$

because the Laplace operator is now equal to the second derivative in x. This is independent of the sign of $\mathrm{i}\omega t$ or $\mathrm{i}\omega x/c$, so that $\phi_2(x,t)$ also provides a valid solution.

It should be obvious that for a given frequency ω the time period of the wave is equal to $2\pi/\omega$. Similar spatial parameters are the wavenumber $k = \omega/c$ and the wavelength $\lambda = 2\pi/k$, which depend both on frequency and on the speed of sound in the medium. Using the wavenumber, (8.33a) can be written as

$$\phi_1(x,t) = \Re\{\phi_1(0)\, \mathrm{e}^{-\mathrm{i}(\omega t - kx + \varphi_1)}\} \tag{8.34}$$

The wavenumber is also defined for the 3-D case. Replacement of the product kx by the vectorial in-product $\mathbf{k} \cdot \mathbf{x}$ gives a plane wave solution for that case.

The wave equations are identical for pressure, temperature, and vector potential, from which vector solutions for particle velocity and displacement can be derived. For a plane wave, the relation between velocity and pressure is even more straightforward, because they are related through the specific acoustic impedance Z_{sa}.

Starting with the plane wave solution for acoustic pressure

$$p(x,t) = A\, \cos(\omega t - kx + \varphi),$$

we obtain for its spatial derivative

$$\frac{\partial p(x,t)}{\partial x} = kA\, \sin(\omega t - kx + \varphi).$$

Application of (8.20) gives as solution for $\partial v/\partial t$:

$$\frac{\partial v(x,t)}{\partial t} = -\,\rho_0\, kA\, \sin(\omega t - kx + \varphi)$$

and integration over time (discarding the possible DC term) gives for the velocity

$$v(x,t) = \frac{1}{\omega}\rho_0 kA\cos(\omega t - kx + \varphi) = \rho_0 cA\cos(\omega t - kx + \varphi).$$

The displacement d is obtained by another integration over time:

$$d(x,t) = \frac{\rho_0 c}{\omega} A\sin(\omega t - kx + \varphi).$$

These equations provide the proof that for plane waves Z_{sa}, which was defined in (8.17), is equal to $\rho_0 c$.

Finally, the acoustic temperature wave is derived from the adiabatic gas law $p^{1-\gamma}.T^{\gamma}$ = constant. This implies that

$$\delta T = \frac{T}{p_{atm}} \cdot \frac{\gamma - 1}{\gamma}\delta p,$$

which gives a factor $\delta T/\delta p = 0.00083$ K/Pa at 101.325 kPa and 20 °C.

A NOTE ON TRAVELING AND STANDING WAVES. The general monochromatic solution of the 1-D wave equation for the acoustic pressure (8.31 where p replaces ϕ) is in complex form:

$$p(x,t) = p_1(x,t) + p_2(x,t) = p_{0,1}\,\mathrm{e}^{-\mathrm{i}(\omega t - kx + \varphi_1)} + p_{0,2}\,\mathrm{e}^{-\mathrm{i}(\omega t + kx + \varphi_2)}.$$

The complex form is used because it simplifies the following analysis. The order of the wave equation (2nd in this case) determines the number of solutions that can exist independently, unless they are constrained by the boundary conditions. The two solutions represent waves that travel in opposite directions: $p_1(x,t)$ in the $+x$-direction, $p_2(x,t)$ in the $-x$-direction. A special case arises when the (peak-) amplitudes of the two waves are equal, i.e., when $p_{0,1} = p_{0,2}$. Let us denote this amplitude by A. Now we use common trigonometrics to analyze the interaction of the waves, to obtain:

$$\begin{aligned} p(x,t) &= A\mathrm{e}^{-\mathrm{i}\omega t}\left(\mathrm{e}^{\mathrm{i}kx - i\varphi_1} + \mathrm{e}^{-\mathrm{i}kx - i\varphi_2}\right) \\ &= A\mathrm{e}^{-\mathrm{i}\frac{\varphi_1+\varphi_2}{2}}\,\mathrm{e}^{-\mathrm{i}\omega t}\left(\mathrm{e}^{\mathrm{i}kx}\mathrm{e}^{-\mathrm{i}\frac{\varphi_1-\varphi_2}{2}} + \mathrm{e}^{-\mathrm{i}kx}\mathrm{e}^{+i\frac{\varphi_1-\varphi_2}{2}}\right) \\ &= 2A\,\mathrm{e}^{-\mathrm{i}\varphi_{av}}\,\mathrm{e}^{-\mathrm{i}\omega t}\cos(kx - \varphi_{diff}) \end{aligned}$$

Note that the time and place solutions are no longer coupled, so there is no longer a traveling wave. The result is now termed *standing wave* because all x-points oscillate in phase, with a cosine-shaped spatial envelope. The location of zeros in the envelope is affected by the difference of the phases of $p_1(x,t)$ and $p_2(x,t)$, the phase of the oscillation or carrier is affected by the average of $p_1(x,t)$ and $p_2(x,t)$.

Also note that any sum of two such traveling waves results in the sum of a standing wave and one remaining traveling wave. Hence, the observer will never perceive two traveling waves moving in opposite directions, but the sum of a standing wave and one traveling wave in the direction of the strongest of the constituents. Although mathematically a standing wave is indistinguishable from the sum of two waves with equal amplitude traveling in opposite directions, physically the term traveling wave has become meaningless as long as it describes (part of) a standing wave. This conclusion applies in particular to monochromatic waves which are analyzed using spectral amplitude analysis. Temporal analysis of click responses can still retrace underlying propagation behavior. With the proper tools, these traces may also be detectable in onset and offset responses of otherwise narrow band signals.

Finally, note that so far this discussion concerned only the behavior of waves traveling in an acoustic medium. As yet it is independent of any boundary conditions or initial conditions.

Another straightforward solution of the general 3-D wave equation is obtained for the sound field generated by a perfect spherical sound source. Note that an acoustic point source is physically unrealizable, but a spherical source behaves similarly (outside of the sphere). We leave the solution to the reader. It has been worked out in most acoustic textbooks. The reader should verify that at large distances the plain wave and the wave from the "point" source become indistinguishable. Near the source, however, you might find some unexpected behavior. Note that this acoustic problem, because of its specific symmetry is most efficiently solved in a spherical coordinate system.

8.2.1 Wave Velocity and Delay

The velocity of a sound wave, or the speed of sound, c in an acoustic medium that has been derived in Sect. 8.2 more precisely is called the *phase speed*, c_{ph}. In the solutions to the general wave equation (8.29), this speed relates the x and t dependance, as in (8.30). Let us consider the forward travelling wave in more detail (the discussion of the backward travelling wave is analogous and is left to the reader). Without too much loss of generality we consider the 1-D case (8.32):

$$\phi(x,t) = f(x - c_{\text{ph}}\,t). \tag{8.35}$$

For a simple, uniform, constant, and linear medium, c_{ph} is constant, and the shape of the wave $\phi(x,t)$ does never change. However, in reality the variables that define c_{ph}, as in (8.28) for gasses, do depend on other physical parameters of the fluid, like temperature, humidity, and viscosity. Some of these parameters are also frequency dependent: e.g., at high frequencies heat exchange between maxima and minima in the pressure wave is negligible, yielding an adiabatic behavior, but at sufficiently low frequencies the heat exchange will be fast enough to generate isothermic behavior, and γ drops from the equation.

When waves propagate through different media or through different mechanical structures, one has to analyze the relevant physical properties of these media or structures. For now, we will focus on the relation between velocity and frequency.

The narrowband version of (8.35), or the wave component at angular frequency ω, can be written as

$$\phi(x,t;\omega) = f\left(\frac{-1}{k}(\omega\, t - k\, x)\right) = g(\omega\, t - k\, x), \tag{8.36}$$

where ω and k, the wavenumber, replace the velocity c. Now we have two variables which are directly linked to c, but note that this formulation does not imply that all frequency dependence is accounted for by ω, because the wavenumber does not have to be a constant.

The formulation in (8.36) is useful as a basis for the classical definition of the group velocity or group speed c_{gr} and the phase velocity or phase speed c_{ph} of the wave:

$$c_{\text{ph}} = \frac{\omega}{k} = \frac{\omega\,\lambda}{2\,\pi}, \tag{8.37}$$

$$c_{\text{gr}} = \frac{\mathrm{d}\omega}{\mathrm{d}k} = c_{\text{ph}} + k\,\frac{\mathrm{d}c_{\text{ph}}}{\mathrm{d}k} = c_{\text{ph}} - \lambda\,\frac{\mathrm{d}c_{\text{ph}}}{\mathrm{d}\lambda}, \tag{8.38}$$

where $\lambda = 2\,\pi/k$. As long as the relation between ω and k is linear, group and phase velocities will be equal, but whenever the relation between ω and k becomes more complex, differences between the two will become apparent. In such a case, the shape of a wave will change during propagation. This phenomenon is known as dispersion.

If a wave propagates through a mechanical structure, then the behavior through this structure is similar to the response of a more general system to its input, as long as that system is characterized by the proper system parameters. Assume that the wave at $x = x_1$ defines the input to the system, and let the transfer function of the system by $H(\omega)$, associated with an impulse response $h(t)$. Now the travel time through the system is usually described by a delay, and one deals with a group delay and a phase delay, which are related to the group speed an phase speed. In general, $H(\omega)$ is a complex quantity, which can be decomposed into a real amplitude part and a phase part, as

$$H(\omega) = A(\omega)\,\mathrm{e}^{-\mathrm{i}\varphi(\omega)}. \tag{8.39}$$

The amplitude term does not affect the delay: all delay generated by the system originates from the phase term. For the system $H(\omega)$ three delays, viz., the phase delay τ_{ph}, the group delay τ_{gr}, and the signal-front delay τ_{fr} are defined by:

$$\tau_{ph} = \frac{\varphi(\omega)}{\omega} \qquad \tau_{gr} = \frac{d\varphi(\omega)}{d\omega} \qquad \tau_{fr} = \lim_{\omega\to\infty} \frac{\varphi(\omega)}{\omega} \tag{8.40}$$

(e.g. Papoulis 1962, p. 134). In principle, the group and phase delays are functions of ω, but the signal front delay is not: it rather is the asymptotic behavior for large ω. Its value might be of particular interest for response delays measured in the ear canal.[6]

Although there are striking similarities between the definitions of phase and group speed an the one hand, and phase and group delay on the other, there also some important differences. Phase and group velocities follow from the setup of the wave equation. They rely on basic differential equations, without restrictions concerning linearity, or constancy.

However, a system which is described by a complex transfer function implicitly assumes linearity of that system, and it assumes one-way traffic through the system, from input to output. This is very useful for the analysis of static responses of linear systems, but it requires a more careful application for the analysis of time domain responses, as is required for click or impulse responses and for stimulus onset and offset responses. Moreover, the definition does not cover the behavior of nonlinear systems (or system elements), thereby leading to unpredictable behavior for such cases. Unfortunately as well as interestingly, the intact auditory system is distinctly nonlinear over most of its input range. Therefore, the use of quantities that have been defined in linear systems analysis is suspect, as long as a full proof of validation of that use is lacking. This also applies to the rather common circular reasoning, which, starting from a linear approximation attempts to converge to *the* solution by making small corrections. Such an approach can easily end at *a* suboptimal solution.

In the current context, this is directly relevant to the signal analysis definitions for time delays τ_{ph}, τ_{gr}, and τ_{fr}. These quantities are not defined for nonlinear systems. Nevertheless, e.g., the group delay is commonly used, because the quantity $d\,\varphi(\omega)/d\,\omega$ is relatively easily measurable.

An indicator of the problem, the presence of a nonlinearity, is the dependence of τ on level.

Currently, the only valid alternative to the frequency domain approach is to just measure delay in the time domain and specify the onset thresholds in proper (in relation to the attained level of accuracy) detail. Two techniques that have been developed along this line are the ABCD technique, developed by Brass and Kemp (1991) and the primary tone phase variation (PTPV) technique, introduced by Whitehead et al. (1996). Both rely on the use of phase effects, in particular the effect of cancellation that occurs if tones of opposite phases, or systematically

[6]The application of the signal-front delay in the analysis of peripheral auditory processing has been promoted by Ruggero (1980), who used it in a discussion of common errors in estimates of BM travel times.

rotating phases, are summed. They are discussed in some detail in Chap. 6, the ABCD technique in Sect. 6.2.1, and the PTPV technique in Sect. 6.3.1.

In other words, although the signal front delay can be of utmost relevance, experimentally one will have to deal with the best achievable time domain alternative. In particular, one should be aware of the fact that the response slope can be steeper than the stimulus onset slope, as well as equal or even shallower. And the onset delay can be anywhere between shorter and longer than the 50% delay.[7]

Apparently, this is an example of a point where the analysis of nonlinear systems is open to improvements.

8.3 Additional Acoustic Standards

In Sect. 8.1 we introduced the primary acoustic variables, pressure, velocity, deflection and temperature. We mentioned in passing the SPL, and the reference pressure at 0 dB SPL, which is 20 μPa. This value is specified in an international standard: ISO-1683 and also in ANSI-S1.1. The general definition of sound pressure, or effective sound pressure, is the root-mean-square instantaneous sound pressure at a point, during a given time interval. A related definition concerns the peak sound pressure: greatest absolute instantaneous sound pressure within a specified time interval. Formally, there is an open end in the specification of duration and time interval. This does not lead to any problems with periodic sounds, where a time window that covers n periods should give a precise result independent of the number of periods. For very short signals, caution is advised. For nonperiodic signals, such as noises, the result will be a stochastic quantity, with a standard deviation depending on the bandwidth of the noise and of the duration of the applied time interval. For a sinusoidal signal, the effective sound pressure is exactly equal to the peak value divided by $\sqrt{2}$, because the average of a squared sinusoid (over 1 cycle) with amplitude 1 is 0.5.

A similar problem can arise for the secondary parameters sound intensity (I) (also termed: sound-energy flux density or sound-power density) and sound power (P or W). The latter is in particular relevant for sound sources (or dissipaters). Intensity is the average rate of sound energy transmitted in a specified direction at a point through a unit area normal to this direction at this point. The unit is W/m^2. For a plane wave, the intensity in the direction of the wave propagation is given by

$$I = \frac{1}{T}\int_0^T pv\,\mathrm{d}t, \tag{8.41}$$

where T is an averaging window duration, which should be long compared to the reciprocal of the lowest frequency of interest; p is the instantaneous sound pressure;

[7]This is the delay between the 50% point in the probe onset envelope to the 50% point in the response onset envelope. Unfortunately, in practice the shape of the response onset can be rather different, in which case this estimate is somewhat arbitrary.

v particle velocity in the specified direction; t, finally, is the running time. For plane waves p and v are in phase, and using the specific acoustic impedance, $Z_{sa} = \rho c$ the result can be expressed as

$$I = \frac{p^2}{Z_{sa}} = v^2 Z_{sa}. \tag{8.42}$$

The large dynamic range of the auditory amplitude scale, in combination with the observation that perceived strength appears to vary more closely with the amplitude ratio than with absolute values, have led to the introduction of the logarithmic measure of the strength, now termed sound level, a dimensionless quantity, but nevertheless expressed in dB (decibel). Sound levels are defined for all acoustic variables, but we only give the two most important ones, the SPL L_p and the sound intensity level L_I. These are defined as

$$L_p = 20 \times \log_{10} \frac{p}{p_0} \qquad \text{where } p_0 = 20\ \mu\text{Pa}, \tag{8.43}$$

$$L_I = 10 \times \log_{10} \frac{I}{I_0} \qquad \text{where } I_0 = 1\ \text{pW/m}^2. \tag{8.44}$$

Historically, the decibel was based on an older definition of the Bel, and 1 Bel was 10 dB. This explains the factor 10 in the definition of L_I. The additional factor 2 for the SPL formula has its root in the fact that for a homogeneous plane wave the values L_p and L_I are almost equal, because (8.43) implies that

$$L_p = 10 \times \log_{10} \left(\frac{p}{p_0}\right)^2 = 10 \times \log_{10} \frac{p^2/Z_{sa}}{p_0^2/Z_{sa}} \tag{8.45a}$$

$$= 10 \times \log_{10} \frac{I}{p_0^2/Z_{sa}} \tag{8.45b}$$

and

$$I_0 = 10^{-12} \approx \frac{400.10^{-12}}{412} = \frac{p_0^2}{Z_{sa}} \qquad \text{W/m}^2. \tag{8.46}$$

The difference for the presented value for Z_{sa} is only 0,13 dB, or negligible in common practice. In other words, the decibel values of L_p and L_I are indiscernible (at least for plane waves).

Three different impedance definitions are commonly used, and standardized, in acoustics, and listed below:

- *Specific acoustic impedance*:

$$Z_{sa} = \frac{p}{v} = \frac{\text{local pressure}}{\text{resulting particle velocity}} \qquad \text{Pa s/m} \quad [\text{ML}^{-2}\text{T}^{-1}]. \tag{8.47}$$

- *Acoustic impedance*:

$$Z_{\mathrm{a}} = \frac{p}{U} = \frac{\text{local pressure}}{\text{resulting volume velocity}} \qquad \mathrm{Pa\,s/m^3} \quad [\mathrm{ML^{-4}T^{-1}}]. \quad (8.48)$$

- *Mechanical impedance*:

$$Z_{\mathrm{m}} = \frac{F}{v} = \frac{\text{local force}}{\text{resulting velocity}} \qquad \mathrm{N\,s/m} \quad [\mathrm{ML^{-2}T^{-1}}]. \quad (8.49)$$

The specific acoustic impedance has been used above in a few other definitions.

These definitions are similar to the probably more common definition of electrical impedance:

$$Z_{\mathrm{e}} = \frac{e}{i} = \frac{\text{voltage}}{\text{current}} \qquad \Omega \qquad [\mathrm{M\,L^2T^{-3}I^{-2}}]. \quad (8.50)$$

8.4 Vibrations in Strings, Bars, and Plates

Although the cochlea is distinctly fluid filled, certain elements of the embedded cochlear partition, in particular of BM and organ of Corti, can have, and show, properties that are more closely related to those of solid structures. Therefore we will give a short description of acoustic properties of these structures.

The best known vibrations of strings are the transversal motions that can be elicited in a string that is under longitudinal stress. In practice, it is more convenient to consider the longitudinal force than the stress, because then we can omit the cross sectional specifications. We assume that the string is very thin in comparison to its length. The longitudinal force in the string is called the tension T. The string can only be kept under a constant T if the ends are clamped, as over the bridges or nuts in a violin or piano. If T is large enough, then effects of gravity on the string motion can be neglected. Otherwise $\mathbf{g}$ and the orientation of the string with respect to the direction of $\mathbf{g}$ have to be taken into consideration. Using the parameter x along the string length direction and y for the transversal direction, it is a good exercise to derive the wave equation that describes the behavior of the string for small amplitudes. Two additional string parameters are required to obtain practical solutions, viz. its length l, and its mass per unit of length m_l. You should arrive at:

$$\frac{\partial^2 y}{\partial t^2} - c^2 \frac{\partial^2 y}{\partial x^2} = 0 \qquad \text{where } c^2 = T/m_l. \quad (8.51)$$

The solutions of the wave equation are the same as before, but now we have clear boundary conditions: $y(0,t) = 0$ and $y(l,t) = 0$. This will limit the stationary solution to a periodic standing wave. The pattern allows a fundamental frequency $\omega_1 = \pi / l \sqrt{T/m_l}$ and all its harmonics; the associated wavelength for the fundamental is $\lambda = 2l$. The actual spectral composition of $y(x,t)$ depends on the initial condition.

The power associated to the string vibration is also worth further consideration. Conservation of power implies, e.g., for a violin, that the amount of energy that is administered must be equal to the sum of the radiated acoustic energy and the losses (damping, friction, viscosity in instrument, air, and room).

For further details, the interested reader is referred to textbooks. An extensive treatment of string acoustics is given in Chap. 2 of Kinsler, Frey, Coppens, and Sanders (1982). A very short string approaches a bar. The cross-section parameters are no longer negligible with respect to the length. Additional material properties and cross-section structures become relevant. Static longitudinal tension is less prominent—if present—than in strings. Both longitudinal waves and transversal waves are important, and the transversal waves are no longer characterized by the common wave equation. Combination of Hooke's law, which relates longitudinal strain and stress, and Newton's law gives again a wave equation for the longitudinal component:

$$\frac{\delta l}{l} = \frac{\partial \xi}{\partial x} = \frac{F}{EA} \qquad \text{Hooke's law,} \tag{8.52a}$$

$$\mathrm{d}F = \rho A \, \mathrm{d}x \, \frac{\partial^2 \xi}{\partial t^2} \qquad \text{Newton,} \tag{8.52b}$$

$$\frac{\partial^2 \xi}{\partial t^2} - c^2 \frac{\partial^2 \xi}{\partial x^2} = 0 \qquad \text{with } c^2 = E/\rho. \tag{8.52c}$$

Parameters: δl (m) is elongation of a bar of length l (m) caused by a pulling force F (N); the bar has a cross-sectional area A (m^2); E denotes Young's modulus (Pa, or N/m^2); $\partial \xi$ describes the stretch over the length ∂x.

Solutions are similar to those of the general wave equation. The velocity of sound is usually one order of magnitude greater than that in gases. But an additional difficulty arises. The transversal wave will no longer follow a common wave equation, but will develop differently, as will be shown below. And then the internal coupling of transversal and longitudinal waves will mix the two to some degree, and also the longitudinal component will be affected by the deviating properties of the transversal wave.

For the bar, the transversal wave involves bending moments and shear forces. These lead to a wave equation with a fourth spatial derivative:

$$\frac{\partial^2 y}{\partial t^2} - c^2 \frac{I_\mathrm{m}}{A} \frac{\partial^4 y}{\partial x^4} = 0 \qquad \text{where } I_\mathrm{m} \text{ is the moment of area.} \tag{8.53}$$

The c-value is the same an above: $c^2 = E/\rho$, but the bending moment of area, and the area of the cross-section also play a role. A general solution can be found by separation of parameters method, or by solving the equation in the frequency domain and transfer the solution back into the time domain. The general form is:

$$y(x,t) = \cos(\omega t + \varphi)\left[y_1 \cosh\frac{\omega x}{v} + y_2 \sinh\frac{\omega x}{v} + y_3 \cos\frac{\omega x}{v} + y_4 \sin\frac{\omega x}{v}\right]. \tag{8.54}$$

Although (8.54) contains a place-independent oscillation term, it does describe traveling waves because the travel speed of the y_3 and y_4 terms depend on frequency, since v depends on frequency: $v^2/\omega = c\sqrt{I_m/A}$, or $v = \sqrt{\omega c \sqrt{I_m/A}}$. This means that wave shapes will change with place. This effect is termed *dispersion*. A detailed description of vibration in bars is given in Chap. 3 of Kinsler et al. (1982).

Plates and membranes are—in good approximation—2-D structures, because the thickness remains small compared to the length sizes. For the often encountered circular shapes, analysis in circular coordinates is most efficient. Membranes and plates often have homogeneous parameters, which leads to circular symmetry. Nonhomogeneous parameters, however, are also possible, in particular in plates.

Transversal vibrations in very thin plates or membranes are described by the regular wave equation, with $c^2 = \sqrt{T/\rho_A}$, with tension T in N/m, and ρ_A the mass per surface area. Vibration modes are determined by the boundary conditions, e.g., the size and shape of the plates, and the value of the surface tension.

Valid vibration modes depend on the shapes of the plates: circular symmetry or antisymmetry for circular plates, and rectangular modes for rectangular plates.

8.5 Theoretical Tools

Many mathematical tools that are used in advanced signal processing and systems analysis are also relevant for the theoretical study of the linear cochlea. The interested investigator should not only be able to use these tools but also have a thorough knowledge of their foundation.

For linear systems, transfer functions $H(\omega)$ are extremely useful, as are their Fourier transform (*FT*) and the impulse responses $h(t)$ of these systems. Efficient application of these functions requires knowledge of practical constraints. Impulse responses, e.g., have to be limited in length if we want to compute their effects. This applies both in the time domain when computing a convolution, and in the frequency domain when using a complex spectrum multiplication and a time response is obtained from a final inverse *FT*.

A general point of consideration regards the *uncertainty relation*, which is a fundamental property from the *FT*. It is properly most clearly demonstrated in Fourier's integral transform. But we start with a simpler elucidation. This goes back to the definitions of the terms frequency and time: the frequency of a signal (or signal component) is the number of cycles, or periods per time unit. In other words, in this context "time" refers to an interval of a non-zero duration, and not

just a time point. The problem then is that the accuracy with which the count can be determined depends on the length of the time interval over which the cycles are measured. Definition and measurement break down if this interval is shorter than one cycle. Next we give a short mathematical specification
Consider a Gaussian window

$$w(t) = 1/\sqrt{2\pi\sigma_t^2}\, \mathrm{e}^{-t^2/(2\sigma_t^2)} \qquad \text{with area 1, or:} \qquad \int_{-\infty}^{\infty} w(t)\, \mathrm{d}t = 1.$$

The Gaussian function is an eigenfunction of the *FT*, which means that the *FT* of $w(t)$, $W(w)$, also is a Gaussian. For the *FT* of this $w(t)$:

$$FT\{w(t)\} = W(\omega) = \int_{-\infty}^{\infty} w(t)\, \mathrm{e}^{-\mathrm{i}\omega t}\, \mathrm{d}t, \tag{8.55}$$

we obtain

$$W(\omega) = \mathrm{e}^{\frac{-\mathrm{i}\omega^2\sigma_t^2}{2}} = \mathrm{e}^{\frac{-\mathrm{i}\omega^2}{2\sigma_\omega^2}} \tag{8.56}$$

or $\sigma_\omega = 1/\sigma_t$ and $\sigma_\omega \times \sigma_t = 1$. In other words, frequency resolution can improve only at the cost of a decreasing time resolution and vice versa.[8]

8.5.1 *Properties of Simple Damped Oscillators*

The damped oscillator, which plays an essential role in Helmholtz's filterbank cochlea model, was briefly introduced in Chap. 1. Some additional information is provided in this section. An excellent treatment of the mechanical and electrical oscillator is presented in Marion and Thornton (1995, their Chap. 3). The full solution of the driven mechanical oscillator equation (1.1), rewritten here in the dot time derivative notation:

$$m\ddot{x} + d\dot{x} + sx = F_0 \cos(\omega t), \tag{8.57}$$

which is equivalent to

$$\ddot{x} + 2\delta\dot{x} + \omega_0^2 x = \frac{F_0}{m} \cos(\omega t), \tag{8.58}$$

where $2\delta = d/m$ and $\omega_0^2 = s/m$. Marion and Thornton address both the complementary and the particular solution of the equation. The latter gives the stationary driven

[8]The general relation is: $\sigma_\omega \times \sigma_t \geq 1$. The minimum is obtained for windows from the Gaussian family, which includes Hermite polynomials.

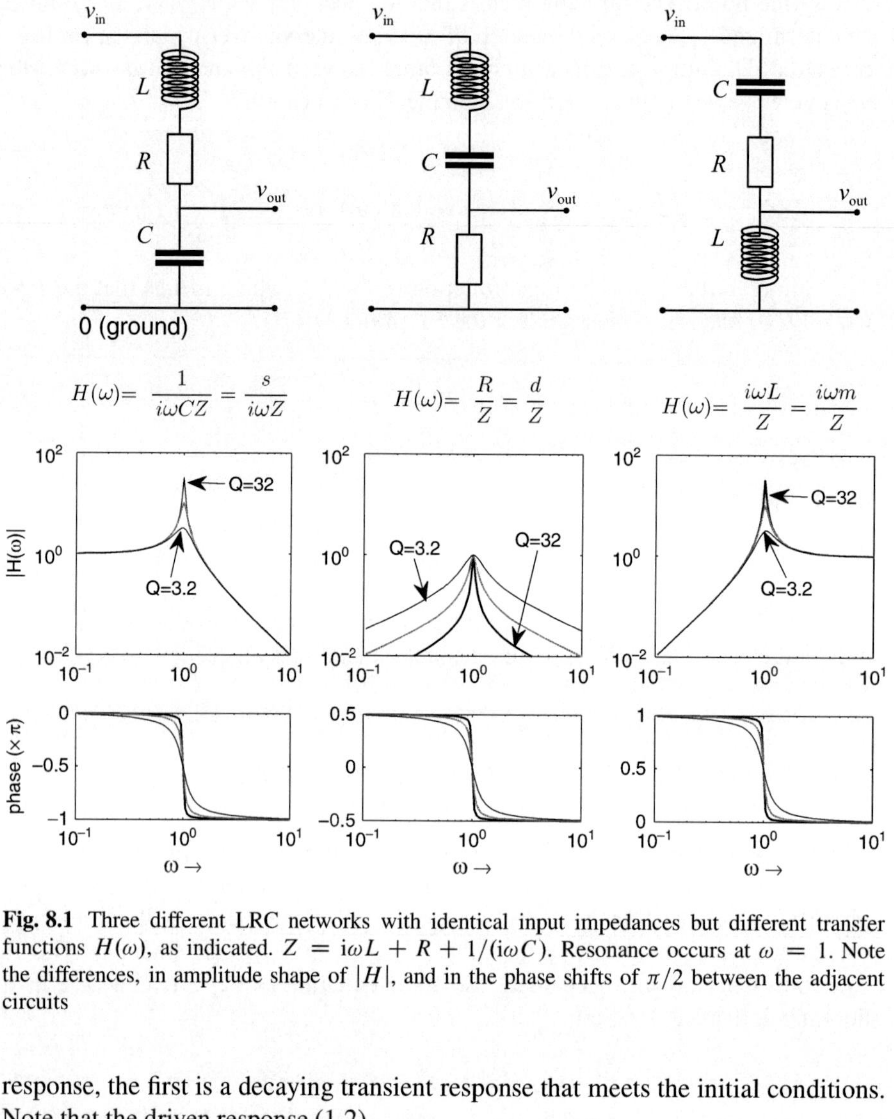

Fig. 8.1 Three different LRC networks with identical input impedances but different transfer functions $H(\omega)$, as indicated. $Z = \mathrm{i}\omega L + R + 1/(\mathrm{i}\omega C)$. Resonance occurs at $\omega = 1$. Note the differences, in amplitude shape of $|H|$, and in the phase shifts of $\pi/2$ between the adjacent circuits

response, the first is a decaying transient response that meets the initial conditions. Note that the driven response (1.2)

$$x(t) = \frac{F_0}{|Z_m|\omega} \sin(\omega t - \phi) = \frac{F_0}{|Z_m|\omega} \cos(\omega t - \varphi), \tag{8.59}$$

$$\text{where } \varphi = \frac{\pi}{2} - \phi$$

has its maximum amplitude at

$$\left. \frac{\partial x(t)}{\partial \omega} \right|_{\omega = \omega_R} = 0. \tag{8.60}$$

This gives for the resonance frequency of driven oscillations (left column in Fig. 8.1)

$$\omega_{\mathrm{R}}^2 = \omega_0^2 - 2\delta^2 \qquad \text{and} \qquad \omega_{\mathrm{R}} < \omega_0. \tag{8.61}$$

The transfer from force F to response R is characterized by the transfer function, defined in the frequency domain: $R(\omega) = H(\omega)F(\omega)$. Obviously, it depends on what is defined as the response. Above it is tacitly assumed that the resulting deflection x is the requested response, but it can also be the resulting velocity or acceleration. In those cases there is an additional term $\mathrm{i}\omega$, or time derivative in the time domain, in the transfer function, and also the location of the wanted ω_{R} with respect to ω_0 changes. This is demonstrated in the electrical analog in Fig. 8.1. In all cases, the input is $v(t)$, the LCR network has the same impedance; in the left circuit, the output voltage is measured over the condenser, in the right circuit, over the resistor. For the middle column, one obtains $\omega_{\mathrm{R}} = \omega_0$, independent of the damping, and for the right column, the shift of ω_{R} is in the opposite direction.

In the text, we used a practical definition for the Q-value of the filter, viz. $\omega_0/\Delta\,\omega$ or $f_0/\Delta\,f$, where the bandwidth $\Delta\,\omega$ is often measure at $1/\sqrt{2}$ below the peak, or at close to 3 dB below the peak on a dB scale. This Q-value bears the simplest relation to mass, damping and stiffness. Other definitions relate to the ratio of the driven resonance frequency ω_{R} and $\Delta\,\omega$, which for high Q-values makes no difference, but for $Q < 3$ begins to show. This value is also related to the ratio of the total amount of power in the system and the fraction that is dissipated.

References

Abramowitz M, Stegun IA (1968) Handbook of mathematical functions, 7th edn. Dover, New York

Beranek LL (1986) Acoustics, 2nd edn. AIP, New York, [1st ed: 1954]

Brass D, Kemp DT (1991) Time-domain observation of otoacoustic emissions during constant tone stimulation. J Acoust Soc Am 90:2415–2427

Davis RS (1992) Equations for the determination of the density of moist air (1981/91). Meterioloia 29:67–70

Kinsler LA, Frey AR, Coppens AB, Sanders JV (1982) Fundamentals of Acoustics, 3rd edn. Wiley, New York

Lamb H (1895) Hydrodynamics, 2nd edn. Cambridge University Press, Cambridge

Landau LD, Lifshitz EM (2004) Fluid Mechanics, vol 6, 2nd edn. Elsevier Buttrworth-Heinemann, Oxford, [1st ed 1959]

Marion JB, Thornton ST (1995) Classical Dynamics, 4th edn. Saunders College Publishing, Fort Worth

Morse PM, Feshbach H (1953) Methods of Theoretical Physics. McGraw-Hill, New York, part I and II

Papoulis A (1962) The Fourier Integral and its Applications. McGraw-Hill Book Company, New York

Rayleigh JWS (1894) The Theory of Sound, vol I, 2nd edn. MACMILLAN, London, [Dover edition: 1945]

Rayleigh JWS (1896) The Theory of Sound, vol II, 2nd edn. MACMILLAN, London, [Dover edition: 1945]

Ruggero MA (1980) Systematic errors in indirect estimates of basilar membrane travel times. J Acoust Soc Am 67:707–710

Whitehead ML, Stagner BB, Martin GK, Lonsbury-Martin BL (1996) Visualization of the onset of distortion-product otoacoustic emissions, and measurement of their latency. J Acoust Soc Am 100:1663–1679

Chapter 9
Nonlinear Tools

Abstract This chapter presents tools that are available for the analysis of nonlinear properties of (some of the constituents of) the cochlea. It starts with a reference to properties of power law devices. In the next section it discusses properties of nonlinear oscillators in more detail than was done in Chap. 5.

9.1 Properties of Power Law Nonlinearities

Power series expansions are useful for the numerical approximation of functions. Probably the most familiar series in this category is the Taylor Series, which is defined both for real and for complex variables. We discuss some general properties of power series along the real x-axis, starting from the Taylor Series. We adjust (shift) the x variable such that we are interested in the development around $x = 0$. In that case, the series is termed the Maclaurin Series, which for the function $f(x)$ has the form

$$f(x) = f(0) + \frac{x}{1!} f'(0) + \frac{x^2}{2!} f''(0) + \cdots + \frac{x^n}{n!} f^{(n)}(0) + \cdots \qquad (9.1)$$

with the remainder

$$R(n) = f(x) - \Sigma_n = \frac{x^{(n+1)}}{(n+1)!} f^{(n+1)}(\varepsilon x) \qquad (0 < \varepsilon < 1), \qquad (9.2)$$

where Σ_n is the partial sum from 0 up to (and including) n. The series converges if R_n tends to zero if $n \to \infty$. This series belongs to the standard power series

$$\sum_{n=0}^{\infty} a_n x^n = a_0 + a_1 x + a_2 x^2 + \cdots + a_n x^n + \cdots \qquad (9.3)$$

H. Duifhuis, *Cochlear Mechanics: Introduction to a Time Domain Analysis of the Nonlinear Cochlea*, DOI 10.1007/978-1-4419-6117-4_9,

Unfortunately, all terms with $n > 1$ are of the expansive type: they grow faster than linearly. If we want to approximate a compressive[1] function, such as a square root, or a hyperbolic tangent, then the radius of convergence of the power series is a very small, limiting the region of convergence to a small range around the point of focus. It also means that the convergence is slow, because it requires many terms in the series before the remainder is negligible.

Compressive functions can be expanded more efficiently with compressive basis functions. A good example is the νth-law device, proposed by Engebretson and Eldredge, and by Smoorenburg (see Sect. 4.1.1). A series expansion in νth-law elements has also been developed in mathematics, but remained less familiar:

$$f(x) = a_0 + a_1 x^{\nu_1} + a_2 x^{\nu_2} + \cdots + a_n x^{\nu_n} \quad \text{where} \quad 0 < \nu_1 < \nu_2 \cdots < \nu_n. \tag{9.4}$$

It was published in 1922 by Bernstein, who proved that an infinite series of νth-law terms also provides a precise approximation of any smooth function (Bernstein 1922). In other words, Taylors expansion with only integer exponents might be considered a special case of Bernstein's.

The properties of νth-law elements that are used in the Bernstein expansion have received considerable attention in electrical engineering, e.g., for the analysis of responses of vacuum tubes and transistors to narrow band signals (e.g. Bennett 1933; Rice 1945; Middleton 1948, 1960; Sternberg and Kaufman 1953; Sternberg et al. 1955; Feuerstein 1957). The expansion is generally given for a half-wave power device

$$y(x) = y+(x) = \begin{cases} c\,(x-b)^{\nu} & x > 0,\ b \geq 0,\ \nu > 0 \\ 0 & x \leq b. \end{cases} \tag{9.5}$$

The general response to the time signal

$$x(t) = \sum_{i=0}^{N} A_i \cos(\omega_i t + \phi_i) \tag{9.6}$$

can be written as

$$y_{+}(x) = A_0^{\nu} \sum_{n_0=0}^{\infty} \cdots \sum_{n_N=0}^{\infty} \frac{1}{2} \mathbf{A}^{(\nu)}_{n_0 n_1 \ldots n_N}(h, k_1, k_2, \ldots k_N) \prod_{i=0}^{N} \varepsilon_i \cos n_i(\omega_i t + \phi_i), \tag{9.7}$$

where ε is the so-called Neumann factor: $\varepsilon_0 = 1, \varepsilon_i = 2,\ \forall i \geq 1$. In this formula, A_0 represents the maximum amplitude: $A_0 = A_{\max} = \max_i\{A_0, A_1, \ldots, A_N\}$. The factors $k_i = A_i/A_{\max}$, the normalized amplitudes of the components are less than or equal to 1, and $b/A_{\max}$ is the normalized bias. Sternberg et al. introduced the term "Bennett function" for the amplitude factors $\mathbf{A}$, which represent the amplitude

[1] The general compressive function has a positive first and a negative second derivative.

factors from multiple Fourier series. The Bennett functions have been expressed in term of contour integrals (Rice and Bennett) and in hypergeometric functions (e.g. Middleton 1960).

This overview is limited to the analysis of a nonlinearity without bias, $b = 0$, and the arbitrary constant c is set equal to 1. The responses to 1- and 2-tone stimuli will be addressed. It is noted in passing that the 1-tone case has been treated in detail by Davenport and Root (1958, Chap. 13).

Duifhuis (1989) proposed to employ the "Modified Bennett function" **B** aiming at decomposing much of the amplitude behavior from the interaction term:

$$\mathbf{A}^{(\nu)}_{n_0 n_1 \ldots n_N}(h, k_1, k_2, \ldots k_N) = \mathbf{B}^{(\nu)}_{n_0 n_1 \ldots n_N}(h, k_1, k_2, \ldots k_N) \prod_{i=1}^{N} k_i^{n_i}. \tag{9.8}$$

B is asymptotically constant if $\forall i, \quad k_i \to 0$.

9.1.1 *Response to a Single Tone*

The response to the single tone $x(t) = A \cos(\omega t + \phi)$ is

$$y_+(x) = \frac{1}{2} A^\nu \sum_{n=0}^{\infty} \varepsilon_n \mathbf{B}^{(\nu)}_n \cos n(\omega t + \phi) \tag{9.9}$$

as follows directly from (9.7) and (9.8). The responses for the full-wave even and odd nonlinearities follow directly from the proper combination of two half-wave rectifier responses. This yields:

$$y_{\text{odd}}(x) = 2 A^\nu \sum_{n=0}^{\infty} \mathbf{B}^{(\nu)}_{2n+1} \cos\{(2n+1)(\omega t + \phi)\}, \tag{9.10}$$

which contains odd harmonics (overtones) only, and

$$y_{\text{even}}(x) = A^\nu \sum_{n=0}^{\infty} \varepsilon_{2n} \mathbf{B}^{(\nu)}_{2n} \cos\{(2n)(\omega t + \phi)\}, \tag{9.11}$$

which contains only even harmonics.

The amplitude factor **B** equals (cf. Duifhuis 1989):

$$\mathbf{B}^{(\nu)}_n = \frac{\Gamma(\nu + 1)}{2^\nu\, \Gamma(1 + \frac{\nu+n}{2})\, \Gamma(1 + \frac{\nu-n}{2})}, \tag{9.12}$$

where $\Gamma(x)$ is the gamma function.

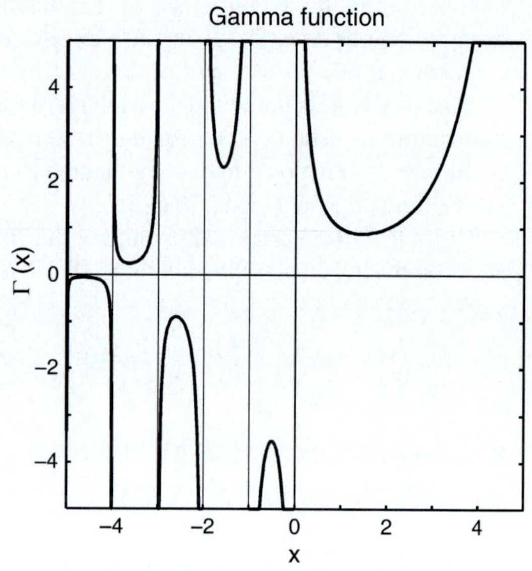

Fig. 9.1 The gamma function over the interval −5 to +5. Note the different signs for the different negative branches of the gamma function. For positive integer values of $x = n$, $\Gamma(n+1) = n!$

The second gamma function in the denominator is of particular interest, because its argument $\{1 + (\nu - n)/2\}$ will be negative for sufficiently large values of n. Figure 9.1 indicates the behavior of the function for arguments below zero. If the argument approaches a nonpositive integer, the value approaches $\pm\infty$, which here implies that the Bennett function will go to zero. This can occur only for integer values of ν, with the consequence that higher harmonics will not be present. For noninteger ν, however, the sign of this term switches every two increments in n, which means that the phase of the distortion product changes accordingly. We come back to the point with the 2-tone stimulus.

Minimum values for $|\Gamma(x)|$ occur approximately at $x = 1.5$ and $x = -0.5 - n$, with $n \geq 0$. Since the increments in the argument are 0.5 at a step 1 in n, these points will not appear in full-wave rectifiers, for the reason given above. Nevertheless the values can serve to establish boundaries. The DC response for the even nonlinearity is equal to the average response of the absolute value of the odd nonlinearity $|y_{odd}|$:

$$< |y_{odd}| >_{av} = < y_{even} >_{av} = A^{\nu}\mathbf{B}_0^{(\nu)} = \frac{1}{\pi} A^{\nu} B\left(\frac{1}{2}, \frac{1}{2} + \nu/2\right), \tag{9.13}$$

where B is the beta function.

9.1.2 Response to Two Tones

For a two-tone stimulus, a product of series of intermodulation terms is obtained. For the stimulus $x(t) = A_1 \cos(\omega_1 t + \phi_1) + A_2 \cos(\omega_2 t + \phi_2)$, the response is

$$y_+(x) = \frac{1}{2} A_1^\nu \sum_{m=0}^{\infty} \sum_{n=0}^{\infty} k^n \varepsilon_m \varepsilon_n \mathbf{B}_{mn}^{(\nu)}(k) \cos m(\omega_1 t + \phi_1) \cos n(\omega_2 t + \phi_2). \tag{9.14}$$

The Bennett function for the 2-tone case is

$$\mathbf{B}_{mn}^{(\nu)}(k) = \frac{2^{-\nu}\Gamma(1+\nu)\,{}_2F_1\left\{\frac{1}{2}(m+n-\nu), \frac{1}{2}(-m+n-\nu); n+1; k^2\right\}}{n!\,\Gamma\left\{1-\frac{1}{2}(m+n-\nu)\right\}\Gamma\left\{1-\frac{1}{2}(-m+n-\nu)\right\}} \tag{9.15}$$

where $|k^2| \le 1$.

${}_2F_1$ is the hypergeometric function. If $|k^2|$ approaches the boundaries 0 or 1, we have the special cases[2]:

$$|k^2| \to 0 \qquad {}_2F_1\{a, b; c; 0\} = 1, \tag{9.16a}$$

$$|k^2| \to 1 \qquad {}_2F_1\{a, b; c; 1\} = \frac{\Gamma(c)\Gamma(c-a-b)}{\Gamma(c-a)\Gamma(c-b)}, \tag{9.16b}$$

where a, b, and c meet the criteria for convergence.

For $k = 1$ it can be shown that $\mathbf{B}_{mn}^{(\nu)}(1) \equiv \mathbf{B}_{nm}^{(\nu)}(1)$, $\forall m, n$, and ν; for $k = 0$, the sign of ${}_2F_1$ can change, because one of the gamma functions can obtain a negative argument. In such a case, the sign of $\mathbf{B}$ will change at least once if k increases from 0 to 1. In practice, this will not occur for the lowest harmonics and intermodulation products.[3]

Note also that $\mathbf{B}_{n0}^{(\nu)}(0) = \mathbf{B}_n^{(\nu)}$.

The response of the full-wave odd nonlinearity follows from (9.14) and is

$$y_{\text{odd}}(x) = 4 A_1^\nu \sum_{m=0}^{\infty} \sum_{n=0}^{\infty} k^n \mathbf{B}_{mn}^{(\nu)}(k) \cos m(\omega_1 t + \phi_1) \cos n(\omega_2 t + \phi_2), \tag{9.17}$$

where $m + n$ must be odd, and $k = A_2/A_1 < 1$.

The Bennett function has the amplitude ratio k as variable, in addition to the power parameter ν and the indices m and n. As long as $k \ll 1$, the dependence of $\mathbf{B}$ on k is rather small. Near $k = 1$, however, its effect can be significant. Examples of the functions are given in Fig. 9.2 for indices m, n up to 4.

As follows rather directly from the expansions, the phases of the components follow the frequency interactions. In other words, the phase of the $2f_1 - f_2$ DP follows directly as $\phi_{CT} = 2\phi_1 - \phi_2$, except for one important point: the sign of $\mathbf{B}_{mn}^{(\nu)}(k)$.

---

[2]See, e.g., Abramowitz and Stegun, 1968 Chap. 15.

[3]Your proof of the above statements is a valuable test of your understanding of the gamma function and the hypergeometric function.

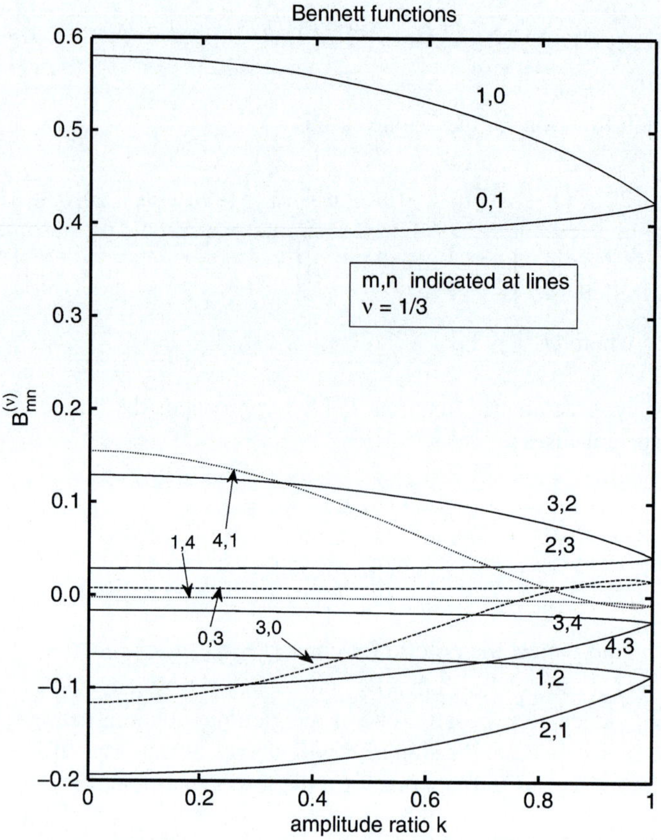

Fig. 9.2 The modified Bennett function for the lowest indices and for $\nu = 1/3$ as a function of the amplitude ratio $k = A_{\mathrm{min}}/A_{\mathrm{max}}$. The value for the CT $2f_1 - f_2$ is negative, $3f_1 - 2f_2$ is positive, etc (not shown beyond 4,3)

9.1.3 The Sign of the Bennett Function

The arguments for the gamma functions and the hypergeometric function are all real, hence the resulting values will be real, but they still can be either positive or negative. This is the basis for an important property of the Bennett functions: its sign. The lower part of Fig. 9.2 shows some negative values, and the lines 3,0 and 4,1 cross zero. Some of these phases flip if ν crosses an integer value. For instance, if ν changes from less than 1 to greater than 1, the phase of $2f_1 - f_2$ flips. Details of this property are presented in Fig. 9.3 for the odd order nonlinearity (even combinations are set to gray). Positive sign is indicated as white, negative sign as black.

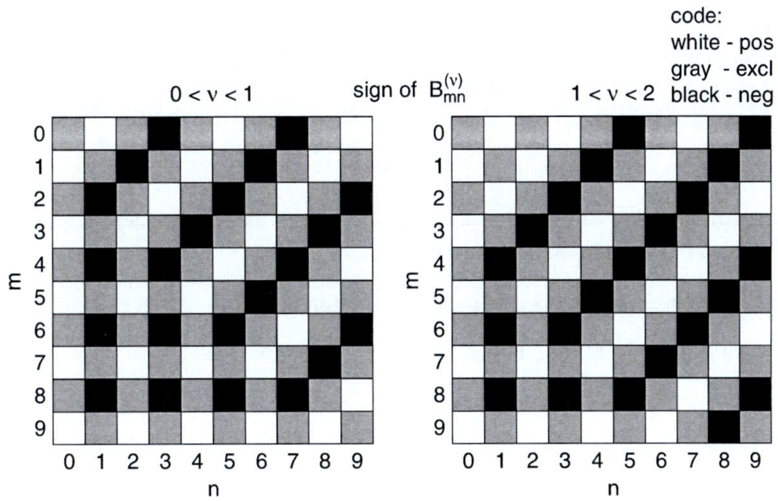

Fig. 9.3 Phase effects (± sign only) of the modified Bennett functions for the indices up to 9 and for a value of ν below 1 (compressive nonlinearity) and for a value above 1 (expansive)

9.2 Properties of Nonlinear Oscillations

Nonlinear differential equations that represent relatively simple physical problems have been known for a long time, even though complete solutions of the equations might have been lacking. A simple one is the description of the pendulum with a fixed but massless bar connection to the rotation point. Another one is the van der Pol oscillator, now well known because of its introduction by van der Pol in 1920 for the theoretical description of the behavior of a triode. However, several decades earlier the same DE had been obtained by Rayleigh who analyzed a bowed violin string. Before dealing with the specific equations, some of the common tools are mentioned. For more information, the interested student is referred to textbooks such as Stoker (1950), Nayfeh and Mook (1979), Guckenheimer and Holmes (1983) and Broer and Takens (2009). The former two have a theoretical physics setup, with many equation examples recognizable from physics. The later two more strictly emphasize the mathematical terminology, and the links with physical examples is less trivial.

9.2.1 Terminology

This section starts with a numbered list of definitions of terms that are used in the field. As yet, several of the definitions lack exclusivity, and where possible alternatives are given. In particular the difference in use in different areas of science can be confusing. The current list will not solve the problem, but aims to be of some assistance.

Definition 9.1. An ODE or *ordinary differential equation*[4] is an equation in one independent unknown which also contains at least one derivative (of possibly more at several orders) of the unknown.

The general shape is

$$\dot{\mathbf{x}} - f(\mathbf{x}, t) = 0 \tag{9.18}$$

because any higher order ODE can be transformed to a first-order by increasing the dimension of the independent unknown(s).

This also applies to the second order oscillation equation.

Definition 9.2. A homogeneous ODE is of the form (9.18).

Definition 9.3. A nonhomogeneous ODE is of the form $\dot{\mathbf{x}} - f(\mathbf{x}, t) = F(t)$.

Definition 9.4. An *autonomous system* of ODEs does not contain the independent variable t in the equation.

[If the timederivative also depends on time the system is called nonautonomous.]

Definition 9.5. A *conservative system* is a system that follows conservation laws. Work can be done by an exerted force. The work is independent of a followed path and matches differences in the energy balance from starting point to end point. The interaction between internal energy (potential + kinetic) and work done is completely reversible.

Definition 9.6. *Dynamical systems* are time-dependent deterministic systems that can be described by differential equations $\dot{\mathbf{x}} = \mathbf{f}(\mathbf{x})$ that originate from the laws of motion. The description is completed by the specification of the appropriate initial conditions. Many examples exist, e.g., in mechanics and fluid mechanics.

[Related definitions: autonomous systems, and conservative systems.]

In current mathematical texts, the relation with the underlying physics may appear to be scaled away (mass, and stiffness, or equivalents, are scaled to unity).

In the following definitions, we drop the vector notation for x and f, even though in general multidimensional situations are implied.

Definition 9.7. *Free oscillations* are solutions to undriven (and undamped) second order differential equations. The mechanical form of the ODE is:

$$m\ddot{y} + s\,y = 0, \tag{9.19}$$

the abstract scaled version:

$$\ddot{x} + x = 0. \tag{9.20}$$

[4]The term "differential equation" goes back to Leibniz, 1676 (EOMlist 2002, Differential equation, ordinary).

Definition 9.8. *Forced oscillations* are solutions to driven (and possibly damped) second order differential equations. A mechanical form of the ODE is (see e.g., Sect. 8.5.1):

$$m\ddot{y} + d(\dot{y}) + s\,y = f(\omega, t, \varphi), \tag{9.21}$$

the abstract scaled version:

$$\ddot{x} + \delta(\dot{x}) + x = F(\omega, t, \varphi). \tag{9.22}$$

Note that Definition 9.7 and Definition 9.8 are 2nd order derivative equivalents of Definition 9.2 and Definition 9.3.

Definition 9.9. *Relaxation oscillations* are nonlinear oscillations that follow a limit cycle path in the phase plane. The limit cycle period is no longer determined by the linear approximation of the resonance frequency, but it is strongly dependent on the extend of the nonlinearity.

9.2.1.1 Additional Definitions for 2nd Order ODEs

Since most of the applications in this book concern dynamical system properties of CMs we are specifically interested in some specific terminology that is in use in this field of mathematics.

Definition 9.10. The *linear mechanical ODE* describing vibrations is

$$m\ddot{y} + d\,\dot{y} + s\,y = f(\omega, t, \varphi), \tag{9.23}$$

which contains only constant coefficients representing mass m, damping d, and stiffness s. Solutions to this equation have been presented in Sect. 8.5.1.

An additional property of this equation is of interest for the undamped case, $d = 0$. Dividing by m, putting $s/m = \omega_0^2$, and considering the driving force to be periodic $\frac{1}{m}\,f(\omega, t, \varphi) = F\cos(\omega t)$ gives

$$\ddot{y} + \omega_0^2\,y = F\cos(\omega t) \tag{9.24}$$

for which the solution has an undriven, free component and a driven component. The first depends on the initial conditions $y(0), \dot{y}(0)$, the second on the driving force. Stoker (1950) showed that for $\dot{y}(0) = 0$ the solution can be written as

$$y(t) = \left(y(0) + \frac{F}{\omega^2 - \omega_0^2}\right)\cos(\omega_0 t) - \frac{F}{\omega^2 - \omega_0^2}\cos(\omega t), \tag{9.25}$$

which gives periodic solutions if and only if $n\omega_0 = m\omega$, and the integers (m, n) are not 0 or both equal to 1. Of particular interest are the cases were m and n are relatively small integers. The solutions for $m = 1$ and $n > 1$ are called subharmonic oscillations. In this case, the oscillation frequency is a subharmonic of the stimulus frequency.

The solutions for $n = 1$ and $m > 1$ have been called ultraharmonic oscillations by Stoker; Nayfeh and Mook use the term super-harmonic. In this case the oscillation frequency is a harmonic (overtone) of the stimulus frequency.

Definition 9.11. The general form of the nonlinear mechanical ODE describing vibrations is

$$m\ddot{y} + d(\dot{y}) + s(y) = f(\omega, t, \varphi), \tag{9.26}$$

which, besides the constant mass coefficient contains in general nonlinear functions in $\dot{y}$ and y.

Some special examples are (scaled, or normalized):

- Equations in which the function $d(\dot{y})$ is nonlinear in $\dot{y}$ (or y), and $s(y)$ is linear in y are termed Rayleigh or van der Pol equations.
- Equations in which $d(\dot{y})$ is linear in $\dot{y}$ (or y) and $s(y)$ is nonlinear in y are termed Duffing equations.

We will primarily consider the Rayleigh or van der Pol equations, which are closely related and readily transposed by differentiation or integration.
For the free (undriven) case, common specific version are given below.

Definition 9.12. *A common version of the Rayleigh equation is:*

$$\ddot{y} - \varepsilon\left(\dot{y} - \frac{1}{3}\dot{y}^3\right) + y = 0. \tag{9.27}$$

Definition 9.13. *A common version of the van der Pol equation is:*

$$\ddot{x} - \varepsilon\,(1 - x^2)\dot{x} + x = 0. \tag{9.28}$$

After differentiation of (9.27), the replacement of $\dot{y}$ by x gives (9.28).

Nayfeh and Mook (1979, p.106) describe a physical basis for the Rayleigh equation in terms of a block (mass) on a rough, rolling belt, where the mass movement is limited by a string attachment to a fixed point. They state (without reference) that Rayleigh used similar arguments to explain the movement of a bowed violin string.

The origin of van der Pol's equation is clearer. It started as the mathematical description of a simple triode network in 1920. Developments have been published extensively, and an extensive review has been presented in van der Pol (1934).

Definition 9.14. The *phase space* is the 2-D plane[5] in which the oscillation velocity is presented against the deflection. Solutions to the oscillation ODEs follow a trajectory in this plane.

---

[5]The 2-D plane applies to the 1-D oscillation. If x is n-dimensional, then the phase space is $2n$-D dimensional.

The trajectories, also called evolution or flow, start at the points representing the initial conditions, and progress with time. The solutions of an oscillator converge to a stable periodic solution or limit cycle.[6]

It is worth noting that the phase plane plots for the "identical" van der Pol and Rayleigh equations can have a very different appearance.

The shapes if the limit cycles depend strongly on the "strength" of the nonlinearity, or the value of ε.

A more general definition of the Rayleigh equation, given by Rozov in the EOMlist (2002), is (for the undriven case) does not limit the damping term to a quadratic factor of $\dot{y}$ in (9.27). It follows the next definition:

Definition 9.15. *A more general version of the Rayleigh equation is:*

$$\ddot{x} + F(\dot{x}) + x = 0, \qquad \text{where} \qquad \dot{x} = \frac{\mathrm{d}x}{\mathrm{d}t} \tag{9.29}$$

and where $F(u)$ satisfies the conditions

$$\frac{1}{u}F(u) < 0, \qquad \forall |u| < \delta \qquad \text{or} \qquad |u| \downarrow 0, \tag{9.30a}$$

$$\frac{1}{u}F(u) > 0, \qquad \forall |u| > \Delta \qquad \text{or} \qquad |u| \to \infty \tag{9.30b}$$

with a similar definition for the van der Pol equation (not shown).
For applications in CMs, a more interesting definition has been given by Broer and Takens (2009), who scaled the linear part in the damping separately from the nonlinear part.

Definition 9.16. The equation

$$\ddot{x} + (x^2 - \mu)\dot{x} + x = 0 \tag{9.31}$$

is a regular van der Pol equation as long as $\mu > 0$, whereas for $\mu \leq 0$ all solutions converge to the zero solution.

The parameter μ is at a Hopf bifurcation at $\mu = 0$ (cf. Definition 9.19). Although the above form may seem different from (9.28), it can be obtained by direct transformation. Substituting $x = \sqrt{\varepsilon}\,u$, where $\mu = \varepsilon$, one goes back to

$$\ddot{u} + \varepsilon(u^2 - 1)\dot{u} + u = 0.$$

We repeat that this formulation of the van der Pol oscillator is very similar to our more recent use of this oscillator in CMs, where we use local variation of the strength of the maximum negative damping.

For completeness, we also mention the Duffing equation, which primarily concerns nonlinearity in the stiffness term of the (mechanical) oscillator.

---

[6]For a more rigorous mathematical definition see Broer and Takens (2009, Sect. 1.2).

Definition 9.17. The *Duffing equation* contains a linear negative stiffness term plus a positive nonlinear component. A driven form is

$$\ddot{x} + \alpha\,\dot{x} - x + x^3 = \beta\cos(\omega t). \tag{9.32}$$

As yet, we interpret the data to indicate that cochlear nonlinear is in first approximation best interpreted as a damping (friction) nonlinearity.

Definition 9.18. If the phase plane trajectory of a solution converges to a periodic and stable path, then that path is termed *limit cycle*.

Definition 9.19. The term *bifurcation* is used in certain areas of mathematics to indicate a point λ on a parameter profile where the properties of a system that depends on the parameter λ changes radically for very small changes of λ.

As indicated above in Definition 9.16, the van der Pol equation (9.31) contains a Hopf-bifurcation at the point $\mu = 0$, where the solutions change abruptly.

9.2.2 Solutions to the Van Der Pol Equation

The van der Pol equation was introduced in Sect. 5.1.3.3 as (5.17):

$$\ddot{x} + \varepsilon\phi(x)\dot{x} + x = s(t),$$

where $\phi(x) = x^2 - 1$. As mentioned above, this equation can simply by transformed to the Rayleigh equation (9.27), with a transformed stimulus term:

$$\ddot{y} - \varepsilon\left(\dot{y} - \frac{1}{3}\dot{y}^3\right) + y = S(t).$$

How do we interpret the difference in terms of physics? Two different answers can be given.

Answer 1: In the above forms, the nonlinearity in the van der Pol equation is an even function in x (in the classical case a simple parabola). In the Rayleigh equation, the parabola is not a function of y, but of $\dot{y}$. In other words, in the Rayleigh nonlinearity depends on the velocity, whereas in the van der Pol it depends on displacement.

Answer 2: Assuming the van der Pol equation to represent the physics most closely implies to interpret x as a displacement, $\dot{x}$ as a velocity and $\ddot{x}$ as an acceleration. If we consider the transformability between the two equations as the link to the other equation, then the y-term in the Rayleigh equation would represent a $\int x\mathrm{d}t$ term, for which we do not have a direct physical terminology. An alternative hypothesis would by to interpret the x-term in the VDP as a velocity, and the higher derivatives accordingly.

Solutions to the van der Pol equation have been around at least since van der Pol himself, alone or with coworkers, and originally as conservator of Teyler's Museum in Haarlem, the Netherlands. He proposed an approximation technique, which is still similar to some of the modern approaches. There is a number of parameters that characterizes the behavior of the oscillator. For the undriven oscillator, the primary parameter is ε. Three important ranges for *varepsilon* are: a) $\varepsilon \ll 1$, b) $\varepsilon \approx 1$, and c) $\varepsilon \gg 1$. These conditions lead to quite different response behaviors.

Although theoretically the response could be locked at $x = 0$, this is not a stable solution. Any deviation from the zero initial condition brings the solution to a limit cycle, the form of which as well as the period duration depend on the parameter ε. For small ε, the limit cycle approaches a circle, and with the resonance frequency $\omega_0 = 1$ the period becomes $T_0 = 2\pi/\omega = 2\pi$. For values of ε around 1, the shape is already distorted, and it gets more distorted with increasing ε. Moreover, the period T_0 begins to depend on ε. A numerical approximation is[7]:

$$T_0 = 1.137\,\varepsilon + 7.0143\,\varepsilon^{-1/3} - \frac{1}{3}\,\varepsilon^{-1}\,\log\varepsilon - 1.3246\,\varepsilon^{-1} + O(\varepsilon^{-4/3}). \tag{9.33}$$

The following section follows a standard approach of dealing with the small ε-case. Usually the large ε-case is treated separately (e.g., Guckenheimer and Holmes 1983).

9.2.2.1 Solutions for Small ε

Now we rewrite this equation, breaking it up in two first order DEs, and introducing

$$\Phi(x) = -x + x^3/3 = \int \phi(x)\mathrm{d}x.$$

or, because $\mathrm{d}x = \dot{x}\mathrm{d}t$:

$$\frac{\mathrm{d}\Phi(x)}{\mathrm{d}t} = \frac{\mathrm{d}}{\mathrm{d}t}\left(-x + \frac{x^3}{3}\right) = -\dot{x} + x^2\,\dot{x} = \phi(x)\,\dot{x}. \tag{9.34}$$

Finally, this leads to:

$$\dot{x} = y - \varepsilon\Phi(x), \tag{9.35a}$$

$$\dot{y} = -x + s(t). \tag{9.35b}$$

[7]This approximation is presented by Nayfeh and Mook (1979, p.146), with references to the sources.

For the undriven case, $s(t) = 0$, and very small ε, the first approximation involves dropping the rightmost terms in (9.35a) and (9.35b) so that the solution of x and y becomes the rotation at $\omega = 1$, which can be frozen in the rotating coordinates u, v, defined as:

$$\begin{pmatrix} u \\ v \end{pmatrix} = \begin{pmatrix} \cos t & -\sin t \\ -\sin t & -\cos t \end{pmatrix} \begin{pmatrix} x \\ y \end{pmatrix}. \tag{9.36}$$

The complete system will yield slowly varying functions u and v. Averaging of the solution over a cycle (or number of cycles) is then justified because $\dot{u}$ and $\dot{v}$ are sufficiently small. It produces

$$\dot{u} = \frac{\varepsilon u}{2}\left[1 - \frac{(u^2 + v^2)}{4}\right], \tag{9.37a}$$

$$\dot{v} = \frac{\varepsilon v}{2}\left[1 - \frac{(u^2 + v^2)}{4}\right], \tag{9.37b}$$

which shows that the condition for small $\dot{u}, \dot{v}$ is met either by sufficiently small ε and/or u, v, or around the limit cycle amplitude $\sqrt{u^2 + v^2} = 2$, in which case the bracket term vanishes. Hence, this limit cycle is an attractor for stable solutions of the form

$$x(t) = r(t)\cos(t + \varphi(t)) \tag{9.38}$$

with an almost constant $r(t) = \sqrt{u^2 + v^2}$ and $\varphi(t)$. The latter is determined by the initial conditions.

In his third *Auditory Physics* paper, de Boer (1991) extends this analysis somewhat further. He also follows Guckenheimer and Holmes (1983, Chap. 2) closely. One of the important points that he addresses is the phenomenon of entrainment, which occurs for stimuli within a certain parameter range: as long as the stimulus frequency is not too far from the limit cycle frequency, and the amplitude not too large compared to ε, the solutions are *entrained*, or drawn to the limit cycle. For large values of ε, the behavior is very different, and the response is a relaxation oscillation, the fundamental frequency of which is no longer independent of ε.

References

Abramowitz M, Stegun IA (1968) Handbook of Mathematical Functions, 7th edn. Dover, New York

Bennett WR (1933) New results in the calculation of modulation products. Bell System Techn J 12:228–243

Bernstein SN (1922) Sur l'ordre de la meilleure approximation des functions continues par des polynomes. Mem Acad Royal Belg Sc 2(IV):1–103

de Boer E (1991) Auditory physics. Physical principles in hearing theory. III. Physics Reports 203(3):125–231

Broer HW, Takens F (2009) Dynamical Systems and Chaos. Springer, New York

Davenport WB, Root WL (1958) Random Signals and Noise. McGraw-Hill, New York

Duifhuis H (1989) Power-law nonlinearities: A review of some less familiar properties. In: Wilson JP, Kemp DT (eds) Cochlear Mechanisms: Structure, Function and Models, Plenum, New York, pp 395–403

EOMlist (2002) Differential equation, ordinary. E.F. Mishchenko (originator), Encyclopedia of Mathematics. URL: http://www.encyclopediaofmath.org/index.php?title=Differential_equation,_ordinary&oldid=13954

Feuerstein E (1957) Intermodulation products for ν-law biased wave rectifiers for multiple frequency input. Quart Appl Math pp 183–192

Guckenheimer J, Holmes P (1983) Nonlinear oscillations, dynamical systems and bifurcations of vector fields. Springer-Verlag, New York

Middleton D (1948) Some general results in the theory of noise through non-linear devices. Quart Appl Math 5:445–498

Middleton D (1960) An Introduction to Statistical Communication Theory. McGraw-Hill, New York

Nayfeh AH, Mook DT (1979) Nonlinear Oscillations. Wiley, New York

van der Pol B (1934) The nonlinear theory of electric oscillations. Proc I R E 22:1051–1086

Rice SO (1945) Mathematical analysis of random noise. IV. Noise through non-linear devices. Bell System Techn J 24:115–162

Sternberg RL, Kaufman H (1953) A general solution of the two-frequency modulation product problem. I. J Math Phys 32:233–242

Sternberg RL, Shipman JS, Kaufman H (1955) Tables of bennett functions for the two-frequency modulation product problem for the half-wave linear rectifier. Quart J Appl Math 8:457–467

Stoker JJ (1950) Nonlinear Vibrations. Interscience Publishers (later: Wiley), New York

Chapter 10
A PTPV Response Collection

Abstract The responses to several PTPV stimulus sets have been obtained with the Groningen NL-time domain model. In most cases we analyzed passive rather than active models because negative damping supposedly would effect the behavior at low stimulus levels only. A nonlinear damping change in the approx. 40 – 60 dB (SPL) interval is present in the NL model versions, however, as demonstrated in Fig. 5.11, and also can be interpreted as "active behavior".

The results to be presented are listed in the Table 10.1. Time plots and spectra are presented at adjacent pages (Figs. 10.3-10.20). But we start with the presentation of the spectra of the responses shown in Figs. 6.6 and 6.7.

All these points were analyzed at primary levels from 30 to 80 dB (SPL), often extended from 0 to 100 dB. Some of those results will be mentioned in the discussion. All results, presented or not, can be easily recomputed from the model.

Table 10.1 List of presented model results in response to PTPV stimulus sets. Presence of d and/or s in the second column indicates a NL version of damping and/or stiffness. The frequencies given in the third group of columns are the f_1 frequencies.

		frequencies (kHz)		
model	d and s	1	2	5
0	s		+	
3a	d		+	
3b	d, s		+	
4a	d	+	+	+
4b	d, s		+	
6	d, s		+	
Hopf	d		+	

H. Duifhuis, *Cochlear Mechanics: Introduction to a Time Domain Analysis of the Nonlinear Cochlea*, DOI 10.1007/978-1-4419-6117-4_10,

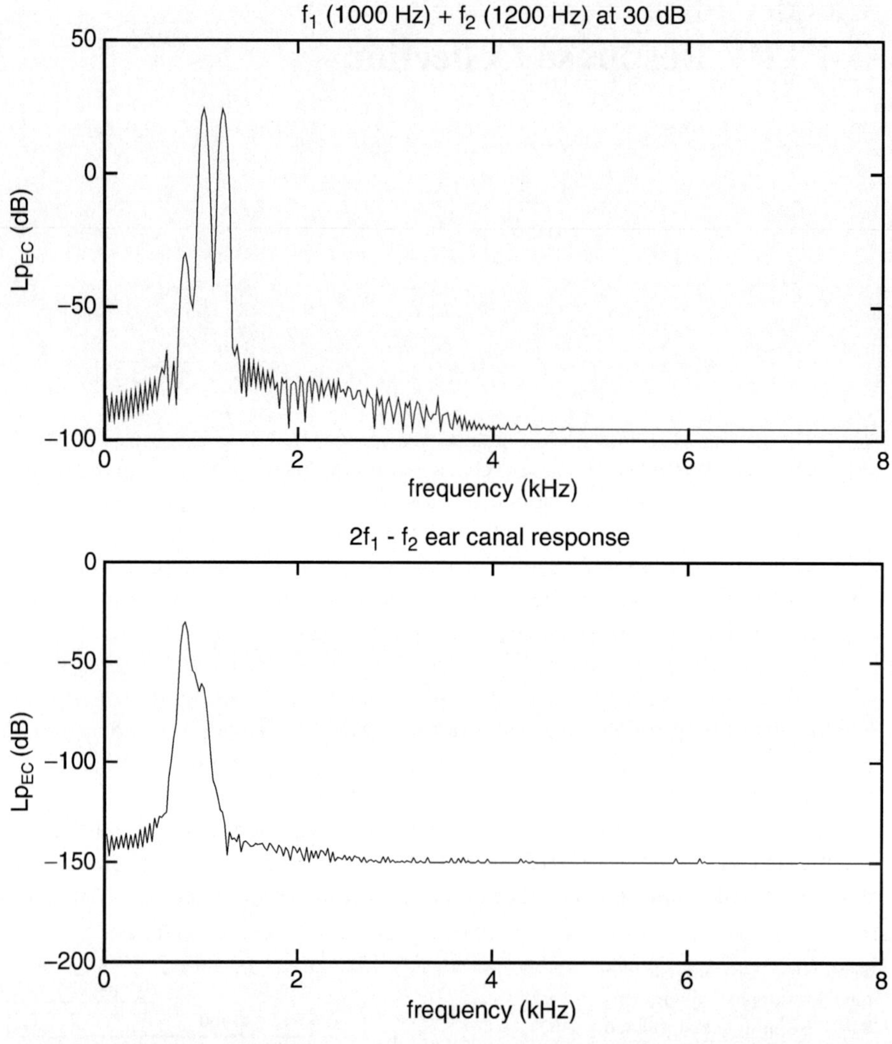

Fig. 10.1 Amplitude spectra of the $f_1 + f_2$-stimulus and PTPV response for model #4 with linear stiffness. Parameters as in Fig. 6.6. The skirts of the peaks represent the time window used for the analysis. At this level $2f_1 - f_2$ is hardly visible in the overall spectrum, and $2f_1 + f_2$ practically disappears in the computation noise. The PTPV result for the DP does show the $2f_1 - f_2$ peaks in detail at ≈ -25 dB. Note that the high-frequency part of the spectrum remains clean.

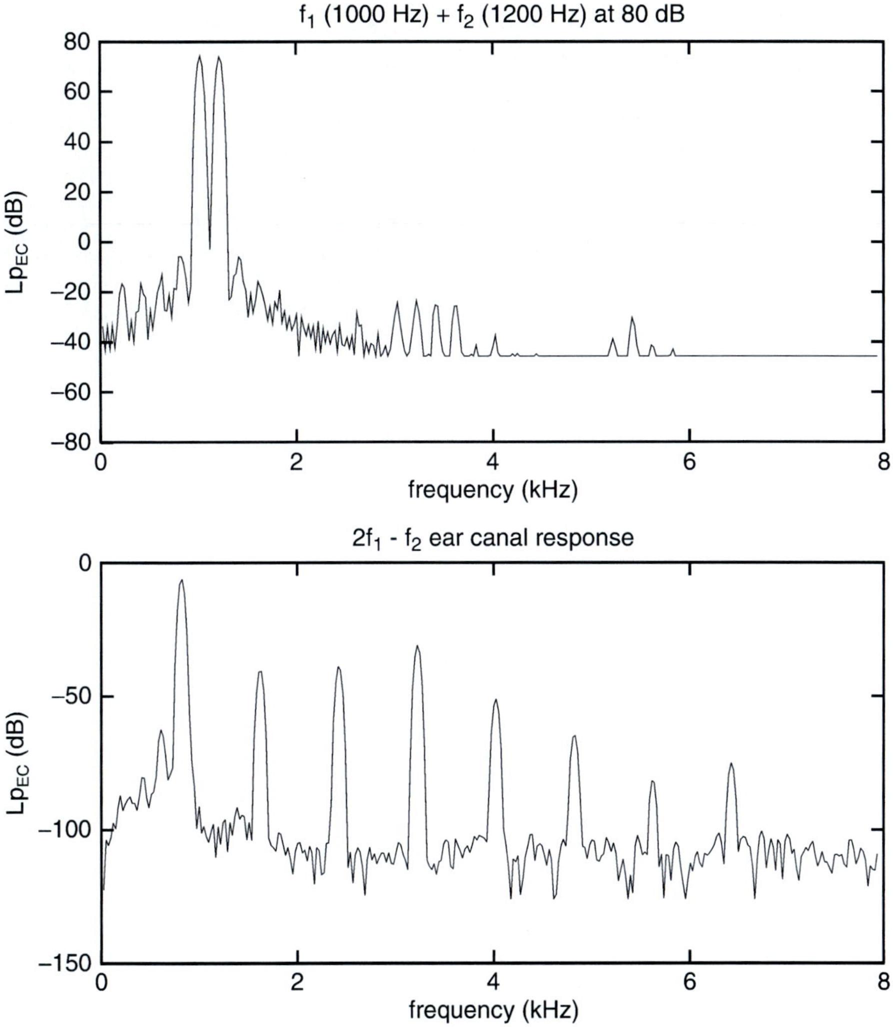

Fig. 10.2 As Fig. 10.1 but at stimulus level of 80 dB. Note that the high-frequency part of the $2f_1 - f_2$ response contains many harmonics.

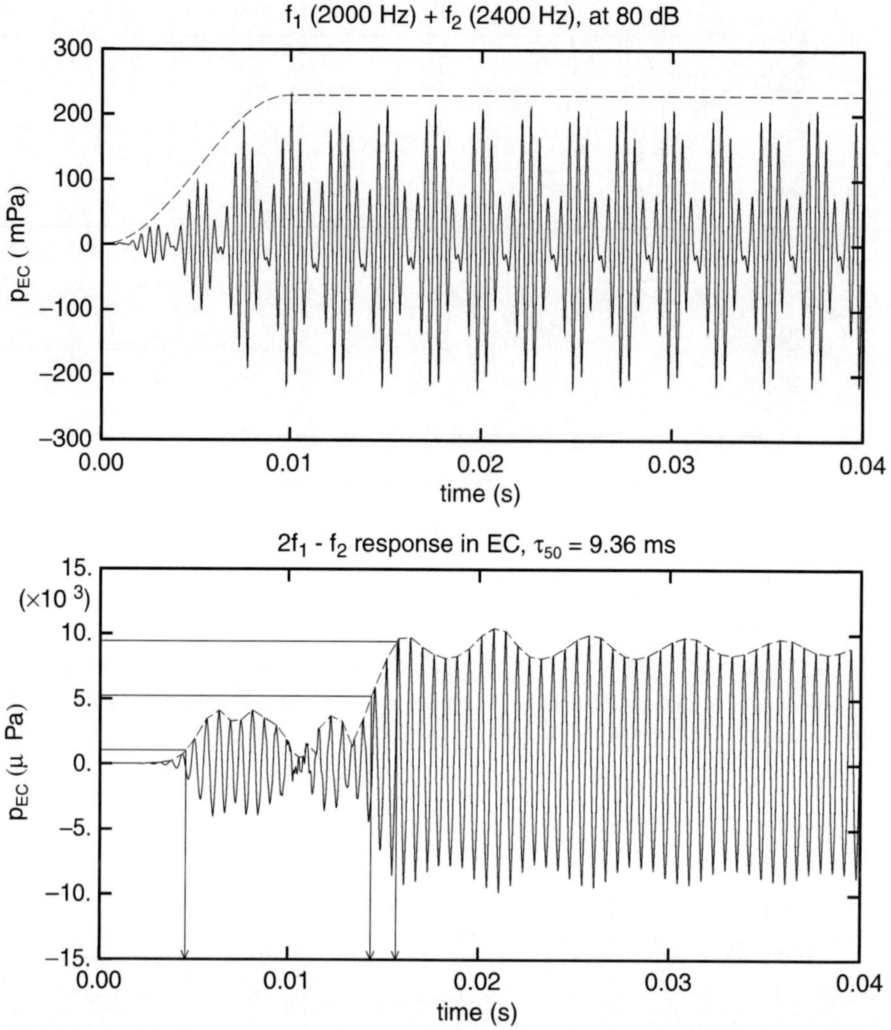

Fig. 10.3 Model 0, linear damping, nonlinear stiffness. Stimulus f_1 and f_2 optimized for the PTPV response to $2f_1 - f_2$ for model #0 with nonlinear stiffness. Stimulus parameters: $f_1 = 2\,\text{kHz}$, $f_2 = 2.4\,\text{kHz}$, $L_1 = L_2 = 80\,\text{dB}$. Time plots over the first 40 ms. The meaning of the response onset delay at 50% (approx. 9.4 ms) is debatable. The asymptotic behavior shows amplitude modulation, or significant spectral sidebands. This is consistent with the actual spectral data.

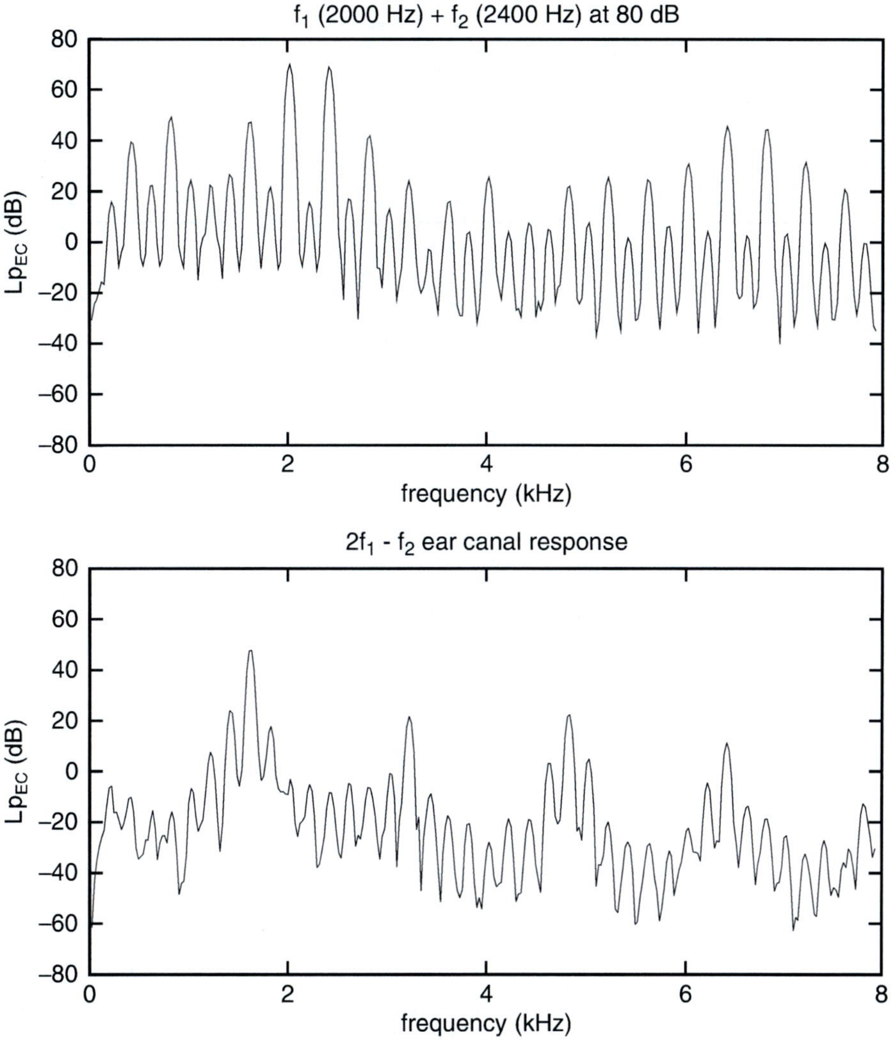

Fig. 10.4 Model 0, linear damping, nonlinear stiffness. Spectrum of the response to the signals shown in Fig. 10.3. The compound spectrum show expected lf-intermodulation and much weaker hf-components. There is fair amount of high-frequency noise, which disappears for the case with a linear stiffness. Note that the NL stiffness generate may intermodulation products, and harmonic in the upper figure, whereas the PTPV response to $2f_1 - f_2$ shows harmonics as well as amplitude modulation components as a primary result.

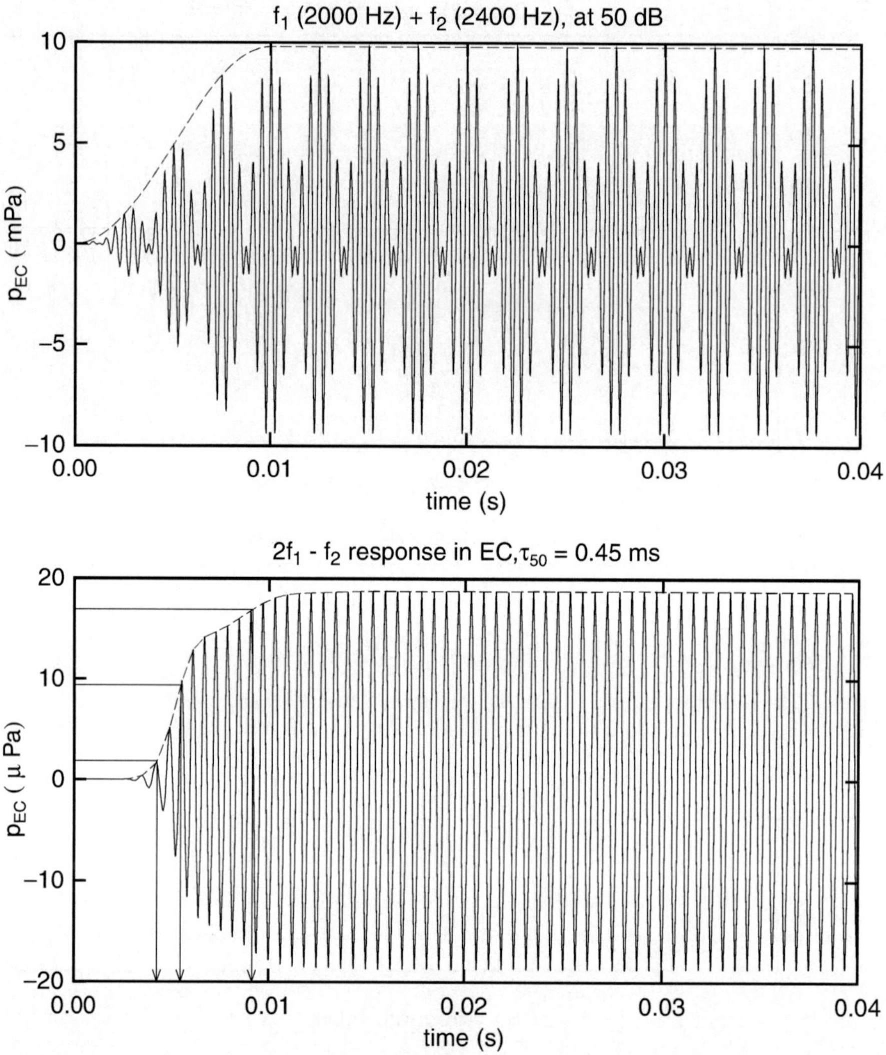

Fig. 10.5 Model 3, nonlinear damping, linear stiffness. Stimulus f_1 and f_2 optimized for the PTPV response to $2f_1 - f_2$ for model #3 with nonlinear damping. Stimulus parameters: $f_1 = 2\,\text{kHz}$, $f_2 = 2.4\,\text{kHz}$, $L_1 = L_2 = 50\,\text{dB}$. Time plots over the first 40 ms. The the response onset shape is clearly deformed, which makes the its relevance unclear. The long term behavior, however, is smooth, again consistent with the actual spectral data.

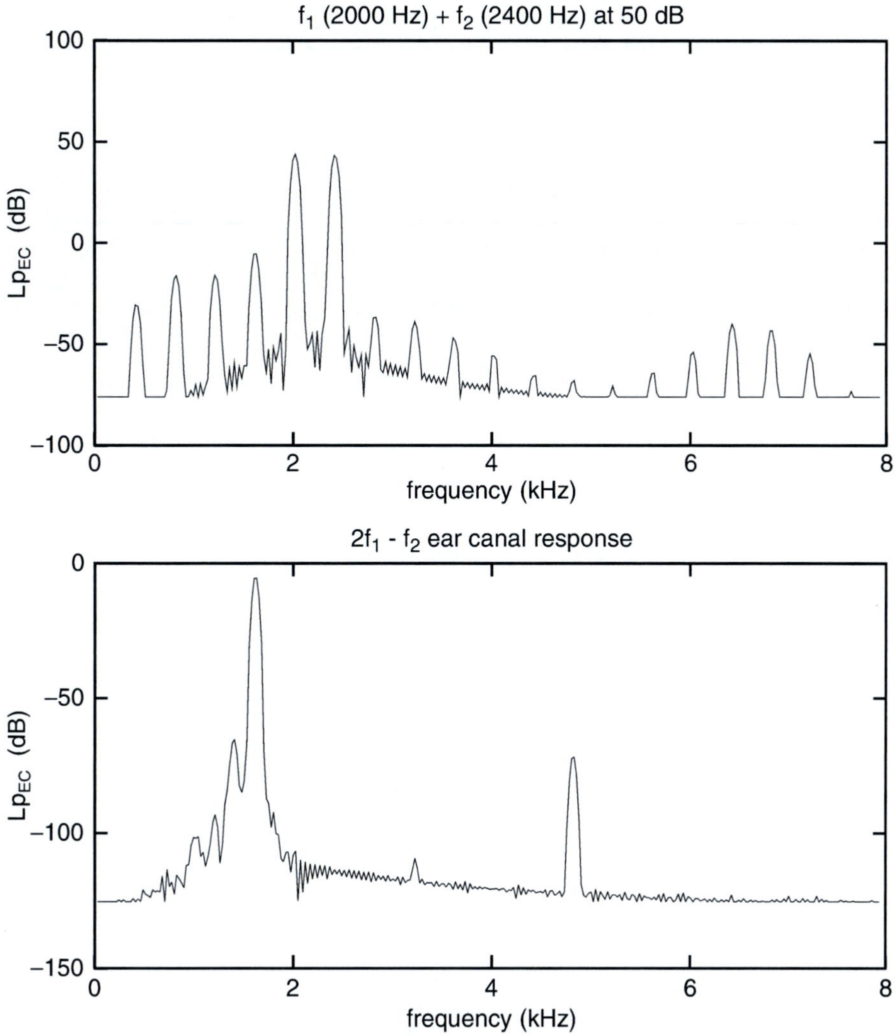

Fig. 10.6 Model 3, nonlinear damping, linear stiffness. Spectrum of the response to the signals shown in Fig. 10.5. The onset effect does not show because the first 10 ms of the response are discarded for the computation of the spectrum. The $2f_1 - f_2$ response is rather clean, with its 3rd harmonic more than 60 dB down.

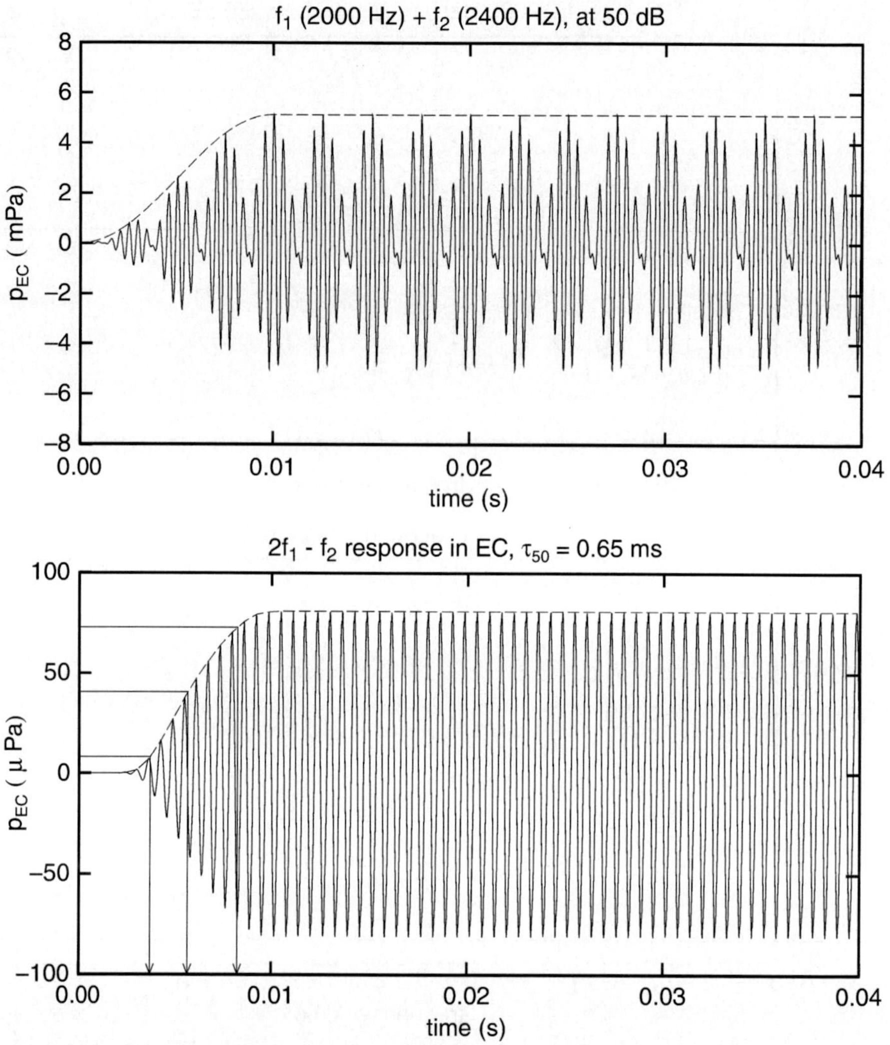

Fig. 10.7 Model 3, nonlinear damping, nonlinear stiffness. Stimulus f_1 and f_2 optimized for the PTPV response to $2f_1 - f_2$ for model #3 with nonlinear damping and stiffness. Stimulus parameters: $f_1 = 2\,\text{kHz}$, $f_2 = 2.4\,\text{kHz}$, $L_1 = L_2 = 50\,\text{dB}$. Time plots over the first 40 ms. The the response onset shape is smooth, but the fine structure of the wave shows multiple peaks, making it impossible for our simple envelope follower to decide the definition of envelope. The long term behavior, however, is stable.

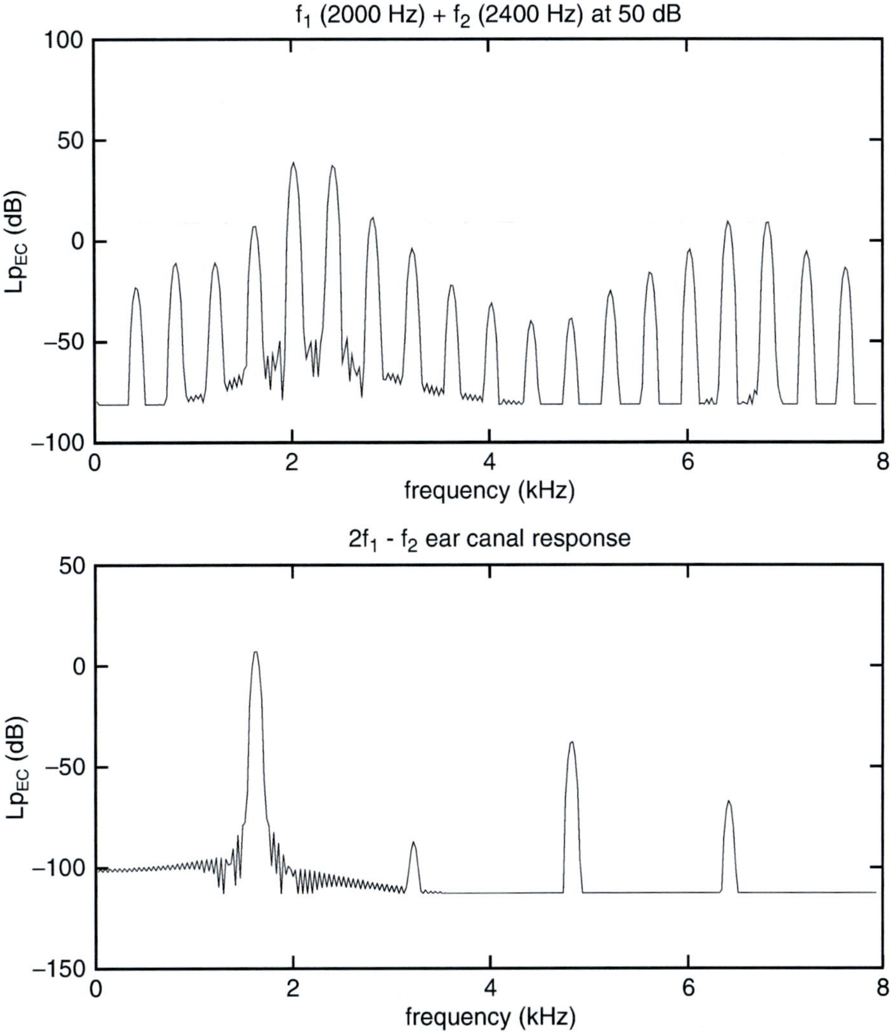

Fig. 10.8 Model 3, nonlinear damping, nonlinear stiffness. Spectrum of the response to the signals shown in Fig. 10.7. The intermodulation products are now symmetrically distributed around the primaries and their 3rd harmonics. Onset effect does not show because the first 10 ms of the response are discarded for the computation of the spectrum. The $2f_1 - f_2$ response is still rather clean, but its 3rd harmonic is now about 40 dB down. The smaller peak at 6.4 kHz is suspect because it is an even harmonics of the DP. Since it occurs in the overall spectrum as a proper intermodulation product, it might be possible that the occurrence in the PTPV response is due to the PTPV-parameter settings.

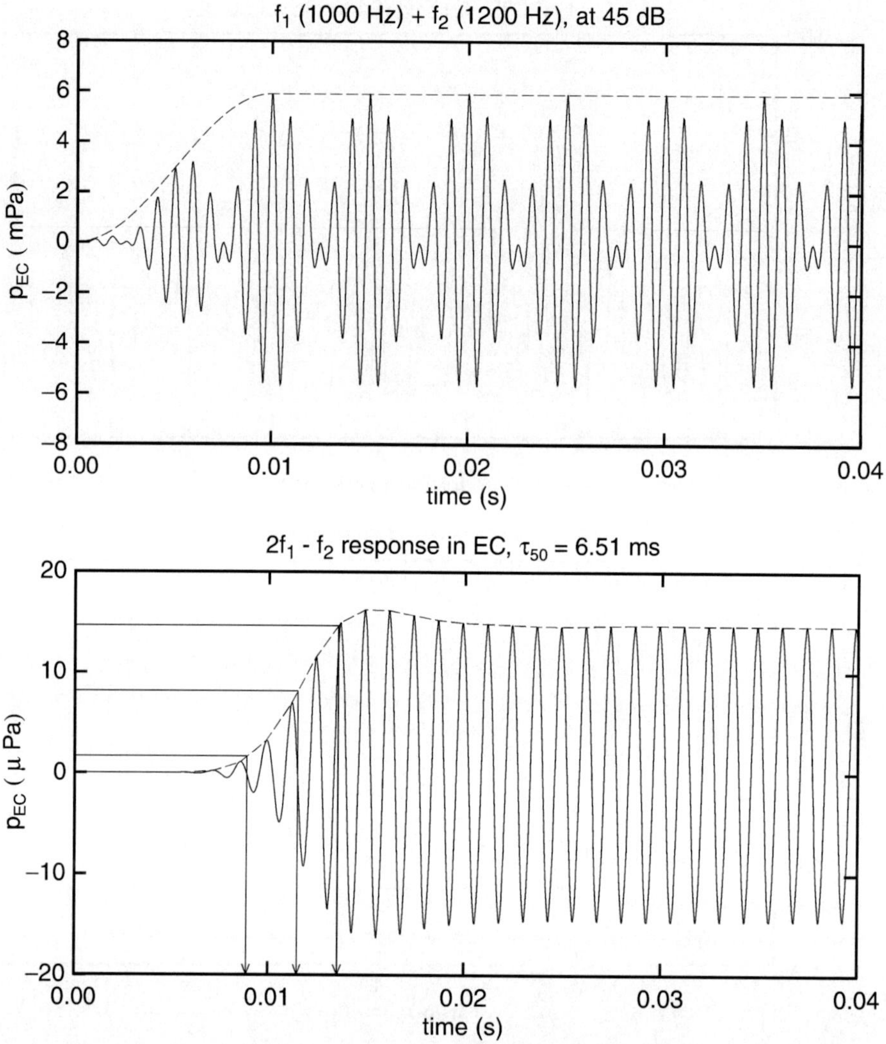

Fig. 10.9 Model 4, nonlinear damping, linear stiffness. Stimulus f_1 and f_2 optimized for the PTPV response to $2f_1 - f_2$ for model #3 with nonlinear damping and stiffness. Stimulus parameters: $f_1 = 1\,\text{kHz}$, $f_2 = 1.2\,\text{kHz}$, $L_1 = L_2 = 45\,\text{dB}$. The the $2f_1 - f_2$-response is smooth over the entire interval. There is a small overshoot in the response onset.

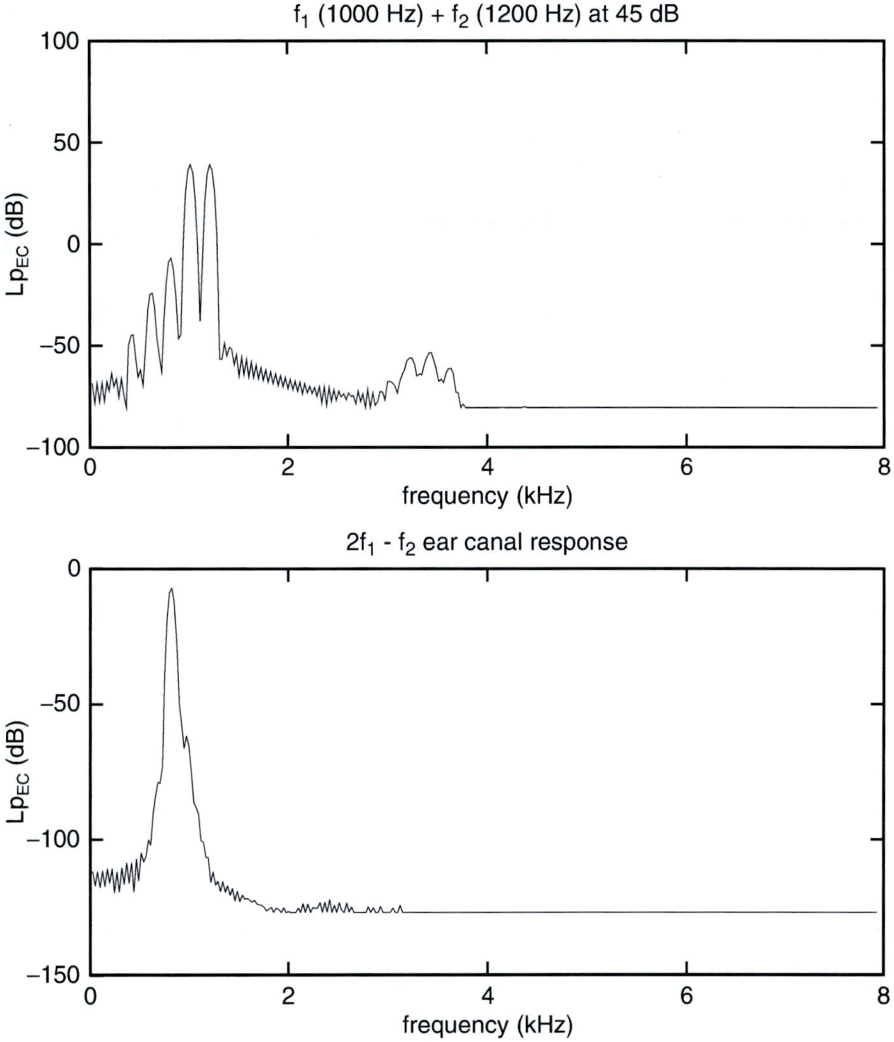

Fig. 10.10 Model 4, nonlinear damping, linear stiffness. Spectrum of the response to the signals shown in Fig. 10.9. The smooth DP time domain response is consistent with the very smooth $2f_1 - f_2$-spectrum.

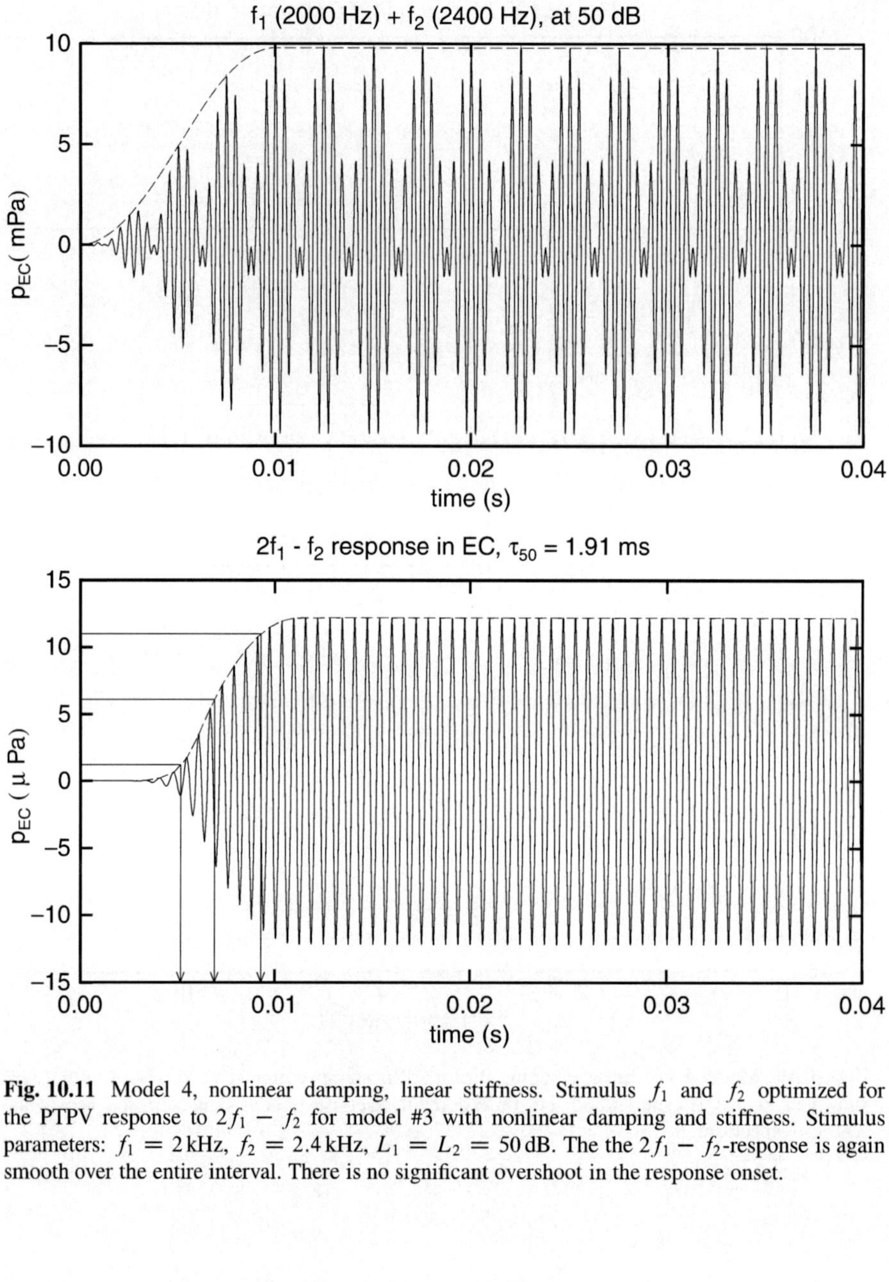

Fig. 10.11 Model 4, nonlinear damping, linear stiffness. Stimulus f_1 and f_2 optimized for the PTPV response to $2f_1 - f_2$ for model #3 with nonlinear damping and stiffness. Stimulus parameters: $f_1 = 2\,\text{kHz}$, $f_2 = 2.4\,\text{kHz}$, $L_1 = L_2 = 50\,\text{dB}$. The the $2f_1 - f_2$-response is again smooth over the entire interval. There is no significant overshoot in the response onset.

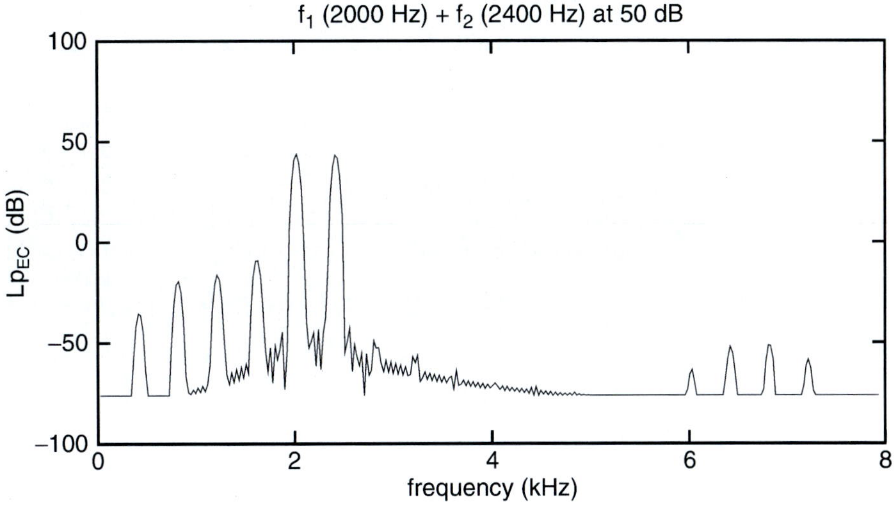

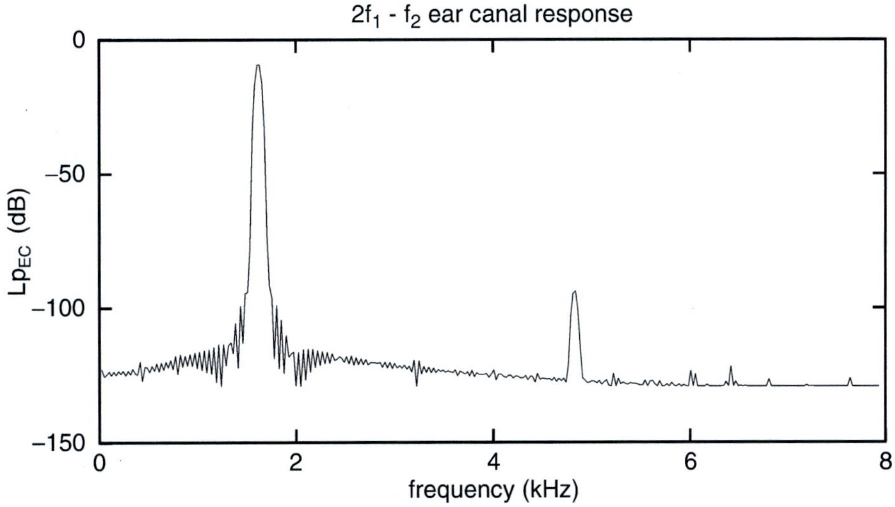

Fig. 10.12 Model 4, nonlinear damping, linear stiffness. Spectrum of the response to the signals shown in Fig. 10.11. The smooth overall spectrum shows lf intermodulation products, the DP response shows significant, but relatively small, odd harmonics: the spectrum shows both even and odd harmonics of $2f_1 - f_2$. The even harmonics (at 3.2 and 6.4 kHz) were not expected. See note at Fig. 10.8.

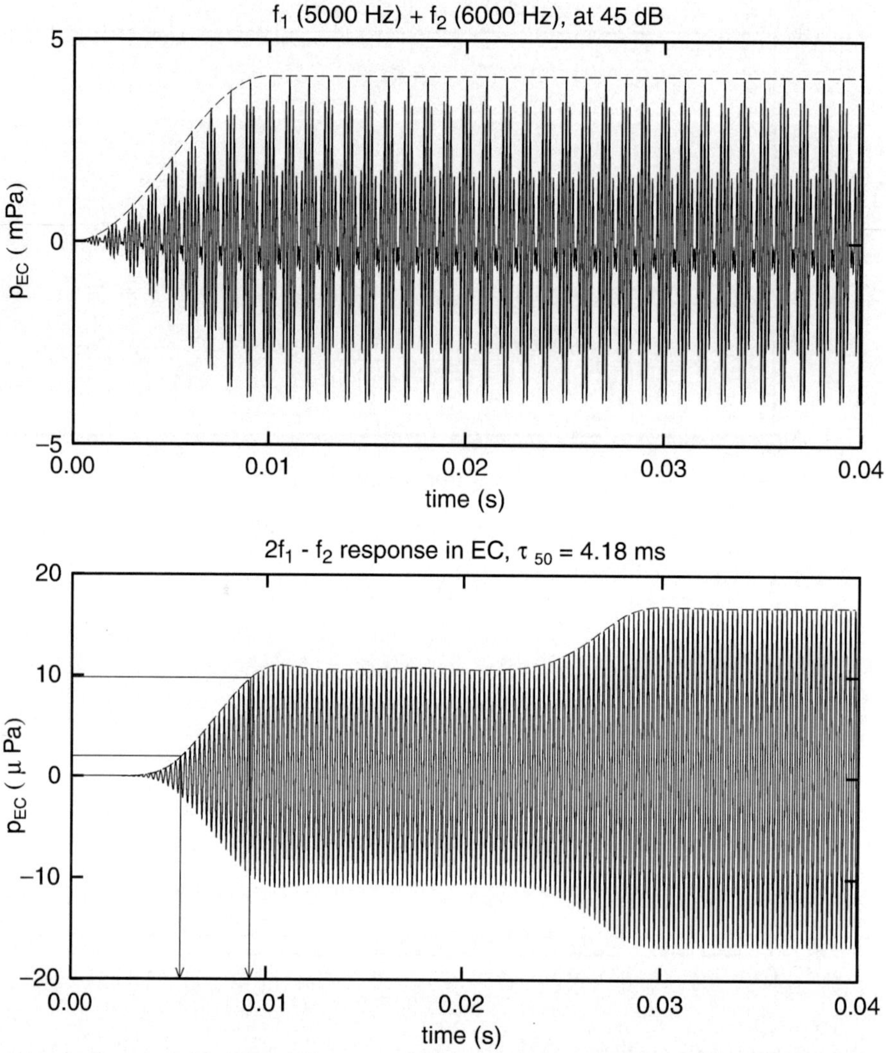

Fig. 10.13 Model 4, nonlinear damping, linear stiffness. Stimulus f_1 and f_2 optimized for the PTPV response to $2f_1 - f_2$ for model #3 with nonlinear damping and stiffness. Stimulus parameters: $f_1 = 5\,\text{kHz}$, $f_2 = 6\,\text{kHz}$, $L_1 = L_2 = 45\,\text{dB}$. The the $2f_1 - f_2$-response, although smooth, comes with a new surprise: after a minor overshoot at the end of the first part of the onset (just after 10 ms) there is a significant second step at around 25 ms. Apparently we see interaction over a broader frequency range.

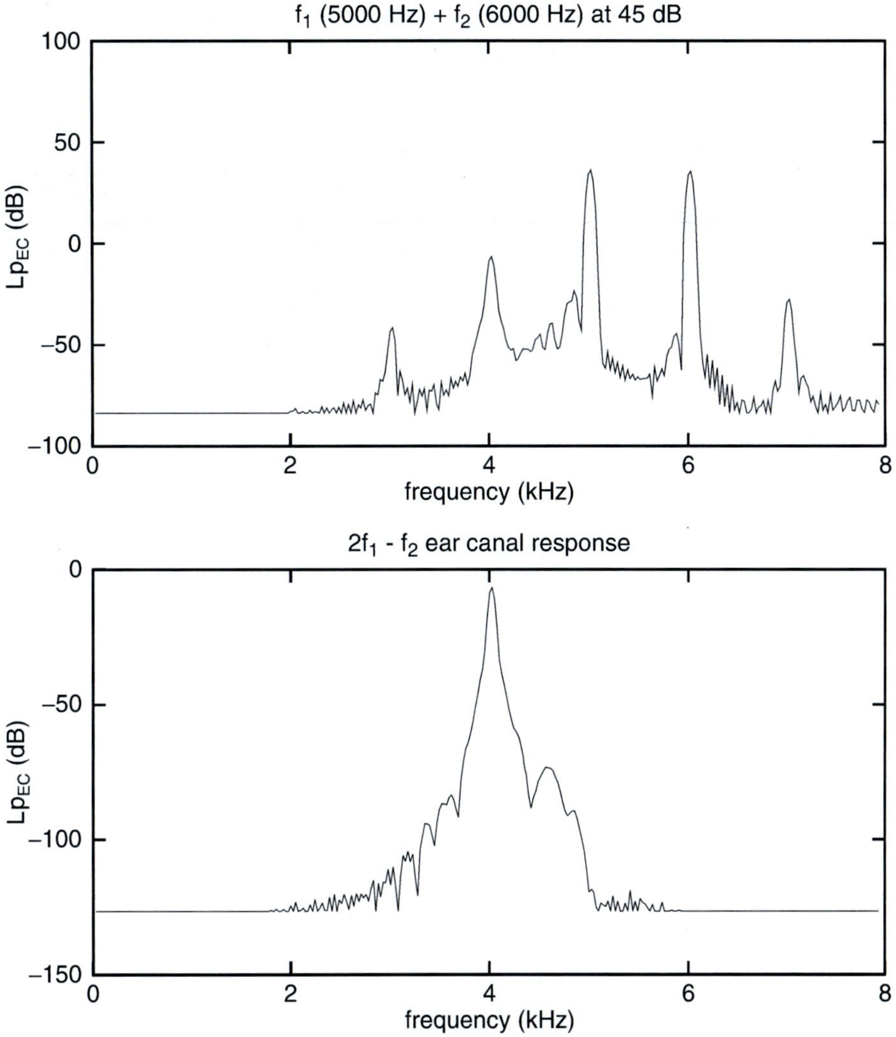

Fig. 10.14 Model 4, nonlinear damping, linear stiffness. Spectrum of the response to the signals shown in Fig. 10.13. The noted special behavior in the time domain only shows in the detailed profile of the main DP component. But harmonics remain below threshold. See note at Fig. 10.8.

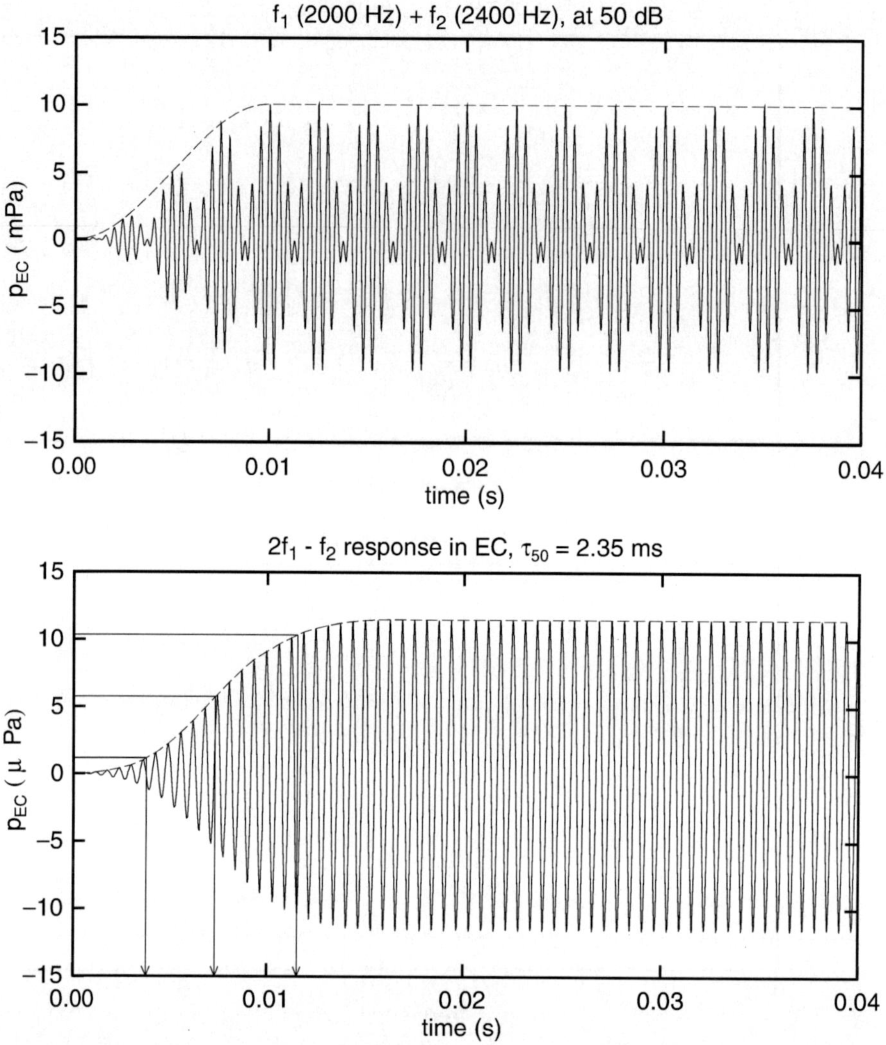

Fig. 10.15 Model 4, nonlinear damping, nonlinear stiffness. Stimulus f_1 and f_2 optimized for the PTPV response to $2f_1 - f_2$ for model #3 with nonlinear damping and stiffness. Stimulus parameters: $f_1 = 2\,\text{kHz}$, $f_2 = 2.4\,\text{kHz}$, $L_1 = L_2 = 50\,\text{dB}$. The the $2f_1 - f_2$-response is smooth over the entire interval but appears to have a shallower slope than seen so far. There is no significant overshoot in the response onset.

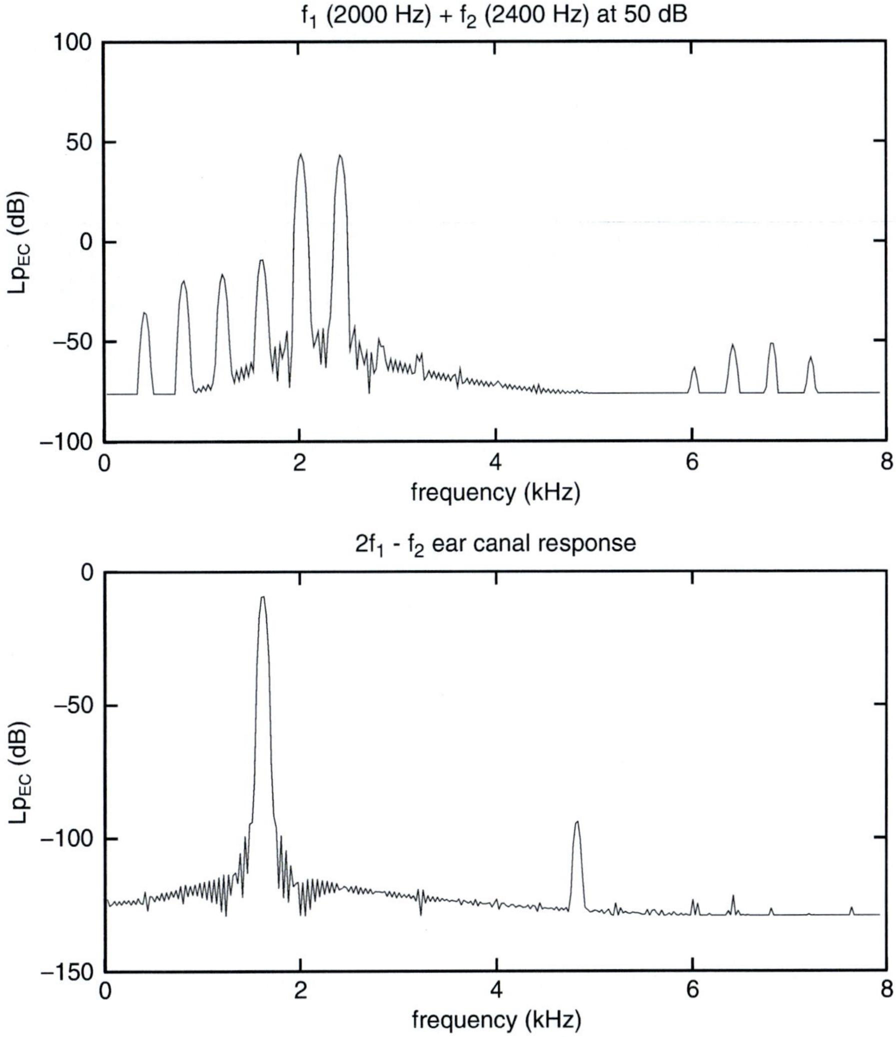

Fig. 10.16 Model 4, nonlinear damping, nonlinear stiffness. Spectrum of the response to the signals shown in Fig. 10.15. The smooth overall spectrum shows lf intermodulation products, the DP response shows significant, but relatively small, odd harmonics: the spectrum shows both even and odd harmonics of $2f_1 - f_2$. The even harmonics (at 3.2 and 6.4 kHz) were not expected.

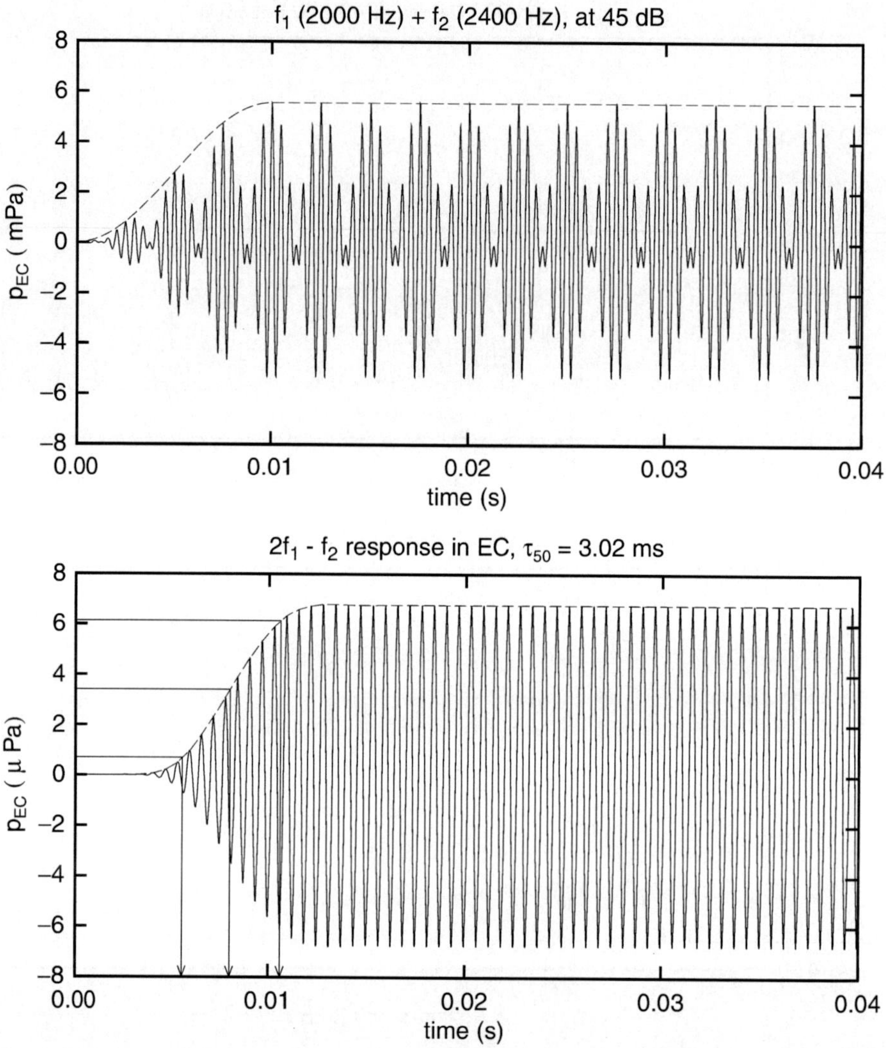

Fig. 10.17 Model 6, nonlinear damping and nonlinear stiffness for Zweig's feedback model. Stimulus f_1 and f_2 optimized for the PTPV response to $2f_1 - f_2$ for model #3 with nonlinear damping and stiffness. Stimulus parameters: $f_1 = 2\,\text{kHz}$, $f_2 = 2.4\,\text{kHz}$, $L_1 = L_2 = 50\,\text{dB}$. The the $2f_1 - f_2$-response is smooth over the entire interval but appears to have a shallower slope than seen so far. There is no significant overshoot in the response onset.

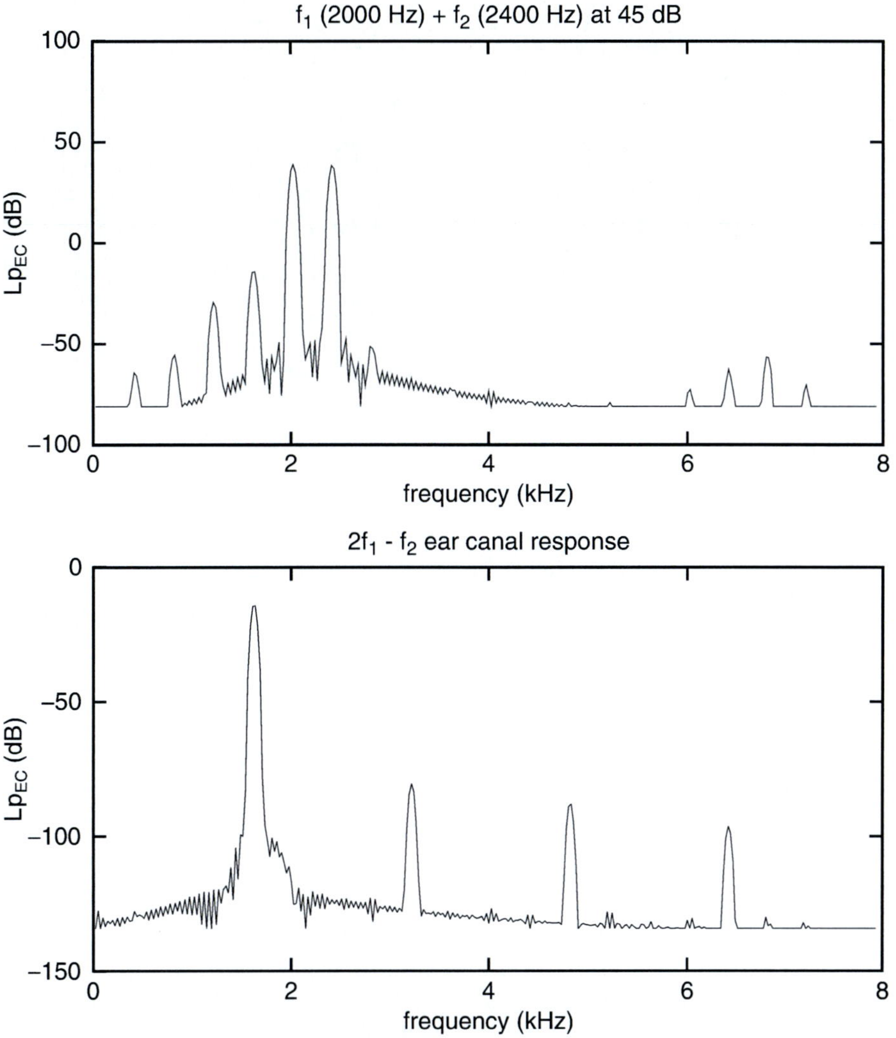

Fig. 10.18 Model 6, nonlinear damping, nonlinear stiffness for Zweig's feedback model. Spectrum of the response to the signals shown in Fig. 10.17. The smooth overall spectrum shows lf intermodulation products, the DP response shows significant, but relatively small, odd harmonics: the spectrum shows both even and odd harmonics of $2f_1 - f_2$. The even harmonics (at 3.2 and 6.4 kHz) were not expected.

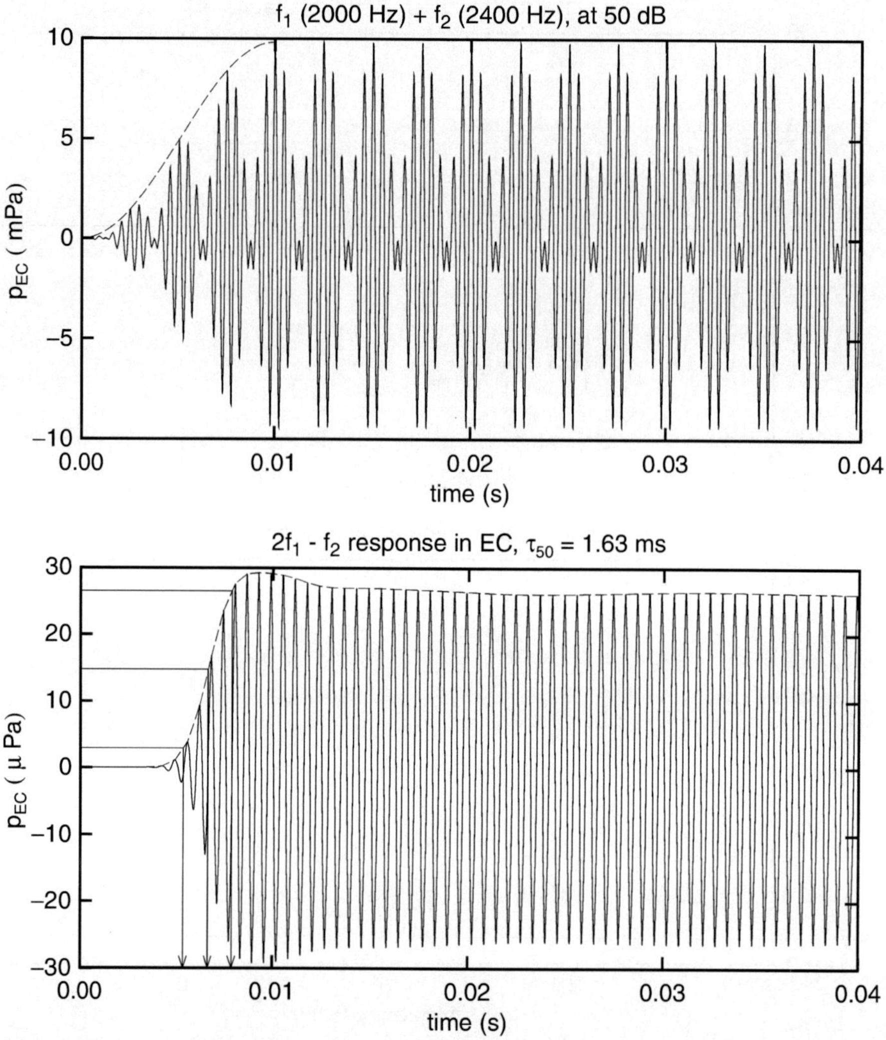

Fig. 10.19 The stimulus and $2f_1 - f_2$ response for Hopf-bifurcation model. It is equivalent to the Van der Pol model but with the difference that the damping does not become negative, but goes to 0 as v approaches zero. Again, the PTPV-responses are rather similar to the responses of most other models. The major characteristics are the steep rise time, causing $\tau_{10} \gg \tau_{50} \gg \tau_{90}$, and the small overshoot at the onset.

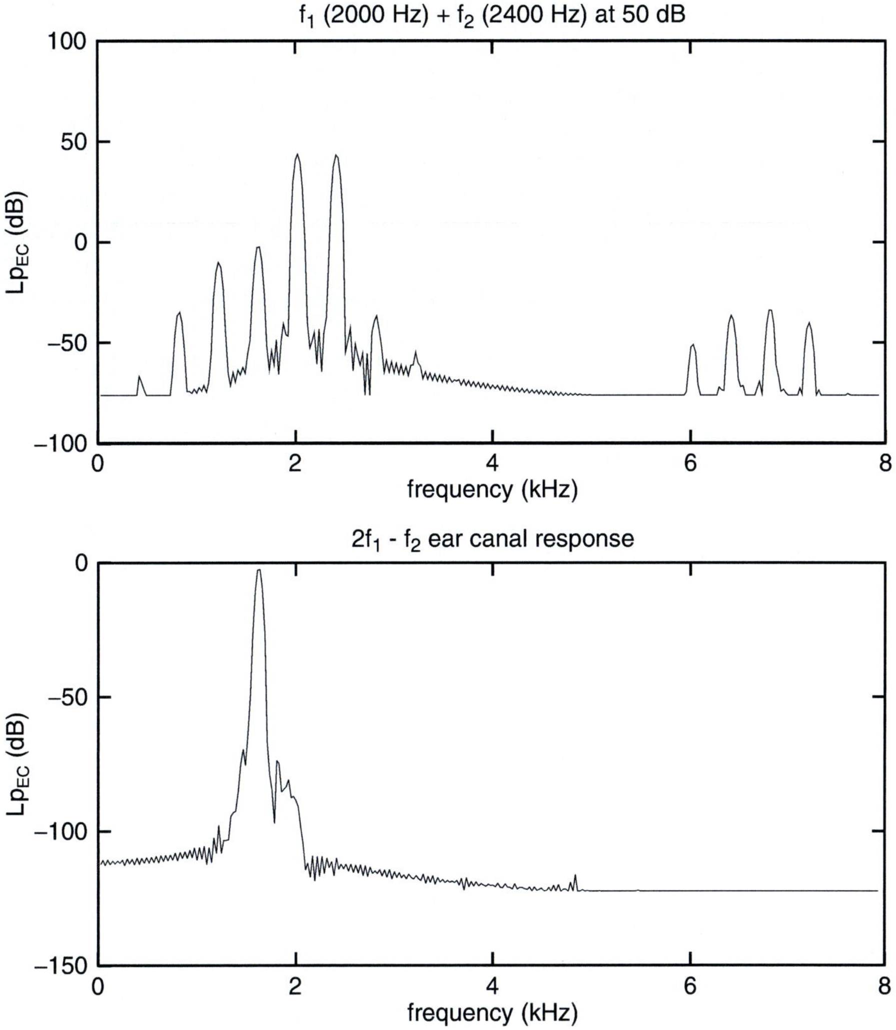

Fig. 10.20 The stimulus and $2f_1 - f_2$ response for Hopf-bifurcation model. Spectra of the responses shown in Fig. 10.19. Again the lf-intermodulation products are stronger than the hf counter parts. Odd harmonic of the primaries are about 80 dB down.

10.1 Some General Conclusions

- NL stiffness only (i.e., with linear damping) generates at high levels (80 dB) an unnatural dose of distortion products (Fig. 10.4). Also the waveform of the $2f_1 - f_2$-response could be termed: aberrant (Fig. 10.3).
- NL stiffness increases the total NL response (Model 3 at 50 dB). This seems most apparent in the spectra (Fig. 10.8 vs Fig. 10.6), but the amplitude results are also interesting. The peak value in the case of linear stiffness is twice the value obtained for the nonlinear situation. At the same time the DPOAE amplitude increases by a factor of 4 (Figs. 10.5 and 10.7).
- A limited amount of NL stiffness, in combination with NL damping has limited effects on the $2f_1 - f_2$ distortion product response at 50 dB (Figs. 10.8, 10.16, 10.18).
- The differences in responses of Models 4 and 6 are "marginal" except for the difference in even harmonics in the $2f_1 - f_2$ response.
- The total time response envelope closely follows the stimulus envelope. The PTPV responses to the DPOAEs can be distorted. The only general trend is that the response is delayed.

Index

H. Duifhuis, *Cochlear Mechanics: Introduction to a Time Domain Analysis of the Nonlinear Cochlea*, DOI 10.1007/978-1-4419-6117-4,